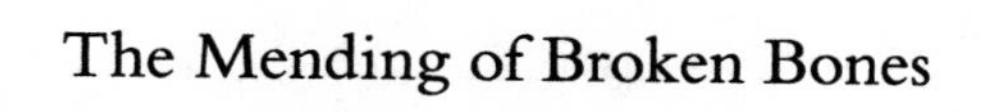
The Mending of Broken Bones

Also by Paul Lockhart

A Mathematician's Lament

Measurement

Arithmetic

The Mending of Broken Bones

A Modern Guide to Classical Algebra

PAUL LOCKHART

THE BELKNAP PRESS OF HARVARD UNIVERSITY PRESS
Cambridge, Massachusetts
London, England
2025

Printed in the United States of America
First printing

Library of Congress Cataloging-in-Publication Data

Names: Lockhart, Paul, author.
Title: The mending of broken bones : a modern introduction to classical algebra / Paul Lockhart.
Description: Cambridge, Massachusetts ; London, England : The Belknap Press of Harvard University Press, 2025. | Includes index.
Identifiers: LCCN 2024045906 (print) | LCCN 2024045907 (ebook) | ISBN 9780674296329 (cloth) | ISBN 9780674300460 (pdf) | ISBN 9780674300477 (epub)
Subjects: LCSH: Algebra.
Classification: LCC QA155 .L63 2025 (print) | LCC QA155 (ebook) | DDC 512—dc23/eng/20241119
LC record available at https://lccn.loc.gov/2024045906
LC ebook record available at https://lccn.loc.gov/2024045907

CONTENTS

The Scribal Art

The scribal art is the mother of orators, the father of masters
The scribal art is delightful; it never satiates you
The scribal art is not easily learned
Strive to master the scribal art and it will enrich you
Do not neglect the scribal art . . .

Sumerian Tablet (c. 600 BC)

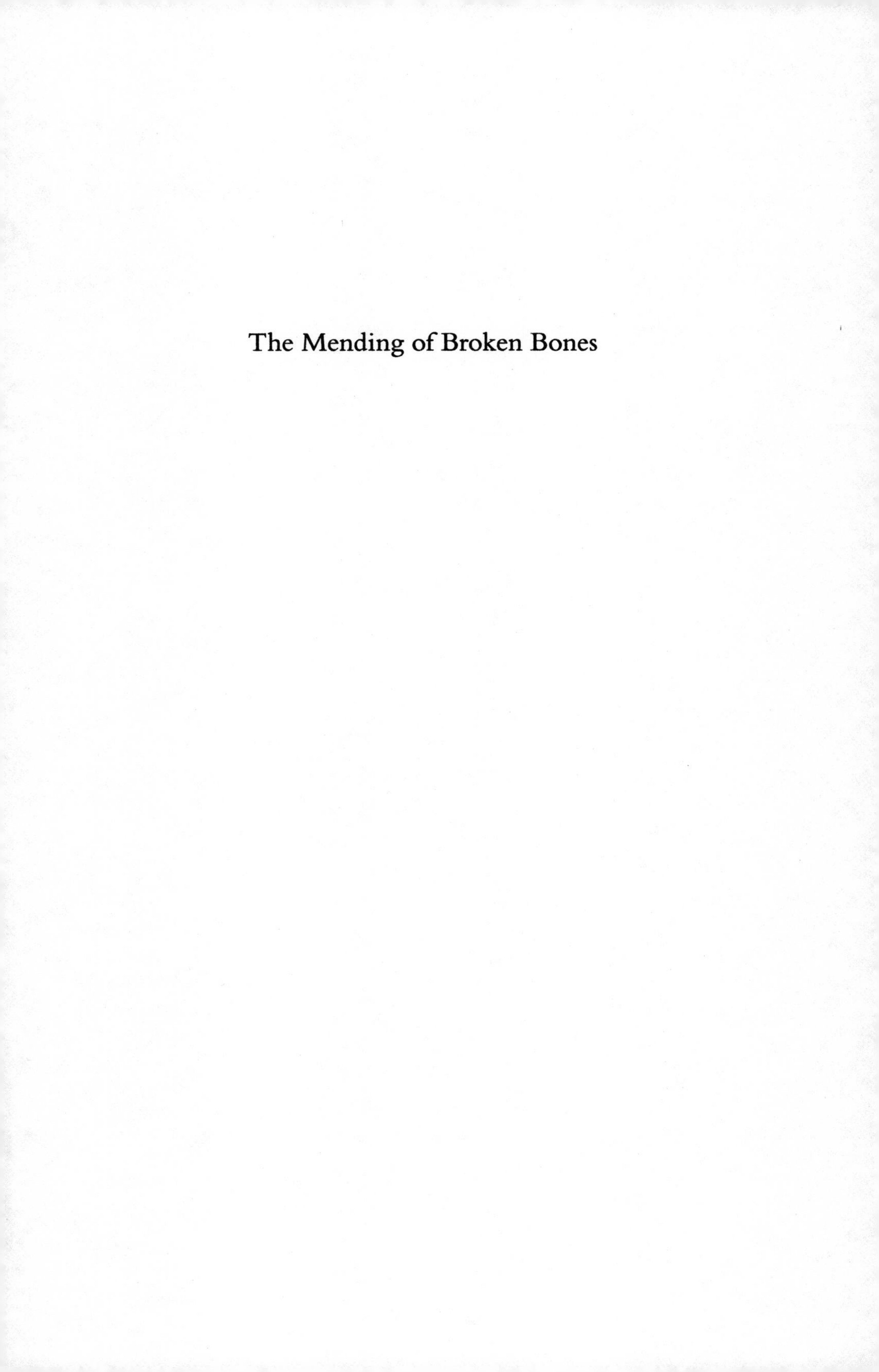

The Mending of Broken Bones

INTRODUCTION

I want to tell you about a peculiar art form that arose in ancient Egypt and Mesopotamia about five thousand years ago. This art is known as *algebra*. Algebra began as the study of a certain type of number puzzle—namely, the kind that arises from practical problems in the realms of engineering and architecture, commerce and taxation, farming and land measurement, and so on. In other words, algebra is a direct consequence of civilization and its increasing complexity and cognitive demands.

Such problems concern quantities and measurements of various sorts combined in different ways depending on circumstances. Typically, one or more of these quantities is *unknown*—a mystery quantity waiting to be determined. Whereas arithmetic concerns the combination and comparison of known quantities, algebra is the deductive craft of determining the values of unknown quantities. In this way, doing algebra is very much like being a detective. We have a mystery number (as opposed to a murderer) along with some clues to its identity that will hopefully allow us to figure out precisely who it must be. As we shall see, there is tremendous flexibility in how we may approach such puzzles. Depending on your taste and level of sophistication, there is ample range to express your own personal style and to invent clever new techniques for yourself.

In many ways, algebra is a lot like tying and untying knots, except that instead of string, we are using numbers, and rather than looping and pulling through, we are performing arithmetical activities such as adding and multiplying. Thus, different problems lead to different types of knots, and part of being a skilled algebraist is recognizing patterns and having familiarity with the various types and the best ways to handle them.

So where do mystery number puzzles come from? Under what circumstances do we find ourselves dealing with unknown quantities subject to constraints? Truth be told, this sort of scenario rarely arises in the arena of the purely practical. If all you want is a quick and dirty solution to a mundane, real-world problem, then common sense and seat-of-the-pants estimation are almost always good enough. It's when we desire more control—not just solving the problem but solving it *well*—that we find ourselves tying numerical knots.

Imagine that you are an ancient Egyptian construction foreman at the site of the new temple. The plans call for sixty-two equally spaced columns, each weighing more than twenty tons. If all that was wanted were some nice decorative columns, we could just eye it and simply accept whatever unequal spacing and asymmetry might result. If, however, we demand accurate equal spacing—simply for the sake of beauty and architectural harmony, an aesthetic rather than structural requirement—we then place an external constraint on the distances.

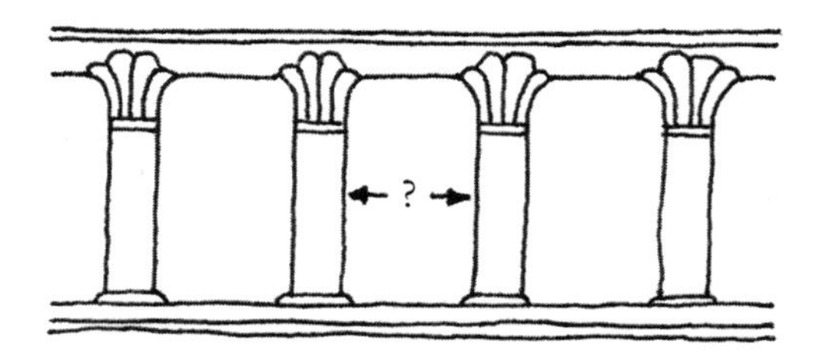

Let's suppose the temple is to be four hundred twenty cubits in length and that each column is two and a half cubits in diameter. The unknown measurement is then the spacing between the columns. This is definitely something you'll want to figure out well in advance, *before* you start hoisting twenty-ton columns into place with levers and pulleys and whatnot. The last thing you need is to finally get these massive objects situated, only to discover that they don't line up right and need to be moved. In other words, you don't want to position them and then see if they fit; you want to position them *so that* they fit.

This is where logic and deduction come in. We require the distance between the columns to be a quantity—one that we do not yet know—with the property that sixty-one copies of it (the total amount of space between columns) added to sixty-two copies of two and one-half cubits (the total width of the columns) make the required total length of four hundred twenty cubits.

So it's not so much necessity as it is design and planning that lead to algebra problems. It is usually not too difficult to get something to work in a rough-and-ready, half-assed sort of way. It's when you want things optimized—either for the sake of art or efficiency—that things get interesting. Desire leads to design, and design leads to constraints. Ultimately, algebra (i.e., the art of solving mystery number puzzles) arises, like most aspects of civilization, from our desire for *control.*

Just as carpenters and cooks develop rules of thumb and simple, easy-to-use methods for commonly recurring situations, the algebraists of the ancient world also began to compile problem collections and systematic methods for their solution. The abstract nature of the enterprise becomes clear as soon as one leaves the world of specifics (e.g., blocks of stone, bushels of grain, plots of land) and begins to study such puzzles in general. Our problems cease to be about mundane, real-world quantities and instead concern the properties and behavior of numbers themselves.

Here is a typical example, taken from a Babylonian problem tablet (c. 1850 BC):

I found a stone but did not weigh it.
After I added one-seventh of its weight,
And one-eleventh of this new weight,
The total was one ma-na.
What was the original weight?

The first thing I want to point out is the extreme artificiality of this question. At first glance, it might seem to be a math problem of the most mundane and practical sort, dealing as it does with stones and weights and all. (The *ma-na* was a Babylonian unit equal to roughly 500 grams.) But on closer consideration, we realize that the scenario depicted here is actually quite absurd. How would anyone come by this ridiculous information in the first place? You mean I somehow have the ability to determine the combined weights but not the weight of the stone I started with?

This is simply not the sort of thing that would actually happen out at the construction site. This is a contrived, made-up problem for the purpose of training young scribes. In fact, such Babylonian cuneiform tablets are nothing other than ancient school textbooks. How ironic that the Hanging Gardens are no more, Nineveh and Nebuchadnezzar are forgotten names, yet amongst the ruins of once-proud Babylon can still be found the math problems at the end of the chapter!

What I find remarkable is that even as far back as 2000 BC, the scribal class was already deeply immersed in a pure mathematical culture—in particular, the study of numbers and number puzzles for their own sake and the invention and dissemination (through schooling) of general methods of solution.

Here is an Egyptian example, taken from the Rhind Mathematical Papyrus (c. 1650 BC):

A quantity, one-half of it added to it, becomes sixteen.

Notice the elegant abstraction—not a quantity of bricks or grain, just a quantity. On the surface, the problem appears to be asking for a number with the specific property that when you increase it by half, you get sixteen. But the *real* question is not so much about the one-half or the sixteen. It's about *any* such question. For instance, the puzzle immediately following this one reads:

A quantity, one-quarter of it added to it, becomes fifteen.

I'm sure you can already feel the rote and repetitive quality (so reminiscent of the exercises in a modern textbook) that is the hallmark of a formal training regimen. Again, the idea was to develop and inculcate general methods, so the exercises were intentionally designed to illustrate and exhibit the various known techniques.

Here's another example, this one taken from the *Greek Anthology* (c. 550 BC):

Make for me a crown of sixty minae, mixing gold and
brass and with them tin and much-wrought iron.
Let the gold and bronze together form two-thirds, the gold and
tin together three-fourths, and the gold and iron three-fifths.

This is a somewhat more elaborate puzzle, involving four mystery quantities—namely, the weights of each of the four metals, presumably measured in *minae* (an ancient Greek unit equal to a little less than a pound).

So the scribes of the ancient world were apparently having a pretty good time—devising fun and intriguing puzzles for themselves and each other, figuring out how to solve them, teaching their students how to do the same, and even writing poems about how cool it is to be literate and intelligent.

At any rate, it is clear that more than four thousand years ago mathematics had already begun to break away from the mundane practical world. A small group—members of the elite priestly and scribal classes, essentially—had the leisure time to investigate numbers and their properties for their own amusement and intellectual satisfaction. The fact that this inquiry also leads to powerful general methods as well as deeper insight and understanding means that the original real-world problems and applications then appear as trivial special cases by comparison.

Thus, algebra becomes the fine art of tangling and untangling abstract numerical information. The setting shifts from the construction site and the counting house to the idealized realm of pure number. Our problems cease to be practical and utilitarian and are instead driven by curiosity and aesthetics. We leave the noisy and complicated world of physical reality behind and move to a quieter, more peaceful realm of abstract pattern and ideal beauty—a place I like to call Mathematical Reality. This is the setting in which algebraists have been conducting their business for the past forty centuries.

Our word *algebra* comes from the Arabic *al-jabr*, meaning "the way of completion" or, more poetically, "the mending of broken bones." Our numbers have been shattered into pieces, strewn about in various ways, and now we want to reassemble them. Just as an archaeologist working with the shards of an ancient vase must be observant and careful, attuned to subtle markings and patterns, so must the algebraist. Untying knots is no joke; anyone who has ever worked with string knows how easy it is to make matters worse. Algebra is a delicate and artful craft; we're not going to untie any knots by brute force. Instead, the idea is to find elegant, graceful, and powerful general methods for untangling so that we may hope to learn something new about both our knots and the nature of our string. (Also, there are no scissors in Mathematical Reality, so get that idea out of your head.)

One thought that may have occurred to you is why not simply use trial and error? We can just guess the solution to our problem and see if it works. If not, we can keep adjusting our guess until we finally get it right. What's wrong with that approach? Actually, it's a pretty good plan so far as it goes, especially with practical, real-world problems that do not require exactitude. There is no point fussing over the precise measurement of a piece of wood, for instance. Carpenters know full well that it will expand and contract with moisture and temperature anyway. So yes, you certainly can guess (and adjust) in a lot of cases, and it'll be perfectly fine.

Of course, this sort of approach becomes increasingly impractical as civilization grows more technologically advanced and higher-precision measurements become necessary. Crude approximations and ballpark estimates may suffice for the ancient world of sandstone blocks and wooden beams, but the design of efficient jet engines and nuclear reactors requires much greater care. Many problems in modern engineering typically demand tolerances of under a micron (one-millionth of a meter). Though still approximate, such precise measurements are practically impossible to guess at, and the arithmetic chore of repeatedly checking and modifying our estimates becomes too laborious to contemplate.

On the other hand, solving specific, approximate problems is not what the art of algebra is about anyway. We don't really care about any individual problem or solution; we care about general types of numerical entanglement and methods for detangling. Any particular example—whether crude or precise—would be just for practice, like the puzzles contained in the Rhind Papyrus.

Many arenas of human creative endeavor possess a high-versus-low distinction: painting a portrait versus painting a desk; symphony versus advertising jingle; literature as opposed to an appliance manual; the use of whatever medium to enlighten, ennoble, and entertain ourselves as opposed to its purely practical utility. This is not so much a value judgment as it is a difference in goals. I love to solve practical problems, actually—building a level staircase, for instance, or designing an efficient electronic circuit. Reality is a fun place to be, don't get me wrong.

But at the same time, I can't help expressing my true feelings and philosophy (much as you might wish otherwise!), and I suppose it's pretty obvious from what I've said so far that I am a pure mathematician by nature. As much as I may enjoy interacting with the physical universe from time to time (e.g., coffee and chocolate), my heart belongs to Mathematical Reality. I spend a good deal of my time there, in fact. If you're talking to me and I'm going "hmm hmm" and not really listening, that's probably where I am.

For me, it's all about seeking understanding. I'm not all that interested in curing diseases or making bridges that don't collapse. I want to understand the creatures that inhabit this weird mental universe in my head. In particular, I want to understand exactly which algebra problems have solutions and which don't, and why. In short, I want to *classify*.

Just as a botanist or zoologist seeks to understand the variation of species in the natural world, mathematicians also face a bewildering jungle full of exotic creatures. The human urge, when confronted with variety, is to sort and categorize. This is in fact the single most fundamental mathematical act: deciding when two things are to be considered the same or not. Even the simple act of counting relies on a tacit agreement as to which items are to be considered distinct, individual units. (The seemingly innocuous question of how many lemons are in the bowl may well depend on your definition of lemon.)

People love to sort and classify—to devise categories and hierarchies and labels and all that—and mathematicians are no different. Mathematics is the study of *pattern*, which means there's quite a lot for us to sort and classify! In the case of algebra, the question is whether there are similarities and analogies (as well as subtle differences) among various general types of problems that can help us to understand better and to see farther, as it were. This is precisely why mathematicians seek generalization and abstraction.

When we find ourselves deep in the jungle, our vision is often impaired—obscured by the very trees and vines we are interested in studying. These are the details and the particulars. To understand the lay of the land, we need to gain altitude—to climb a hill of some sort. This hill is *generality*. The more abstract and general our patterns and ideas, the more they encompass and the more they allow us to see. What we gain is perspective, and that is a rare and precious commodity. This can sometimes be a fairly difficult climb, but the view is always worth it.

The point being that the method of trial and error, as easy and understandable as it may be, is simply not going to cut the mustard. Not for a scribe with any serious pretensions toward understanding, that is. Trial and error is unsatisfactory for a number of reasons. First of all, we learn almost nothing about the general pattern. Okay, you've tried some numbers and eventually you find some that happen to work. But that gives you no information about the solutions to any other problem. This is an ad hoc approach, rather than a systematic treatment. Second, in the context of pure mathematics, our quantities are not fuzzy, approximate, real-world measurements; they are absolutely precise conceptual quantities—exact numbers, in other words. You might guess different values until the cows come home and still never hit the mark.

Another reason why trial and error is insufficient is that even when successful, it simply produces *a* solution, not necessarily all of them or even the ones that may be the most interesting. The goal is complete understanding, and that includes knowing how many solutions a given type of problem tends to have and what might be the exceptions, if any. Finally, who really wants to do all that arithmetic? You could get a computer to do it, I suppose, but even then, you still don't achieve any real cognitive benefit. Again, it's not about solving this or that problem; it's about understanding numbers and their behavior.

In this way, algebra—and mathematics in general—moves increasingly further away from the physical reality that inspired it. Specific problems in physics and geometry yield to vast generalizations, revealing unforeseen connections among disparate subject areas and leading to elegant and far-reaching unifications. Abstraction heaped upon more abstraction! Mathematicians are addicted to altitude. The greatest height reveals the greatest depth, and the feeling is intoxicating.

But even for a pure mathematician like myself, it's not just about breathing the rarified air of abstraction for its own sake. The high-altitude perspective is also the most practical in the long run. Mathematics is the study of pattern in the abstract, so of course it applies to any human activity that involves patterns, from quilt making to music theory to signal processing. And this wide-ranging utility in no way detracts from the beauty and splendor of the mathematics itself. In fact, the more applications a particular result has—both practical and theoretical—the more testament to its power and significance. Just as a profound work of art is known for its many admirers and imitators, the beauty and importance of a mathematical structure can be judged by its wide variety of uses. The system of counting numbers is just the earliest example of this kind of multipurpose abstraction. The thing is, once you have put in the work to climb higher, it is always relatively easy to come back down.

So let's try to get into the scribal mindset, where the goal is to find general methods, not just ad hoc solutions. As a start, you can try your hand at a few of the ancient puzzles I mentioned, or even design your own. How are you going to solve these problems? I don't know, and maybe you don't either. Fantastic! Here's an opportunity for you to be creative and clever, to notice patterns and to delight in them. Maybe you will come up with an idea, maybe you won't. Let's say you fail to solve a problem. Even better! Now you have a mystery and maybe even some curiosity.

Or maybe you don't. Perhaps this airy world of abstract quantities bouncing around in a fictitious realm of pure beauty is not your cup of tea. Maybe you couldn't care less about the scribal arts and the development of systematic methods for the solution of number puzzles. I certainly couldn't blame you. I myself am utterly uninterested in ballet, for instance. I understand that it's beautiful, creative, extremely challenging, and all that. I get that there is a world there, full of richness and culture, history and personality. It's awesome, and I'm glad that it exists. But it's just not my thing. So it's fine if you have no interest in mathematics (it appears that most people don't, in fact). But I am fascinated by it and can't seem to stop thinking and writing about it. You, however, are under no obligation to do so.

On the other hand, maybe I should be more open-minded and learn more about ballet. Maybe I would really like it—probably not enough to put on tights, but maybe enough to gain a richer and fuller appreciation of the art form. So that would be one proposal for how to read this book. Forget about mastering anything or studying deeply; just enjoy learning more about the algebraist's craft, as well as getting a taste of the modern mathematical aesthetic.

In any event, I will assume (if you are still reading) that you are curious to find out what these ancient scribes were up to and that you want to hear me blab about it. I'll try my best to explain the development of these ideas in a way that will satisfy both the casual as well as the more studious reader. (You should feel free to skip over any details that strike you as overly gory.)

Algebra, of course, continues to develop and to expand its purview. Classical algebra, which I will define loosely as the period from roughly 2500 BC to about 1800 AD, is primarily concerned with the study of number puzzles arising from arithmetic—that is, the knots we tie are created using the operations of addition and subtraction, multiplication and division, applied to numbers both known and unknown. Gradually, mathematicians came to recognize that many other sorts of operations and activities are capable of similar treatment. For instance, geometric transformations such as rotation and reflection can be combined to form closed systems that strongly resemble number environments.

Modern algebra is the study of systems of activities and operations in general—numerical, geometrical, physical, whatever. At this level of generality, what matters is not so much the string as it is the ways knots are formed—the various moves and maneuvers and how they combine. A good example is the Rubik's Cube, a closed system of physical turns that combine in a subtle and beautiful way. The Rubik's Cube is a problem in modern algebra and can be solved with techniques that are a direct outgrowth of those developed in the classical period. And, of course, modern algebraists are busy sorting and classifying—investigating the behavior of all manner of algebraic structures and trying to categorize and understand their various interrelationships.

The modern viewpoint in mathematics is abstract and structural, just as it is in music and art. Simplicity, generality, and structural unity are the guiding aesthetic principles. I want to introduce you to this way of looking at mathematics while at the same time taking our subject to be classical algebra. In particular, I have no qualms about injecting modern notions into discussions of ancient thought. This is not a book about history; it's a book about math and art and ideas.

With any luck, it's also a book about me (which no doubt means you'll have to put up with a few of my inevitable rants). Thus, along with the smatterings of history and the lucid explication of ideas, expect a fair amount of abstract blither-blather and romantic mumbo-jumbo. It is important to me to try to get across the way that modern mathematicians view their subject, and I also want to earn your trust as a reader by being as honest (and opinionated) as I can be. Sound good? Then I will leave you with this famous puzzle, the epitaph of Diophantus of Alexandria:

This tomb holds Diophantus. Ah, how great a marvel!
He was a boy for the sixth part of his life,
adding a twelfth part to this, he clothed his cheeks with down;
He entered wedlock after a seventh part,
and five years later the gods granted him a son.
Alas, late-born wretched child; having attained the measure
of half his father's life, chill Fate took the boy.
After consoling his grief for four years, he ended his life.

NUMBERS

Before we start designing knots and attempting to untie them, it might be a good idea to take a closer look at what constitutes our string: *numbers*. This is a fraught subject, it seems. There are few more effective ways to cause anxiety than to shove a bunch of numbers in someone's face and demand that they perform a calculation. Lots of people are terrified of numbers and want nothing to do with them. Numbers can be used to trick and delude; numbers are strange and eerily abstract. Numbers are fundamentally *alien*.

But there are mean, scary aliens and there are friendly, helpful aliens. We won't try to kid ourselves that numbers aren't abstract and otherworldly —they most certainly are. But we can adopt a less bigoted attitude. For me, as a pure mathematician of the most idealistic stripe, it is precisely this abstraction—as well as the sheer simplicity—that draws me to things like numbers and other perfect, Platonic realms, such as the Euclidean plane.

I absolutely adore simple things. To me, *simple is beautiful*. In my romantic and idealistic view, something like seven lemons or seven dollars is dirty and distasteful, corrupted by a physical reality full of messy atoms and obnoxious humans. Seven, on the other hand—that linguistic mental encoding of quantity in the abstract—is pure and pristine, clothed in the spotless raiment of Mathematical Reality. (I guess it's a good thing I went into math instead of poetry!)

So what are numbers, exactly? The notion has evolved a great deal, especially among mathematicians and philosophers. From the practical point of view, the most important developments concern the expansion of our concept of number to include fractional and negative quantities, leading to a more flexible, multipurpose number system that can be used to hold and manipulate numerical information easily and effectively. The utility of arithmetic in everyday life is obvious and evident.

Mathematically, however, the more important step was the move away from the concrete notion of number as quantity and toward the more abstract, modern view of number as *entity*. To an ancient Egyptian scribe, numbers represented actual (or at least potential) *amounts* of something

—bushels of grain or weights of stones, say. Even when dealing with numbers in the abstract (with no mention of objects or units), the concept was fundamentally quantitative. To a modern algebraist such as myself, however, this outlook is much too narrow and confining.

I understand that concreteness is comforting, and I am in no way against it (especially at first), but if the history of mathematics has shown us anything, it is that abstraction equals power and depth. If a mathematical structure carries unnecessary baggage, then it needs to be jettisoned, however difficult that may be conceptually. In the case of numbers, it turns out that requiring them to represent quantities (whether discrete or continuous) is unnecessary. The modern mathematical notion of number is far more abstract and general. Numbers are creatures that engage in behavior, and it is the behavior that matters more than the creatures. It turns out that we can capture the behavioral properties of numbers—for all our algebraic intents and purposes—without having to commit to what they actually are.

What this means is that a modern algebraist does not study numbers in the usual sense but rather *algebraic structures*—systems of entities and operations, together with specific demands on their behavior. This greatly expands the algebra project to include the study of pretty much any way of combining anything, especially activities such as permutations, geometric transformations, and sliding block puzzle maneuvers. In this way, algebra grows to encompass more and more phenomena. New, more abstract notions of negation or multiplication allow for wider scope; the higher we climb the more we see.

Mathematics is sometimes called "the science of necessary consequences." The idea is that we make our various assumptions (for instance, the behavioral demands we place on numbers) and then we deduce the consequences. Whenever an unnecessary assumption can be removed or a new, exciting consequence added, this is progress. The goal of the mathematician is to understand—as simply and as elegantly as possible—which assumptions lead to which conclusions. In order to have the clearest possible view of this complex logical hierarchy, we want our imaginary structures to be as tight and minimalist as they can be, with no excess fat.

Returning to earth, let's think a little bit about counting and measuring in the ancient world. What exactly are we doing when we are counting something? When my daughter was a toddler, she used to count her toys by picking up each of them in turn and saying, "a-nuh" (her word for "another"). Although she did not arrive at a final total this way, she clearly had the idea of associating words and actions with individual items in a collection. Counting is essentially *matching*. In the typical case of a collection of rocks (or coins, or marbles, or whatever), we match each item in turn with a previously established, societally agreed-upon list of number words. At the end of our counting process, we arrive at the word indicating the quantity in question.

The result is that we are now giving names to the *sizes* of collections, independent of the specific nature of their contents. We have this word *three* (the third word on our tribal list), which now has the feel of an abstract noun such as *hope* or *joy*. It is not referring to anything or anybody in particular; this is a general, all-purpose name that we get to apply to a specific situation (e.g., three lemons) or use in its abstract, noncommittal form, as when we do arithmetic. At some point in the history of our species we moved from concrete, real-world quantities to abstract, linguistically encoded quantities. Signs and symbols then followed naturally, once we had the idea of writing.

There is a long history of human number languages and symbolic encoding systems that I will not belabor you with here. (See *Arithmetic* for details.) Mathematically, none of that matters very much. How our creatures *behave* is what's interesting; what names you choose to give them is more of a cultural matter. The number four is neither the word *four* nor the symbol 4. Neither is it *quatre*, *vier*, IV, nor 四. The number four doesn't have a look or a sound; it is an abstract concept. What we can do, if we choose, is encode that abstract idea symbolically and verbally. There are all sorts of ways to do this, and we get to choose the ones that are most convenient for our purposes. For example, tally marks might make more sense for scoring a game than a decimal place-value encoding.

Notice how counting relies on an underlying notion of same and different. Not only is there the issue of making sure you don't count things twice, or even miss a few entirely, but there is also the question of what you even mean by "thing." Suppose I am asked to count a bag of jellybeans and I discover a strange sort of half-sized one, perhaps the result of a manufacturing error. Does this count as a jellybean or not? What exactly is a jellybean?

This is yet another instance of the mucky, nose-wrinkling complexity of physical reality. The replacement of the actual sticky bag and its contents by the ethereal purity of a single number allows us to leave this godforsaken world of grease and grime and enter the simpler, quieter realm of Mathematical Reality, where abstract patterns live in peace and harmony.

Actually, I don't mean to be so pejorative. Both realities—the mathematical and the physical—are exciting and wonderful places full of mystery and breathtaking beauty. And there's lots of fun to be had moving between the two realms as well. But of course I can't hide my aesthetic preferences. I can understand having a fear of the abstract and a desire for concreteness and solid ground beneath one's feet. The challenge for me is to convey the opposite feeling—the desire for complete abstraction and utter removal from the world of everyday life. I guess that's not so hard to get across: it's called *escapism*. My project, then, is to exhibit mathematics as a psychedelic drug. This means we're going to be talking about new ways of seeing and the profound effect such epiphanies have on us.

We have then both the practical realm of approximate real-world physical quantities, as well as the abstract, idealized numbers that we use to represent them. Another version of this dichotomy occurs in the act of measurement. In the ancient world, as well as now, the measurement of continuous, smoothly varying quantities (e.g., liquid, cloth, or tracts of land) was performed by comparison with a culturally determined standard known as a *unit*. Such a unit (e.g., a cup or a yardstick) is then used repeatedly to count off parts of the whole, until the remainder is smaller than the unit. As I'm sure you're well aware, the idea is then to chop up or subdivide our unit into smaller pieces (e.g., ounces or inches) and to continue until the remainder no longer matters. Depending on how accurate our measurements need to be, we may end up requiring half units, or possibly twelfths, or even millionths.

The abstraction that arises from these sorts of practical activities is the notion of a *fraction*. These numbers form a larger, more intricate system of idealized quantities, but the idea is the same: to each approximate, psychologically and culturally determined physical measurement we can assign a perfect, permanent, and exact fraction. Thus, the prosaic "three and a half or so, give or take" inspires the Platonic and romantic ideal of $3\frac{1}{2}$, the number lying exactly halfway between three and four.

An algebraist may prefer to think of this number as $\frac{7}{2}$, meaning the number that when doubled is exactly seven. The value of this description is that it indicates behavior rather than proximity. If you care about the *size* of your number, then $3\frac{1}{2}$ (or 3.5, if you prefer that format) might be convenient for comparison purposes, but this is not really an algebraist's concern. We are more interested in the ways our numbers engage with each other and the patterns of behavior they exhibit.

Counting and measuring thus inspire a new craft: *arithmetic,* the art of combining and rearranging symbolically encoded quantities. Whatever system of notation and nomenclature you choose to employ, the fundamental act of addition—the combining of two or more quantities—had better be reasonably simple to perform symbolically or your system is not going to stand the test of time. Ideally, one would like a symbolic number language to include all numbers whole and fractional, with simple patterns that encode the basic numerical operations such as addition, multiplication, and comparison. Over the centuries, a number of such systems have been developed. None of them is anything to write home about, but they do get the job done.

The system that eventually caught on (and has been in continuous worldwide use for the past century or so) is the vaunted Hindu-Arabic decimal place-value system. It's fine, I suppose. The place-value thing is convenient, but having ten as a grouping size is a bit cumbersome. Anyway, it doesn't really matter. Our questions about numbers are independent of how we choose to speak and write them. Similarly, a research scientist may choose to give pet names to her lab rats, but this in no way affects their behavior.

Even within our conventional Hindu-Arabic notation system, every number has infinitely many names. The number twelve, for instance, can be written as 12 (which is really a kind of shorthand for ten plus two) as well as $5 + 7$, $18 - 6$, 3×4, and so on. The value of being skilled at arithmetic is that we become facile and flexible and can move easily among these different representations. It's like being an interpreter. It's not that one language is the real one and all the others need to be translated into it; rather, each is a valid means of communication, and we want the ability to move smoothly and freely from one to the other without getting confused or having to work too hard.

How many ways can the number 72 be written
as a product of two whole numbers?

Arithmetic can be viewed as a set of basic dance moves. The reason dancers practice and rehearse such maneuvers is so they can combine and arrange them fluidly to help express what they are trying to convey and to make beautiful, nuanced works of art, such as (*ahem*) ballet. Therefore, we will not do arithmetic mindlessly; we will rearrange numerical information when necessary, in order to say something interesting and meaningful.

The modern view is that we have these creatures (to whom we have given pet names like *twelve*) that engage in various activities (such as multiplication) that cause them to become entangled with each other. We discover that each of our entities can be produced in a variety of ways. We find, for example, that six sixes is the same as four nines. There is a subtle and amusing interplay among our creatures, and the patterns created by this interplay are interesting for their own sake, quite independent of any practical utility they may possess. So we study and amuse ourselves with these patterns—and others inspired by them—just as the ancient Egyptian scribes did so many centuries ago. (The Egyptians, of course, had their own way of writing and communicating numerical information and used a completely different set of symbols and cultural conventions. Their symbol for twelve was ∩||, for instance.)

Classical algebra concerns numbers in the usual sense, and so we will need to develop a reasonable familiarity with numbers and their properties in order to create and solve interesting and intriguing algebra puzzles, as well as to gain a deeper understanding of numbers for their own sake.

Of course, if we're going to be thinking and talking about numbers—in particular, if I'm going to be writing and you reading about them—we're going to require some sort of system of nomenclature and notation.

Since we are conversing in the English language (or possibly another of the world's natural languages, if you are reading this in translation), we have already made the unconscious decision to refer to quantities in the ten-centric fashion. We may as well employ the customary Hindu-Arabic decimal place-value system while we're at it.

There is also the social and cultural aspect to consider. It may not matter which system we choose, but there are advantages to all of us agreeing to use the same one. Just as with the metric system or music notation, if we all share the same language it makes everything simpler and easier.

I want to be very clear that although we will be discussing notational conventions from time to time (just as a music teacher must), this is not what math (nor music) is really about. A notation system is merely the messenger, not the message. Though we certainly will adopt what has become the conventional notation system among modern algebraists, I don't want you thinking this is any sort of big deal. We have grammatical rules and conventions in spoken language as well, and occasionally we make mistakes and cause momentary confusion, but we still manage to make ourselves understood. As we go along, if the need to introduce new symbols arises, then we can discuss it at the time. Perhaps I can even inject a little history.

So let's assume that we are reasonably familiar with the Hindu-Arabic decimal place-value system and the usual methods of calculation with whole numbers (otherwise known as pencil-and-paper arithmetic). Truth be told, lack of fluency in arithmetic will not be much of an obstacle. For one thing, algebra is fairly theoretical; we will not require much explicit calculation. And if the need for practical, concrete computation should arise, then we can always grab a calculator.

Which is larger, 14×18 *or* 16×16*?*

Fractional quantities (e.g., two-thirds) are also easy to denote. Here, we imagine our unit not as an indivisible object like a rock or a person, but more as a smooth, easily divided quantity, such as a bolt of cloth or a pail of milk. We can then specify how many equal parts we wish to chop it into, as well as how many of these parts we wish to consider. Thus, the number two-thirds is usually written $\frac{2}{3}$ (or sometimes ⅔), meaning that we wish to break our unit of measurement into three equal parts (i.e., thirds) and take two of them.

The bottom number, known as the *denominator* (Latin for "namer"), thus refers to the number of parts, whereas the top number, the *numerator* (Latin for "counter"), indicates how many of them we want. A symbolic encoding such as $3\frac{5}{7}$ tells us that we have three whole units, plus an additional fractional quantity amounting to five copies of one-seventh of a unit. This is not quite the way that fractions were represented in the ancient world, but it is more or less the same principle: we chop things up and take the pieces we need.

Already we are in fairly abstract territory here, actually. Although fractional quantities frequently appear in the workaday world, it is not often that you find yourself dealing with elevenths or seventy-thirds. We choose our monetary and other measuring units for convenience, after all. Sure, we may require a half inch, or a quarter pound, or even a sixteenth note from time to time, but you don't see too many carpenters measuring out a length of $28\frac{13}{47}$ centimeters. For one thing, a large denominator means extremely high precision, and that is not typical of most practical everyday measurements. In truth, the real-world utility of fractions is rather limited. Mostly it's a lot of chopping things in half.

Of course, there is an enormous difference between the practical and mathematical uses of fractions. In the real world, when we ask the deli guy for a half pound of roast beef, or we measure out 2¼ cups of flour, we are dealing with approximate quantities. Not only is half a cup never exactly half of a cup, but even a cup itself is not a precise, sharply defined quantity. This is simply the nature of measurement in physical reality.

Mathematically, however, our fractions are *perfect*. The number $\frac{1}{2}$ is precisely who it claims to be: the number that when doubled is exactly equal to 1. We do not need to think in terms of real-world subdivisions such as breaking a stick or chopping with a knife. These activities are crude and clumsy, leaving shards and splinters behind, and never truly yielding identical, equal-size pieces. Instead, we can chop up our imaginary idealized numerical entities *linguistically*. A fraction such as $\frac{3}{8}$ can be described as "three copies of the thing that if it were multiplied by 8 would be 1." Perhaps an even simpler and more elegant description would be "the number that when multiplied by 8 is 3."

In fact, this gives us a uniform way to think of all fractions. Even a complicated entity such as $\frac{355}{113}$ is simply telling us what it does for a living: "I'm the guy who gets up in the morning, has a cup of coffee, and then spends my day being the number that when multiplied by 113 makes 355. It's a good, steady job, and the benefits are terrific—especially getting to be an abstract concept with no expenses."

In this way, all of our numbers—whole and fractional, large and small—can be represented in a uniform fashion as a numerator and a denominator. The number 6, for instance, can be thought of as $\frac{6}{1}$ (i.e., the thing that when multiplied by 1 is 6), and $4\frac{1}{2}$ can be rephrased as $\frac{9}{2}$, if we so desire. The reason why we might so desire is that whereas the representation $4\frac{1}{2}$ tells us the *comparison* news (namely, that our number is halfway between 4 and 5), it does so at a slight cost of simplicity, as well as burying the behavioral information. This is simply the number that when doubled makes nine. From an algebraic point of view—that is, in terms of operations and behavior—this is a much more succinct and useful description.

There is one small annoyance with this representation scheme, however. Although every one of our abstract numerical entities now has a uniform name (i.e., a numerator and denominator), this naming system is in fact highly redundant—every number has an infinity of names of this form. For example, the number two-thirds can be written $\frac{2}{3}$ as well as $\frac{4}{6}$, $\frac{6}{9}$, $\frac{8}{12}$, and so on. This is a direct consequence of our choice of meaning. If I am the number that when tripled makes 2, then I am necessarily also the guy that when multiplied by 6 makes 4.

At first, it may seem a bit unwieldy to employ a notation system where any given individual has infinitely many names. But on the other hand, this means we gain flexibility—we get to choose the name that best suits our purposes. The typical example is with comparison: Which is larger, $\frac{8}{3}$ or $\frac{5}{2}$? One approach would be to rewrite these numbers as $2\frac{2}{3}$ and $2\frac{1}{2}$, and then know from experience that $\frac{2}{3}$ is larger than $\frac{1}{2}$. But a more general technique is to rewrite the two numbers so that their denominators match: $\frac{16}{6}$ and $\frac{15}{6}$. Now the numerators are counting the same things (namely, sixths), so it's easy to tell which is larger and by how much.

Is $\frac{1}{2} + \frac{2}{3} + \frac{3}{4}$ larger or smaller than 2?

So just like the Babylonians and Egyptians before us, we now have a symbolic representation system for idealized whole and fractional quantities. If we have chosen well, then our system should be reasonably simple and easy to use, and the patterns inherent in the numbers themselves should be reflected in corresponding symbolic patterns. We want our symbols and representation systems to help us understand the ways that numbers behave, not to get in our way and distract us with hieroglyphic nonsense. There is nothing at all sacred about the choice of ten as a grouping size, nor the crossbar format for fractions. These things evolve and are subject to the whims of history and culture; they are in no way built in or even preferable. The only real question is whether your system is so awkward and overly complicated that you can no longer even tell what you are saying. So long as we can easily read and understand our notation—the simpler the better—not much else really matters.

There is one interesting proviso regarding our fraction notation, and this concerns the unusual number *zero*. The number zero itself can be represented as $0, \frac{0}{1}, \frac{0}{2}, \frac{0}{3}$, and so on. No matter how many pieces you chop your unit into, if you don't take any of them then that's exactly what you've got: nothing. So when zero is in the numerator of a fraction, the denominator is irrelevant.

Zero in the denominator, however, is an entirely different kettle of fish. In fact, it never occurs—for the simple reason that there is no number that when multiplied by zero makes anything other than zero. The symbol $\frac{3}{0}$, for instance, would supposedly indicate a number that when multiplied by 0 is 3, and there simply is no such quantity.

What is your feeling about the symbol $\frac{0}{0}$?
Does it have any useful meaning?

So we find that not every conceivable name necessarily corresponds to an actual, legitimate entity. Just because we have a notational format doesn't mean that every possible instance of it is meaningful. Besides, even in the prosaic practical sense, what would chopping a quantity into zero equal pieces supposedly mean? (Even doing nothing at all still leaves us with one piece!) Other than this very sensible restriction, any number can be used as a denominator. We may chop our units into one, two, three, or however many parts we choose, and then we can take as many such parts as we please—including none of them.

Once you have such a representation system in place, the question then becomes how the basic operations of arithmetic (e.g., addition, multiplication, comparison) appear symbolically. The numbers themselves engage in these activities whether or not we choose to notate them: two plus two is four independent of anyone's names and symbols. The way these various interactions and relationships *appear* to us will then very much depend on the manner in which our numbers have been encoded.

Arithmetic is the skillful craft of using a symbolic representation system to perform computations and comparisons. A fundamental question of arithmetic would be how to take two numbers (encoded symbolically) and determine which is larger, or whether they are equal. Apprentice scribes in any culture would need to be able to do this; the details would depend on the system. The ancient Egyptians were perfectly capable of doing arithmetic using their notation system, just as we are more or less comfortable with ours. There are two very separate issues here, however.

The first is becoming highly proficient at such a skill. There certainly have been times in human history when this has been valuable. Now, however, is not such a time. Of course, I don't mean it's a good idea to be completely clueless and innumerate—that would be difficult socially and commercially, if nothing else. All I mean is that there is not much value in being able to quickly multiply three-digit numbers together in your head or even with a pencil and paper. Everyone has access to a calculator these days, so let's use them and have one less bit of drudgery in our lives. Of course, if you *want* to get good at it, then great! Arithmetic can actually be pretty fun, if you like that sort of thing.

The more important aspect of arithmetic—not only theoretically but also for the attainment of true proficiency and sophistication—is the patterned behavior of numbers in the abstract. What I mean by being skilled at arithmetic is having a firsthand, personal relationship with numbers and their properties and the ability to move information around smoothly and easily according to these patterns.

Thus, in addition to referring to specific numbers, we are also going to want to talk about number patterns in general. For instance, one of the first discoveries ever made about counting is the fact that multiplication is symmetrical: three copies of five is the same as five copies of three. One nice way to see this is to imagine a 3×5 rectangular array of rocks. Viewed in one way, it is three rows of five; turn your head and it is five rows of three.

Suppose you wanted to convey the general pattern. One way, of course, is to spell it out in words: "multiplication is symmetrical," or "whenever you multiply two numbers together, the result is independent of their order." This is, in fact, the way a great deal of mathematics was conducted and conveyed over the centuries.

Another reasonable approach would be to provide a set of concrete instances illustrating the general principle:

$$3 \times 5 = 5 \times 3$$
$$7 \times 8 = 8 \times 7$$
$$4 \times 14 = 14 \times 4$$

The idea being that one is supposed to then say to oneself, "Oh, I see what's being said here; the order doesn't matter." This was the approach taken by the Babylonian and Egyptian scribes and their students. Ancient problem tablets and papyri are essentially long lists of similar puzzles illustrating a particular method of solution. This method would then be inculcated by repetition, as well as explicit instruction.

The modern approach is to take the best of both schemes. We will describe the general pattern succinctly and precisely in symbols, avoiding long and complicated (and possibly ambiguous) sentences and paragraphs. In place of the repetition of specific numerical values, we will adopt a more flexible, abstract notation that captures the idea of "any old number, no one in particular."

For example, one way to express the symmetry of multiplication is to write something like

$$\square \times \triangle = \triangle \times \square.$$

Here, the box and the triangle act as placeholders for any two numbers whatsoever. So we aren't wasting our time making lists of examples and urging our readers to "get the pattern," and we also aren't crafting some convoluted and torturous phrasing in our native tongue. We have a simple and concise way of describing numerical behavior without having to name names.

Depending on what we are doing, our numbers may have different levels of concreteness. Some numbers may be known, others unknown. Some may be general and noncommittal, others specific and particular. We need our system of notation to be able to deal with the varying status of our information.

Of course, this pertains to any linguistic framework. This is why we have pronouns and indefinite articles and all the rest—so as to indicate the difference between *this* thing, *some*thing, and *any*thing. We will find ourselves wanting to say things like "there is a number that does this" as well as "every number does that." The use of generic names like □ and △ is one way to get these sorts of sentiments across concisely.

So a modern mathematician will often write something like "for any numbers a and b, we have $a \times b = b \times a$." Here I have named my general, nonspecific numbers a and b for simplicity, but any two (not utterly confusing) names would be fine, including Harold and Barbara, 中 and Я, or whatever you like. The point is to be generic, so it literally doesn't matter. The only thing that does matter is that these names be meaningless —if they already carry too much cognitive baggage one might get easily confused. (If you want to name your generic number *o*, be my guest, but don't blame me if something goes haywire!)

The symmetry of multiplication is a beautiful and important feature of our arithmetic system. But let's understand that it is a feature of the *numbers themselves* (along with the multiplication operation itself). This is not a statement about mere symbols but rather about the idealized quantities to which they refer. The symmetry of the symbols reflects the symmetry of the behavior.

Addition, of course, is also symmetrical: we have $a + b = b + a$ for any numbers a and b. Not that this is in any way surprising, given that pushing two collections together is a symmetrical activity, but it is an attractive feature nonetheless.

I suppose I've been a little sloppy in assuming that these various symbols are all well known to you, but I think they probably are. The plus sign (+) signifies addition. Like the ampersand (&), it began as a simple abbreviation for the Latin word *et*, meaning "and." Its use in arithmetic came later. In the Renaissance period, for instance, it was still quite common to write *plus* (Latin for "more") or to abbreviate using *p*. The times sign (×) is often used for multiplication as well as to denote rectangular dimensions, such as an 8½ × 11 sheet of paper.

The equals sign (=) is another nice symbol. The two identical sticks say *equal* in such a simple and elegant way. It's really just efficient, well-designed sign language—like you sometimes see at international airports. Anyway, its meaning is the important thing. An *equation* (that is, a statement of equality) is simply an assertion that two things are identical—exactly one and the same object. This means that any statement made about one is also a statement about the other.

In particular, the equals sign has no directionality; it doesn't move information from left to right or the other way. It simply encodes the fact (or the question or the hope) that two expressions are merely different names for the same thing. Thus, $5 + 7$ is not a problem or a question, and 12 is not its answer. It is simply that $5 + 7 = 12$, meaning that these two names refer to the exact same entity. There is no mathematical context in which 12 appears where it cannot be replaced by its alternate name $5 + 7$. Because we are really talking about a numerical entity in and of itself, any distinctions among its various names are beside the point. (I know I'm sounding like a broken record about this.)

An algebraic equation is thus a statement asserting that one result of entangling a set of numbers is exactly the same as another. Depending on context, this might be a general pattern, or we may be seeking specific numbers that exhibit this behavior. The statement $a + b = b + a$ is a general assertion valid for all numbers, whereas the equation $3 \times w = w + 1$ is expressing a very peculiar property of an as yet unknown number w.

Can you figure out who w must be?

We will often write our pattern information in the form of algebraic equations, and context will (hopefully) make clear what is meant by our symbols and the scope of their generality.

One convenient innovation is to dispense with the times sign altogether and to signify multiplication by simple juxtaposition. Thus, in place of $a \times b$ we can simply write ab (or sometimes $a \cdot b$, if more typographical clarity is needed). This pretty much forces us to choose our generic names to be single characters, or things will get too confusing. The convention is to choose single letters from the Roman and Greek alphabets, avoiding conspicuously ambiguous letters like *o*. But any reasonable set of symbols will do. The symmetry of multiplication can then be expressed by the generic equation $ab = ba$. However, I would not recommend this scheme for explicit products like 13×5, lest we find ourselves writing $135 = 513$. Notational abbreviations and simplifications are great—right up to the point where they fail to be conveniences and start causing trouble.

The usual way to avoid such readability problems is to use grouping symbols like parentheses and brackets. This can be a good way to clarify exactly what you mean. If, for example, we wish to take a generic number x, add three to it, multiply that result by seven, and then subtract two, we could indicate that precise sequence of operations by writing $[(x + 3) \times 7] - 2$. Had we simply written $x + 3 \times 7 - 2$, there's no telling what sort of confusion might arise. (One possible misinterpretation might be $(x + 3) \times (7 - 2)$, for example.) Simplicity and economy of notation are a definite plus, and we should always strive for brevity and typographical elegance, but the paramount consideration is always clarity. Ambiguous, confusing notation is not going to help us discover or invent anything.

The convention that has become standard practice is to use parentheses (and other such delimiters) when necessary to avoid confusion but otherwise to use as few symbols as possible. The elimination of the times sign is one example. (Besides, it looks too much like the letter x.) Another is to prioritize multiplication over addition. This convention is mostly due to the relative simplicity of addition—we don't usually mind a set of parts being assembled with plus signs; that's always easy to read.

Thus, in our previous example, $[(x + 3) \times 7] - 2$, we can remove the times sign and the brackets and simply write $7(x + 3) - 2$. Of course, we need to understand that the 7 is being multiplied by the $x + 3$ and *then* the 2 is subtracted. So there is a (very minor) literacy issue involved in navigating such notational conventions, just as there is when learning music or Spanish grammar. Thankfully, algebraic notation is far simpler than most of the symbolic languages that one learns to use in daily life.

Notice that I have chosen to write $7(x + 3)$ instead of $(x + 3)7$, even though both are perfectly understandable and (of course) exactly equal. This is more of a stylistic choice; I tend to prefer putting known quantities before unknowns, for instance. After a while, you may find yourself starting to prefer certain formats and typographical conventions as well. My advice would be to write whatever you like but also to consider any potential ambiguities and misinterpretations. Feel free to make liberal use of parentheses, especially at first. If we're going to be tying and untying knots, we need to be able to specify which loop is going around which and what strand is passing through what, exactly when. So our language must be precise enough to describe our moves in complete detail, without ambiguity. Parentheses essentially control the timing, indicating which operations need to be performed before which.

Any complex maneuver, such as a ballet movement or a knitting pattern, typically consists of dozens of basic steps, combined in sequence to create a rich and expressive statement that is greater than its individual parts. A knot may consist of nothing more than a sequence of tucks and overlaps—each alone not much to speak of—but the convoluted manner in which they are combined can create tangles of unlimited subtlety and depth.

So we want our notation system to reflect the hierarchical nature of compound operations. We want to be able to see at a glance what is being done to whom and when. This is as true for knots in string as it is for knots in numbers. If I can't tell what move comes when, it will be hard for me to know what knot I'm really talking about.

Suppose we have two simple numerical operations—adding 1 and doubling, say. If we do the adding first and then the doubling, a given number y would become $2(y + 1)$, whereas if we double first we get $(2y) + 1$. Here, the minimalist convention would then suggest $2y + 1$, the idea being that since we are choosing to favor multiplication (meaning that all things being equal, we perform multiplications first), the parentheses become redundant.

Putting matters of notation aside for the moment, let's just pause for a second to appreciate how weird this whole mathematics business is. It's one thing to fantasize and to create fictional characters—I can imagine a purple unicorn who loves to eat silver tomatoes, for example. But these sorts of creatures do not really possess any definite features or behavior. Does my unicorn enjoy playing tennis? Sure, I guess . . . Or maybe not? Whatever. We can make up such things as we go along. There are no necessary consequences to any of our decisions because such fictive realms are not required to satisfy the demands of logic. Stories are dreamscapes and are subject to an entirely different set of aesthetic criteria. The literary aesthetic demands plot twists and characters that develop and grow—in short, *complexity*. The mathematical aesthetic, on the other hand, demands simplicity above all else.

What is funny about our mathematical creatures—such as numbers or perfect circles—is that they are both fictional as well as logical; that is, they do not (and cannot) exist in the real world, yet they have definite properties and possess eternal truths that are beyond our control. Mathematical entities are every bit as fantastical and unreal as my purple unicorn but not so arbitrary and random. Mathematical structures are at the same time more functional and utilitarian—in that we create them to hold pattern information—yet far more abstract and imaginative. After all, a purple unicorn is not really all that farfetched; it's basically a horse with a couple of minor adjustments. Mathematical abstractions such as infinite-dimensional spaces and non-Euclidean geometries are in an entirely different league of imagination and creativity.

We thus find ourselves confronted with an amazing fictional universe full of abstract patterns and pattern-holding structures, and although none of it is even the slightest bit real—in the sense of being part of the physical universe—it nonetheless possesses definite, knowable properties and timeless, eternal truths. Mathematicians are actively engaged in the creation, exploration, and study of this realm of rational fiction. And to think that people have been amusing themselves with such investigations for so many thousands of years! I suppose the need to escape is as old as civilization itself.

In any event, we have these imaginary entities known as numbers (both whole and fractional), and we also have a reasonably convenient way to notate them. Now we need to think about the various activities we can perform on them. The simplest is probably addition: putting two or more quantities together. For whole numbers written in the Hindu-Arabic fashion, there are familiar procedures (e.g., carrying) for producing the total, also expressed in this form. (There is also the + button on your calculator.)

Adding fractions presents a more interesting challenge. Given two fractions expressed in the numerator/denominator format, how do we produce their total? For example, what fraction represents the total of $\frac{3}{7}$ and $\frac{2}{5}$? Here we have a problem in linguistics. One of our creatures is described as the thing that when multiplied by seven is three, and the other is the thing that when quintupled is two. Put them together and what do we get: The thing that when multiplied by what is what? Another way to look at this is that we are using different measuring units—a seventh of a cup and a fifth of a cup, say—and we are taking three scoops of one and two of the other. So we're adding apples and oranges (that is, sevenths and fifths) and finding it difficult to perceive and understand.

Once again, the clever idea is to use the fact that each of our numbers has a variety of possible names. Specifically, we can take any given representation, such as $\frac{3}{7}$, and scale both the top and bottom by the same factor without changing the actual value of the fraction in any way. Thus, we can write the number $\frac{3}{7}$ as $\frac{6}{14}, \frac{9}{21}, \frac{12}{28}$, and so on. Similarly, the number $\frac{2}{5}$ can also be written as $\frac{4}{10}, \frac{6}{15}, \frac{8}{20}$, et cetera. The trick is to find alternate names for your numbers so that they *do* speak the same language—in other words, rescale everything so that the denominators match. Then it will be easy to add (as well as subtract and compare) them.

Some people like to be extra clever and find the lowest common denominator of two fractions (in other words, the smallest denominator that works for both), but I find this rarely to be of much value. If we like our denominators to be as small as possible, we can always adjust them later, after our computations are completed. The main thing is to understand the pattern, and in this case, there is a much simpler, more general approach: let each fraction be rescaled by the denominator of the other. To add $\frac{3}{7}$ and $\frac{2}{5}$, we rescale $\frac{3}{7}$ by a factor of 5 to get $\frac{15}{35}$, and we scale $\frac{2}{5}$ by 7 to get $\frac{14}{35}$. Now both names share the denominator 35, which is simply the product of our two original denominators. Their total is now easy to obtain: $\frac{15}{35} + \frac{14}{35} = \frac{29}{35}$; in other words, the number that when multiplied by 35 is 29.

Can you represent the number $\frac{1}{2} + \frac{1}{3} + \frac{1}{5} - 1$
as a single fraction?

Now let's see what this pattern looks like in the abstract. Instead of explicit, particular fractions, let's use generic fractions $\frac{a}{b}$ and $\frac{c}{d}$. Here, my symbols a, b, c, and d refer to arbitrary whole numbers (the denominators b and d presumably being nonzero). I'm not committing to their values because I do not want to commit. I want to understand how *all* numbers behave, not just $\frac{2}{5}$ and $\frac{3}{7}$. So I keep my options open by giving my symbols *roles* rather than values. For instance, the symbol a refers to the numerator of my first fraction—regardless of what actor you might later get to play the part.

Now, using our clever rescaling method, we see that $\frac{a}{b}$ can be rewritten as $\frac{ad}{bd}$ (multiplying both top and bottom by d) and $\frac{c}{d}$ can be given the stage name $\frac{bc}{bd}$. So in general, we see that

$$\begin{aligned}\frac{a}{b} + \frac{c}{d} &= \frac{ad}{bd} + \frac{bc}{bd} \\ &= \frac{ad+bc}{bd}.\end{aligned}$$

Here is an example of a numerical pattern succinctly expressed in the modern algebraic idiom. We can, of course, convey the same idea in words: "The sum of two fractions is again a fraction, its denominator the product of the two original denominators and its numerator the sum of the products of each numerator with the denominator of the other." I guess I don't really have a problem with this in terms of comprehension, though such descriptions do tend to get rather lengthy and wearisome. Not only is the algebraic code more concise, but it's also actually easier to read and understand, once the symbolic language becomes reasonably familiar.

This same rescaling technique can be used for subtraction and comparison as well. In particular, we have the general pattern

$$\frac{a}{b} - \frac{c}{d} = \frac{ad-bc}{bd}.$$

Thus, we find that the addition and subtraction of fractions can be reduced to ordinary whole number arithmetic.

Which is larger, $\frac{7}{9}+\frac{1}{8}$ *or* $\frac{4}{3}-\frac{5}{12}$*?*

Something odd happens when we express our ideas symbolically. Once we start writing things down—algebraic formulae and equations, for instance—the very symbols themselves begin to take on a certain tactile and kinetic quality. Symbol choices may be a minor issue in the grand scheme of things, but once you do select a notation, your primate brain will then go to work, and your symbols will begin to connote other things by association. Perhaps you already feel this way in regard to digit symbols and letters of the alphabet. (I'm always worried that the letter P is about to fall over, being so top-heavy.)

So let's look at our fraction addition pattern again:

$$\frac{a}{b}+\frac{c}{d}=\frac{ad+bc}{bd}.$$

When I read this, I can't help but feel my left thumb and forefinger wanting to squeeze the a and the d on the left side together to form the product ad in the numerator on the right. So I imagine this little finger dance in which I snap the a and d together, snap the b and c together, and throw them both into the top to be added. Then I grab the bottom numbers b and d and squeeze them into bd and shove it down below.

Perhaps this strikes you as juvenile or insane—I don't know. It's pretty unconscious; I think it's simply the way humans tend to remember things. The point is that this pattern matters to me (as all patterns do), so I want to understand it, remember it, and incorporate it into my arsenal of techniques. If I'm exploring in the jungle and I manage to build a raft out of logs and vines, then I'm certainly going to want to remember exactly what I did and in what order.

As we discover more and more such patterns (and invent increasingly clever ways to use them), the value of the symbolic approach over the more literary descriptions becomes evident. It is much easier to remember a sequence of arithmetic operations encoded symbolically—with whatever strange psychological and kinesthetic associations that may entail—than to remember a long mantra involving numerators and multiplication.

On a more practical note, these patterns allow us (or our calculators) to add and subtract fractions simply and easily, using only addition and multiplication of whole numbers. We see that $\frac{19}{3} + \frac{7}{2} = \frac{59}{6}$, and $\frac{19}{3} - \frac{7}{2} = \frac{17}{6}$, for example.

Multiplication of fractions presents us with a somewhat different issue: What do we want $\frac{3}{7} \times \frac{4}{5}$ to mean, exactly? For whole numbers such as 5 and 8, we already know what we want multiplication to mean: 5×8 means 5 copies of 8. But making copies is an activity, and although I am perfectly capable of performing an activity once, twice, or five times, I simply do not know how to repeat an action $\frac{3}{7}$ times. So we must decide on a meaning for multiplication of fractions if we want them to engage in such a practice.

On the other hand, our experience with practical reality suggests that since we can put $4\frac{1}{2}$ muffins on a plate, we should be able to speak of $4\frac{1}{2}$ sixes or $4\frac{1}{2}$ sevenths, or $4\frac{1}{2}$ anything. In other words, the idea is to interpret multiplication as a *proportion*. To multiply by $\frac{1}{2}$ is to take half. The result is that multiplication becomes a kind of enlarger or copy machine. We can double a number by multiplying it by 2, and we can cut it in half by multiplying by $\frac{1}{2}$.

How can we write $4\frac{1}{2}$ sevenths as a single fraction?

Multiplication by a number like $\frac{2}{3}$ can then be viewed as a doubling followed by a shrinking down by a factor of 3, or, if you prefer, you can first shrink down by 3 and then blow up by 2. Either way, we end up with a quantity that is two-thirds of what we started with. Now it is easy to see the general pattern: to multiply by $\frac{a}{b}$ is to blow up by a factor of a and shrink down by a factor of b.

One way to think of a fraction like $\frac{a}{b}$ is to imagine that we start with our unit (i.e., the number 1), then put it through the copier, enlarging it by a factor of a and then shrinking it down by a factor of b, to arrive at the fraction $\frac{a}{b}$. To multiply this by $\frac{c}{d}$ is to enlarge it again by c and then shrink by d. So, all told, our unit was placed on the copier, and two blowups (by factors of a and c) occurred, as well as two shrink-downs (by b and d).

Thankfully, since order doesn't matter, we see that the result must be the fraction $\frac{ac}{bd}$—that is, the general multiplication pattern for fractions is to simply multiply their numerators and denominators separately:

$$\frac{a}{b} \times \frac{c}{d} = \frac{ac}{bd}.$$

Are your thumbs and forefingers feeling anything? Snap-snap, now that's a pattern I can remember! It's also one I can readily understand —not merely as a rule or a formula but as a necessary consequence of a natural choice of meaning.

Notice that to multiply two fractions, there is no need to rescale them to a common denominator—we just multiply the tops and bottoms, that's all. Why does multiplication have such a simpler pattern than addition and subtraction? The reason is that fractions are multiplicative by nature. The fraction $\frac{a}{b}$ literally means "the number that when multiplied by b is a." Another way to say it is that $\frac{a}{b}$ is a divided by b.

Just as subtraction can be viewed as the opposite of addition (i.e., subtracting 3 undoes the act of adding 3), we can also think of a division into 7 parts as being the opposite of multiplication by 7. Thus, $a - b$ refers to "the thing that when added to b is a" in the same way that $a \div b$ (or $\frac{a}{b}$) means "the thing that when multiplied by b is a." Once again I have accidentally introduced symbols without explanation: the minus sign (–) is used in place of *minus* (Latin for "less"), and the division symbol (÷) is simply the fraction crossbar with dots indicating where the numbers go.

The point being that $\frac{a}{b}$ can be viewed as $a \div b$ if we wish. Whether you think of $\frac{3}{8}$ as three of those things called eighths, or as the result of dividing 3 into 8 equal parts, either way you've got the thing that when multiplied by 8 is 3. So every fraction is an implicit statement about multiplication, which is why the multiplication pattern is so simple and direct.

Division of fractions is also rather pretty. Since multiplication by a fraction $\frac{c}{d}$ performs a blowup by a factor of c and a shrink-down by d, the way to undo this would be to blow up by d and shrink down by c—in other words, to multiply by $\frac{d}{c}$. Thus, every division is actually a multiplication, only by an inverted, *reciprocal* version of the divisor: to divide by 2 is to multiply by $\frac{1}{2}$, and to divide by $\frac{3}{7}$ is to multiply by $\frac{7}{3}$. Thus, we have the general pattern

$$\frac{a}{b} \div \frac{c}{d} = \frac{a}{b} \cdot \frac{d}{c} = \frac{ad}{bc}.$$

Notice, by the way, that we could also choose to write this division in the form

$$\frac{a/b}{c/d}.$$

That is, we can express such a ratio as a fraction both of whose numerator and denominator are themselves fractions! Of course, we can always rescale this (rather ugly) compound fraction by bd to obtain $\frac{ad}{bc}$ as before. The point being that representations such as $\frac{7}{3}/\frac{8}{5}$ are perfectly sensible and can always be converted into whole number fractions (in this case, $\frac{35}{24}$), if desired.

Which is larger, $\frac{3}{4} \div \frac{2}{3}$ *or* $\frac{2}{3} \div \frac{3}{5}$*?*

The upshot is that we now have a symbolic arithmetic system capable of representing all of our creatures (namely, fractions) together with the four basic operations of addition, subtraction, multiplication, and division. The Babylonians, Egyptians, and classical Greeks all had similar systems as well. They used their notation to encode ideas, pose and solve problems, and invent methods—in short, to do algebra—and were every bit as clever and creative as we are. They were not, however, as comfortable with abstraction. The history of mathematics is essentially the history of abstract thinking. The more time that goes by, the higher we climb and the wider our perspective becomes.

A modern algebraist works in a considerably more abstract environment than that of an ancient Egyptian scribe. Not that the scribes weren't doing honest-to-goodness mathematics—of course they were. Their abstract quantities are our abstract quantities; their interest in numbers and number puzzles is the same as our own. The difference is that their mental models were firmly rooted in practical reality, whereas modern mathematicians have found it both simpler (from the logical and aesthetic points of view), as well as more powerful and effective (in the sense of allowing us to see farther and to understand more deeply), to detach the objects of our study from physical reality entirely.

Of course, we can't help grasping onto concrete mental images, but we also understand fully (and often painfully) that these mental and physical models are not the actual objects under consideration. Three lemons is a representation of three, but it is not three. A drawing of a circle is not a circle. Try as we might, we cannot produce a truly accurate mental or physical model of a perfect mathematical circle. On the other hand, we can certainly hold the *idea* of a circle quite precisely using language.

So mathematics is essentially a language game, in that our objects of study are ultimately defined linguistically. However, it's not an arbitrary, random language game because the objects we are encoding via language actually mean something to us—namely, they are structures we design for the express purpose of holding abstract pattern information.

IDEALIZATION

One way to think about the mathematics of quantity is that we are making an *improvement* of physical reality by abstracting our ideas about counting and measuring to a nicer, more ideal realm. We humans are always imagining utopias: better versions of reality that do not actually exist. (The word *utopia* is Greek for "no place.") Mathematical Reality is the ultimate example—the ideal environment for impractical dreamers who wish things were simpler and more beautiful than they really are. I've always used math as a refuge from the often ugly and depressing world of ordinary reality, just as painters and poets use their alternate realities to soothe their souls and to express their feelings and desires.

My desire is to *understand*. I want to explore the imaginary realm of mathematics and see all of its beauties and learn all of its secrets. I want to climb the highest mountains and see the farthest horizons that I am able. I want to understand both the grandest possible vistas as well as the tiniest, most intimate details. This is fascinating terrain indeed, teeming with abstract imaginary life.

Numbers, in particular, are out there doing stuff. What are they up to? We've been investigating these creatures for more than five thousand years, and we still barely know anything about them. Even though (in a sense) we invented them, we are powerless to control them. It's a lot like Frankenstein, actually. We have the ability to dream up and create these entities, and even to endow them with operations and behavior, but once these are specified and our number system is up and running, then our little monsters will do whatever it is they do, whether we like it or not. Their properties are dictated by logic alone, not by us.

It is a strange feeling to create something—a child, an invention, or a movie franchise—and then have it disobey, go haywire, or escape from your control. I'm sure many artists have felt that way at times about their paintings or their hit singles. "What have I done? Why did I bring this monstrosity into the world?" On the other hand, how fascinating and wonderful when something we've made takes on a life of its own.

This is certainly the case with numbers, no doubt. How long did it take, once the idea of representing quantities with rocks was invented, before we discovered prime numbers? Arranging rocks in rows and columns is fairly natural; at some point somebody is bound to notice that certain numbers (e.g., 5, 13, 41) cannot be laid out in nice rectangles. These are the prime numbers (or *primes*, for short). Now we have a mystery that we cannot control. We didn't ask for these weirdos; they just *appeared*. The primes are an unexpected consequence of our concept of quantity, a side effect of our linguistic and aesthetic decisions.

Any intelligent beings that play around with collections and counting will eventually make this same discovery. The prime numbers are a permanent feature of the system of counting numbers, and their properties are already etched in stone. Every well-defined mathematical structure already knows its own truths; the question is whether *we* will ever know them. At the present moment we have hundreds of unsolved problems and unanswered questions regarding arithmetic, and many of these concern prime numbers.

So yes, we have the freedom to create and invent new structures in mathematics and to demand various features and properties of our creations in order to define them and to fix their meaning, but after that we lose control and must accept whatever consequences our choices entail. It's not so much that there are rules so much as there are creative decisions that either do or do not please us as mathematical artists. If you can create something lively and interesting—or even (Lord help us) *useful*—then your work will be appreciated and valued. If you create an abstract structure of some kind—however logically coherent it may be—that is boring and does not reveal nor enlighten nor connect to anything interesting and meaningful, then it will fail as a work of mathematical art. Numbers—even the ordinary counting numbers—constitute a timeless mathematical masterpiece of profound depth and understated elegance. (One of the chief benefits of being a mathematician is getting to rub shoulders with such intensely beautiful creatures.)

To an ancient Egyptian, numbers were really just real-world quantities dressed in abstract clothing. Though it was understood that one could work with the numbers 3 and 5 independently of whatever was being counted, there was nothing like the modern conception of numbers as imaginary entities engaged in patterned behavior. Numbers in the ancient world, though abstracted and idealized, were still concrete in the sense of representing quantity and being imagined and visualized in quantitative as opposed to purely behavioral terms.

Negative numbers, for instance, held a very different status in the early days of arithmetic. The idea most likely comes from the world of commerce, where numbers are used to record transactions—both profits and losses. At some point someone must have noticed that you can combine both income and expenditures in the same column, as long as you indicate when you have gone "below zero" and are in debt. Negative numbers are a clever bookkeeping trick that keeps all the arithmetic consistent so that no matter in what order the profits and losses are tallied, the net result is the same.

So we can write things like –3 to indicate that we owe three dollars, or that it's three below zero on the thermometer, but that's really just a *marked* 3—it's still three of something, just an inverted something. It's a much greater leap of imagination to accept –3 as a legitimate number in its own right.

From the point of view of counting collections, of course, negative quantities make no sense—there simply is no pile of rocks that when we add three to it becomes empty. No wonder the ancients were in such a muddle about negative numbers! If numbers must be quantities—prosaic or idealized—then the notion is clearly quite meaningless, whatever the accountants might say.

There are other possible interpretations, however. Let's consider our system of counting numbers: 0, 1, 2, 3, and so on. These entities are also known as *natural numbers*, meaning the numbers immediately available to our intuition. The system of natural numbers is usually denoted by a stylized capital $\mathbb{N}$ (naturally).

One way to get a feel for the additive structure of $\mathbb{N}$ is to imagine our numbers as an endless row of dots:

$$\begin{array}{cccccccccccc} \bullet & \bullet & \bullet & \bullet & \bullet & \bullet & \bullet & \bullet & \bullet & \bullet & \cdots \\ 0 & 1 & 2 & 3 & 4 & 5 & 6 & 7 & 8 & 9 & \end{array}$$

This sort of structure is often referred to as a *number line*. One nice feature of this imagery is that it provides us with an alternative view of addition: adding a number is the same as sliding (or hopping) that many places to the right. Thus, if we start at the dot labeled 3 and move five places over, we end up at the dot marked 8, illustrating the fact that $3 + 5 = 8$.

If we like, we can even go a step further and reconceive what a natural number is. Instead of 5 being the number of fingers on one hand, we can think of it as being the act of shifting every dot five places over. In other words, we can reinterpret our numbers as *actions* rather than quantities. In this way, the operation of addition simply becomes the act of combining two motions together: the equation $3 + 5 = 8$ expresses the fact that moving three places over and then moving an additional five is the same as sliding by eight. The point being that we have multiple ways to think about numbers—what they are and what they do—and that provides us with a certain amount of mental flexibility.

What move does the number 0 *correspond to?*

One immediate consequence is that we get an entirely new window on subtraction. If adding 3 is shifting three places to the right, then subtracting 3 must be the opposite: a shift of three places to the left. Now we have a very clear and striking visual illustration of what goes wrong with subtraction: whereas any dot can move any number of spaces to the right, this is not the case in the other direction. The trouble is, we run out of dots.

Here is our opportunity to be mathematical artists. We feel an obstacle—an unpleasant asymmetry in our system—and our aesthetic of simplicity and structural elegance is nagging us to do something about it. So we create a piece of mathematical art and improve our system by extension—in other words, we add more dots!

. . . • • • • • • • • • . . .

Now our line of dots goes on forever in both directions, making it more symmetrical and hence *simpler*. Having an endpoint is a complication, an unnecessary and unattractive obstruction to our desires. Now we have no endpoints to worry about, and every dot feels the same as every other. As before, the number 3 still means shifting three places to the right, but now we can also shift three to the left, and if we combine the two moves (thus adding them) we return to where we started, as if we had done nothing.

In this way, we can bring −3 into existence very naturally. It's not some spooky pile of ghost rocks or a half-baked mystical notion of potential being, nor is it an indication of debt or freezing temperatures. It's simply a shift in the opposite direction and is therefore every bit as natural and deserving of attention as its right-handed cousin. This means we can expand our number line image as well:

. . . • • • • • • • • • . . .

−4 −3 −2 −1 0 1 2 3 4

To add and subtract is simply to move forward (i.e., to the right) or backward as you please. So this mental model helps us to see what ℕ really wanted: to be improved and completed and made more symmetrical, thereby allowing it to fulfill its role as a pattern-carrying structure more fully.

It is not hard to extend our notion of multiplication to these new numbers as well. Since multiplication by 2 is the same as adding a number to itself, we see that $2 \times (-1) = -2$. So doubling has the effect of blowing up, or magnifying, our number line so that the spacing is twice as wide.

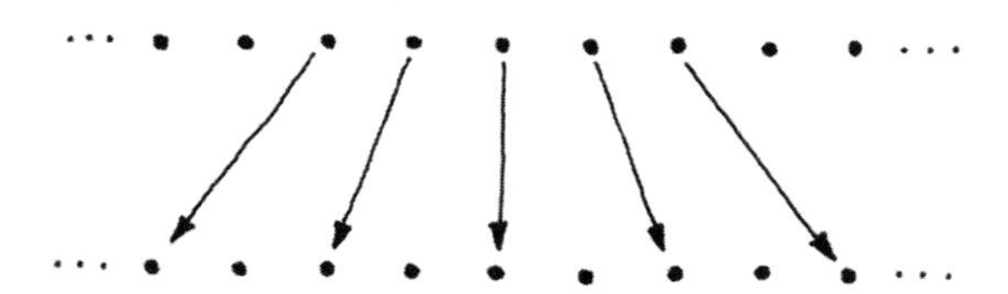

The point 0 stays where it is, while the point 1 moves to position 2, the point −1 moves to −2, the points 2 and −2 move to positions 4 and −4, and so on. The same goes for multiplication by 3 or any other positive whole number: we are simply *stretching* the line, pulling each point away from 0.

But what about multiplication by a negative number? What, if anything, do we want multiplication by −1 to mean? It certainly doesn't mean making that many copies! So we get to *choose* our meaning, and the usual idea is to choose it so that all of our previous patterns remain intact. In particular, we would like our multiplication operation to be symmetrical, if at all possible.

This suggests that we define (–1) × 2 to be –2 so that it agrees with 2 × (–1). In other words, the demand for symmetry forces us to take multiplication by –1 to mean *negation*—the act of exchanging a number with its negative. This clearly has the visual effect of *flipping* (i.e., reversing) our number line, sending each point to its reflection on the other side of 0.

If we like, we can even incorporate this into our copy machine metaphor: multiplication by a negative number both enlarges and reverses, just as many real-life copiers are built to do.

The upshot is that all of the various relationships and truths of whole number arithmetic now have visual, geometric realizations. For example, multiplication by –3 would perform a stretch-flip, so it is easy to see that (–3) × (–4) = 12, for instance. (See *Arithmetic* for more details.)

This new, improved number system (including both positive and negative whole numbers) is known as the *integers*. (The word *integer* is Latin for "whole.") The system of integers is usually denoted ℤ (short for *Zahl*, the German word for "number"). So the move from ℕ to ℤ is really an aesthetic decision, albeit one with a great deal of practical utility as well. So it's a win-win situation all around, I guess.

As an algebraist, the way I feel about it is that ℕ is sort of pent up and frustrated, and ℤ is its way of breaking loose and breathing free. Negative numbers are ℕ's way of coming out of the closet and proudly announcing to the world, "I want to freely add and subtract with no limitations!" The row of dots image is a nice way for us to come to grips with this unfamiliar (and possibly unsettling) behavior. Numbers less than zero? Yes, indeed! It's what ℕ wants, and I, for one, am happy to give her whatever she needs—no matter how hard I may have to work to adapt and conform my intuition.

Although we can add and subtract in ℤ without restriction, the multiplicative structure of the integers is in fact quite intricate and deep, pertaining as it does to divisibility and primes and so forth. Algebraically, we are in the same position as before. We have certain actions—such as multiplication by 6—that cannot necessarily be undone. Again it turns out that we are missing some numbers and will need to expand our system. In this case, it means throwing in all of the fractions, as well as their negatives, so that we have an improved environment in which we can add and multiply, as well as unadd and unmultiply, with complete freedom (or as much freedom as logic will allow, at any rate.)

Thus we arrive at the system of positive and negative fractions, otherwise known as the *rational numbers* and denoted $\mathbb{Q}$. (The word *rational* pertains to the fact that each of our numbers can be represented as a whole number ratio; the symbol $\mathbb{Q}$ stands for "quotient.") Thus, $\mathbb{Q}$ contains not only all of the whole numbers such as 4 and –3 but also fractions like $\frac{8}{5}$ and $-\frac{3}{4}$.

From an algebraist's point of view, the progression from $\mathbb{N}$ to $\mathbb{Z}$ to $\mathbb{Q}$ is one of continual improvement and expansion, enlarging our realm of numbers so as to allow our operations to act as freely as possible. Since we are planning to tie and untie algebraic knots, we want our string to be as flexible as it can be. Working with $\mathbb{N}$ or $\mathbb{Z}$ is like trying to tie knots in dry spaghetti—it's just too stiff!

This is not to say that $\mathbb{N}$ and $\mathbb{Z}$ aren't interesting environments full of beautiful patterns and unsolved mysteries—of course they are. It's only from an algebraic perspective that they demand improvement. If your interest is in divisibility and prime numbers, as opposed to untying algebraic knots, then $\mathbb{N}$ (or $\mathbb{Z}$, if you prefer) is the perfect setting for that study. Indeed, the entire notion of divisibility pretty much evaporates in $\mathbb{Q}$, in the sense that everything is divisible by everything else. So different patterns require different structural housings, and part of being a modern mathematician is being sensitive to which structural setting is natural and appropriate to a given problem.

For an ancient algebraist, that place is $\mathbb{Q}$. Or rather I should say the *positive* half of $\mathbb{Q}$. The equation $z + 2 = 0$ would, in their eyes, have no solution. Again, this is due to their quantity-based mindset. Once you free yourself from those shackles, it becomes much easier to create and invent. Abstract imaginary structures are far more easily stretched and modified, and are thus more amenable to our artistic desires. Even changing our perspective with some dots allowed us to see better. So the flexibility of the abstract viewpoint makes it far easier to expand and improve our mathematical structures. In particular, if we choose to live by the modern algebraists' creed—*a number is what a number does*—it becomes almost effortless to make these sorts of extensions.

The modern view is that since the number 3 is imaginary anyway, it may as well have an imaginary friend, –3. The defining feature of their relationship is that they add up to 0. Thus, the number –3 is simply defined by what it does: it's the thing that when added to 3 makes 0. In this way, each number has its own partner; the imaginary friend of –3 is then 3. (Poor little zero must console herself by being her own friend, however.) The point is, we don't necessarily require the mental image of dots and shifting in order to get negative numbers off the ground logically. It's more of a question of whether a given mental model provides any useful intuition.

Fortunately, the number line concept is fairly robust and easily survives the passage from $\mathbb{Z}$ to $\mathbb{Q}$. We simply chop up the space between the dots in the same way that we chop up our units. In this way, we obtain the *rational number line*:

Now we have a line instead of a discrete row of dots, and the hash marks and numerical labels are in fact infinitely dense: between any two fractions, no matter how close, there are always infinitely many more. It's a serious mental image, actually. Every positive and negative fraction makes an appearance at the exact right spot, and in this way addition and subtraction again correspond to shifting, and multiplication and division to stretching and contracting (as well as flipping). We thus obtain a geometric perspective on rational arithmetic: it's all about sliding and stretching lines.

Show that between any two distinct rational numbers there exist infinitely many others.

In any event, we now have a more symmetrical environment of idealized numerical entities (namely, $\mathbb{Q}$) in which we are free to add, subtract, multiply, and divide—provided, of course, that we do not attempt to divide by zero.

With regard to tying and untying algebraic knots, this is very good news. Whenever we have an interesting action or operation of some kind—whether it be twisting a loop or adding six—we will eventually want to undo it. The inclusion of negative quantities in our number system allows us to undo any addition with the appropriate subtraction. Moreover, since we will often be working with unknown quantities, it is nice to be able to write something like $q - 3$ and not have to worry about whether q is large enough; the subtraction makes sense regardless.

The modern way to think about subtraction is to view it as *addition*: subtracting 7 is the same as adding –7. (This can also be seen in commercial terms: you can pay the seven dollars now, or we can put it on your tab.) In other words, $a - b$ is exactly the same as $a + (-b)$. The point being that the expansion of our system to include our negative friends has also rendered subtraction obsolete and redundant. Why carry around an extra operation when we can let the numbers themselves do the work? This is essentially the reason why the minus sign (–) is used for both subtraction and negation: –2 is just shorthand for $0 - 2$, and $5 - 3$ is an abbreviation for $5 + (-3)$.

We can do the same thing with division as well. Dividing by 2 is the same as multiplying by $\frac{1}{2}$. To each (nonzero) number $\frac{a}{b}$, we can associate its reciprocal, $\frac{b}{a}$. This is simply the number that when multiplied by $\frac{a}{b}$ makes 1. We can undo multiplication by any number (other than 0) either by dividing by that number or multiplying by its reciprocal.

Thus, every number has an opposite in both the additive and multiplicative senses. From the additive point of view (i.e., thinking solely about the way our numbers behave with respect to addition), we see that 0 is the number that does nothing, and two numbers are opposite if they add up to 0. The opposite of $\frac{3}{8}$ is $-\frac{3}{8}$ and the opposite of $-\frac{3}{8}$ is $\frac{3}{8}$. On the multiplication side, we see that 1 is the inert element and that two numbers are multiplicatively opposite if their product is 1. So here the opposite of $\frac{3}{8}$ is $\frac{8}{3}$ and vice versa.

What is the multiplicative opposite of $-\frac{3}{8}$?

We now see that our two basic arithmetic operations, addition and multiplication (from which subtraction and division may be derived), have a number of important similarities. They are both symmetrical binary operations (meaning they act on two numbers at a time), and whatever can be done can also be undone—with the notorious exception of multiplication by zero. They resemble the most basic moves of knot tying; each on its own is simple and reversible, but combined together in sequence, they form intriguing puzzles of infinite variety and subtlety.

And thus our utopian desires have led us from the practical, real-world acts of counting and measuring to the idealized realm of pure number we call $\mathbb{Q}$. Of course, our idealism doesn't stop there. Just as we feel the need to abstract (and to some extent romanticize) our notion of quantity, it is also natural to dream of a simpler, more perfect world of objects and measurements. A real tabletop is an insanely complicated mass of interacting molecules that we can never hope to fully understand. A perfect imaginary rectangle, on the other hand, offers us the possibility of exact measurement and permanent knowledge. Why bother having such a pristine and elegant system as $\mathbb{Q}$ if we're only going to end up measuring some clunky, approximate, atom-based monstrosity? Surely we want a perfect imaginary world of objects to go along with our perfect imaginary world of numbers. The interesting aesthetic question is what we want this perfect spatial realm to look like. What properties do we demand that it have?

Humans have been idealizing space and the world around them for quite some time. Not only do we replace an actual group of lions with a single counting number, we also replace the actual moon with a perfect circle and an actual garden plot with an imaginary rectangle. These sorts of radical simplifications of reality are pretty much built into our brains. The fact is, if you're going to go around encoding the world into language, then you're going to have to make some compromises. Sure, I can tell you about my trip to Hawaii—it was great, thanks—but a few descriptive sentences are never going to capture all of the detailed sense memories. Whatever words we choose, we always end up describing a class of things more than a specific thing, and we simply omit the details and distinctions we don't care about. So right next door to the notion of assigning words to objects is the concept of the *ideal* object—the purest possible incarnation of the word you are using.

In our minds we can then replace a wooden crate with a perfect cube and a bowling ball with a perfect sphere. We move from literal *geometry* (Greek for "earth measurement") to Geometry, the mathematics of idealized space. Of course, such objects as perfect cubes and spheres cannot exist in the harsh climate of physical reality; they are idealizations and must remain imaginary. Sure, there are some fairly round objects in our universe—especially the man-made ones—but they still consist of atoms, and their measurements vary from moment to moment. A perfect sphere, on the other hand, is a permanent resident of Mathematical Reality and thus possesses permanent, exact truths. In particular, perfect mathematical shapes have perfect mathematical measurements.

Let's try to imagine an ideal straight line.

Of course, this smear of ink isn't it. In order for you to see it, this ink line has to have some thickness, which means it's really more of a long, skinny ink *rectangle*. To qualify as a perfect line, it would need to have no thickness at all, yet still "exist" in some sense. Such a line cannot consist of ink molecules or even individual atoms. That means we will have to get comfortable with the idea of a dimensionless point—what Euclid referred to as "that which has no part."

An idealized point would have no size or extent, no left or right sides, and would take up no space. And yet we do not wish it to be nothingness. So that's a bit of a mental hurdle in that we have no experience (nor will we ever) with any such material. We are inventing a new spatial universe, and we get to make whatever decisions we want, regardless of whether they are achievable in physical reality. The whole point is that we are trying to improve upon the place, not copy it in all of its grisly details.

So let us mentally bring into existence an imaginary spatial realm consisting of perfect, idealized points. No matter what images you may have in your head right now, I can guarantee you they're wrong. We simply do not possess the right optical equipment—not in our eyes and not in our brains—to handle it. Instead, we have to live in a sort of netherworld, where we can pin things down with language and logic, but we can't quite see them due to the blur created by our own physical existence.

Well, too bad! We don't get to live in mathematical paradise, and we don't even get to have a clear view from outside the gates. We just have to think and reason and imagine as best we can and—like the scientists—build whatever tools we can build (in our case, purely mental ones) to help us in our blindness. A great mathematical idea (e.g., shifting dots) is like a telescope, allowing us to see just a bit more clearly.

Now let's try to imagine a perfectly flat plane.

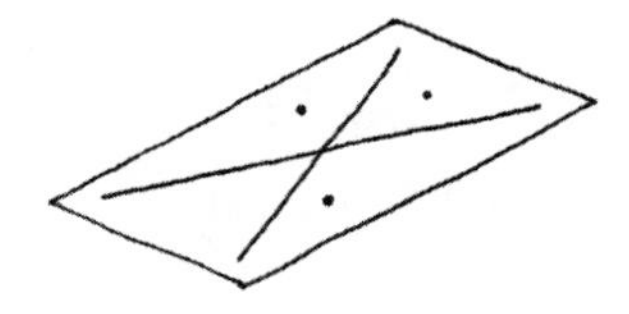

Here is my mental image of a (zero thickness) plane floating in space, with some points and lines on it. Well, perhaps I shouldn't say on it; it's more that they are *of* it, if you get my drift. The plane consists of certain points in space, and the lines are certain subsets of these. There's not really any *on* about it. When we make drawings, we are adding ink to paper, laying it on top. But in our fantasy world of perfect shapes, it is much simpler. A line in a plane is simply a *selection* of some of the points, not an addition. This means that if you were to look at the back (or underside) of the plane, you would see the same dots and lines; there are no front and back faces to a perfect, zero-thickness plane.

Similarly, if I select a point and a distance and I summon those points in the plane that are at that distance from the selected point, then a circle appears—all of its points called into being simultaneously by that one mystical incantation. There is no need for careful, steady-handed drawing or carving; our pattern of perfect points is made with a patterned verbal description instead.

The ancient Greeks were the first to systematically study such a realm and to explore its logical consequences. The most daunting feature of idealized geometry is that in order to be perfect, our objects must also be fictitious. So how on earth are we supposed to measure them? It's not like we can bring a wooden yardstick into Mathematical Reality and start laying off units.

This means we need an entirely new way to measure. We need a notion of "pure" measurement to match the idealized objects we are measuring. In particular, there won't be any external, real-world measuring units dirtying up the place. We will instead measure our perfect idealized objects in the only way we can: against each other! In this way, all of our measurements will be *relative*. On the other hand, isn't all measurement relative anyway? We are always measuring something against something else; all we are doing here is eliminating the unit middleman.

For example, if our bowling ball and crate are arranged so that they have the same width (as if the bowling ball were sitting in its packing case), then we can ask for their relative volume measurement. How much of the box does the ball occupy?

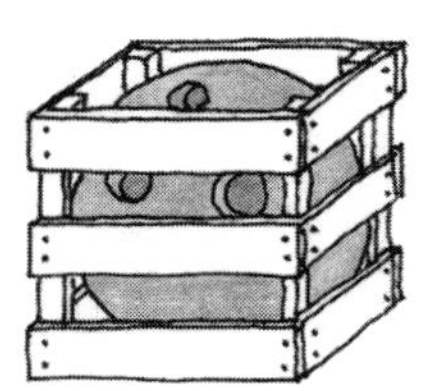
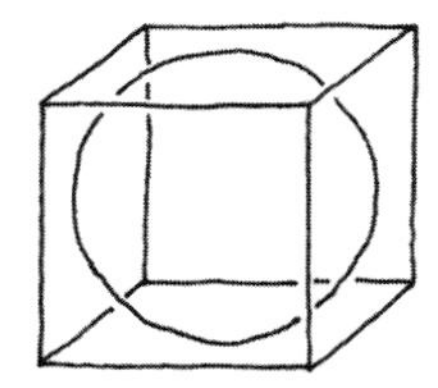

In the case of a real bowling ball sitting in a real wooden crate, any answer to this question would clearly be approximate—there is no such thing as an exact measurement. But there does happen to be a pretty nice rule of thumb: it's about *half*, actually. The ball takes up about half the volume of the crate. There certainly was a world of ancient practical geometry and engineering that included such knowledge. But this is not the same as asking for the perfect truth about perfect objects. The difference becomes quite glaring when one sets about trying to make such a measurement.

The practical measurement, though only approximate, does have the advantage that it is relatively easy to make. We could fill the box with water, say, and then remove the ball and measure the remaining water level. (As I said, it will come up about halfway.) To measure a tabletop, you can get out a tape measure. I'm not saying that all real-world measurements are always so easy to obtain (certainly not the astronomical or subatomic ones!), but the difficulty with measurement in physical reality is merely logistical, whereas in mathematical reality, it is philosophical.

The question facing a scientist or carpenter is: How am I going to make this measurement, and to what degree of accuracy? The mathematician's question is rather different: Is this measurement knowable? Does there exist a piece of logical reasoning that will explain exactly how much of a cube a sphere takes up? It's not so much a practical thing. There's nothing you can do with your hands, and no equipment will help—except perhaps a pencil and paper to write down your ideas, I suppose.

The point is that the idealized perfection of our objects actually makes them much, much harder to measure. To measure the kitchen floor, I just need a ruler, but to measure a perfect mathematical rectangle, I need an *idea*. Measurements in geometry are not made by repeatedly laying off culturally determined units; they are achieved by reason alone.

So we invent an idealized realm of perfect shapes, and the first thing we discover is that we don't know how to measure them. Or rather, we do—it's just that it happens to be extraordinarily difficult and creatively demanding. I call it the price of perfection. We get to have perfect spheres and perfect cubes, and therefore perfect measurements; the only trouble is that we may never know what they are, unless some clever scribe somewhere comes up with an idea and an explanation.

That means all of our knowledge about Mathematical Reality is going to have to come from our imagination, creativity, ingenuity, and careful reasoning. For example, we know that a triangle always takes up half its box:

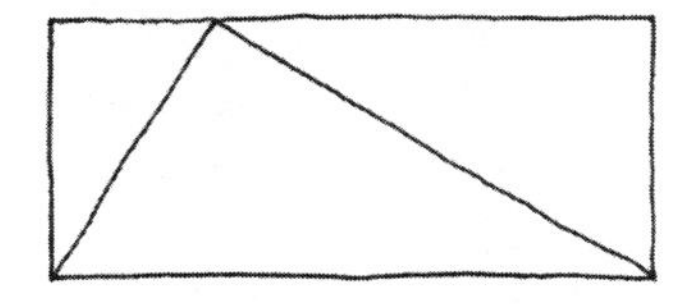

That is, if a triangle sits inside a rectangle of the same height and width, then the triangle will occupy exactly half of the space inside the rectangle.

Of course, we could draw pictures on paper (in fact, I just did!) and roughly measure the approximate area and see that it comes out to about half. That is certainly encouraging, but it doesn't really say anything about an ideal triangle sitting in a fictitious rectangle.

What we need is an *argument*—a piece of logical reasoning that will not only tell us that it is exactly half but also *why* it must be so. There is no "why" about an approximate measurement of an ink smear. We are being dreamy philosophers here: we don't need any approximations; we seek beautiful explanations instead.

Suppose we chop the rectangle in two, from the tip of the triangle straight down to the bottom.

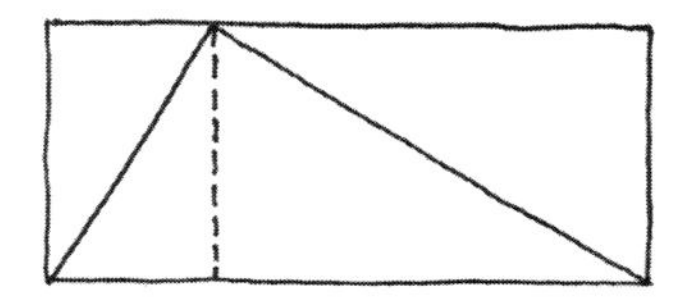

Now it's easy to see that each of the two smaller rectangles has been divided diagonally in half, so there is exactly as much space inside the triangle as there is outside. Notice how we don't have to worry about any inaccuracies due to the thickness of our chopping line—our lines have no thickness at all! (This is one of the great advantages of working with imaginary shapes.)

Of course, this is a ridiculously simple example, but at least it shows that such knowledge is possible. It turns out that we can know things about Mathematical Reality, but in order to do so, we must be creative. We need to create a way of seeing the truth, and this means crafting an explanation—what mathematicians call a *proof*. Our proof of the triangle-box relationship basically comes down to drawing one clever line and then using the symmetry of diagonally cut rectangles. This would be an example of "low-hanging fruit" in the mathematical jungle. The fact that a triangle occupies half its rectangle is a relatively elementary discovery, easily explained by a simple and direct argument. (The measurement of the sphere inside its cube is significantly more difficult and requires a correspondingly more elaborate and ingenious technique.)

As another example, we have the famous *theorem* (the fancy math term for a statement that has been proved) that says the sum of the angles of any triangle is always a half turn.

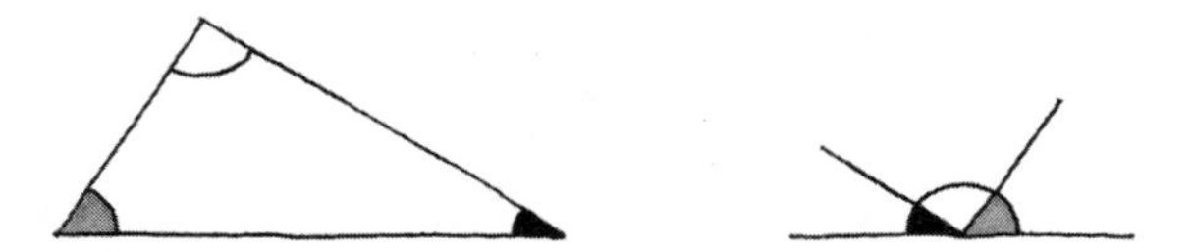

As you are probably aware, it is customary among carpenters and engineers to measure angles in *degrees* (i.e., 360ths of a full turn). Thus, a half turn would be 180 degrees. The classical Greeks preferred to use a *right angle* (that is, a quarter turn) as their unit, so a half turn would be two right angles. I see no reason to choose any units at all; we're really just saying that the three angles of a triangle fit together to make a straight line.

So why should this be true? Even if you satisfied yourself somehow that it were true, that still wouldn't explain why. For mathematicians, the why *is* the what. It is also the how, the when, and the where. The only remaining question is *who*: Who will be the lucky devil who finds the gorgeous explanation that will cause future mathematicians to weep for joy?

Well, I don't know who first invented this argument, but I was fortunate enough to discover it for myself. The idea is to make a triangle sandwich by imagining a line through the tip of the triangle parallel to its base:

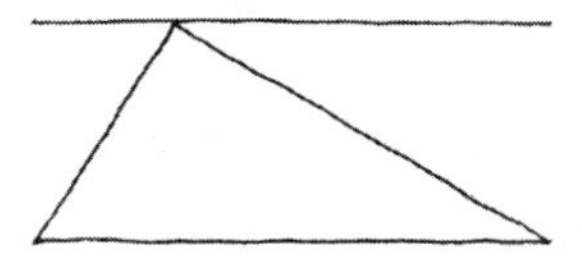

Notice the Z shape that is formed by this line together with the left side and base of the triangle. There is also a (backward) Z shape on the right as well. What makes Z shapes so nice is that they are *symmetrical*.

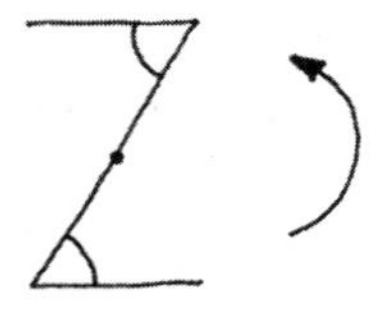

A Z shape has a center point, and because the top and bottom sides are parallel, when it is rotated halfway around (180 degrees), it makes the exact same shape. Since each angle is brought to the other by this rotation, the two angles of a Z must then be equal. This allows us to transfer the angle information from the bottom of the triangle to the top.

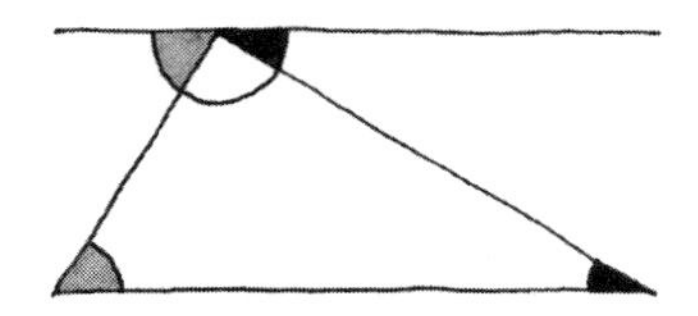

Now we can clearly see the three angles conspiring together to form a straight line, or half turn. Isn't that pretty?

So this proof required a bit more ingenuity and imagination. As our questions and discoveries become increasingly nuanced and sophisticated, our arguments and techniques must also. No matter when you join the mathematical jungle cruise, one thing is always certain: it's only getting harder. The low-hanging fruit has already been picked.

Of course, if you are an amateur—a lover of mathematics and mathematical ideas—then you are in great shape. There is far more than a lifetime's worth of beautiful problems, awe-inspiring abstractions, and insanely clever arguments for you to appreciate and enjoy. But if you want to be a pioneer and solve an unsolved problem and devise an argument that has never before been made, then you're in for one seriously demanding mental rafting trip. On the other hand, there is no greater feeling in the world.

The amazing thing is that we actually are able to attain knowledge about perfect idealized objects using reason alone, and the medium of this artful science is proof. Our goal then, as geometers, is to understand the relationship between shapes and their measurements. The classical geometric measurements are angle, length, area, and volume. Rather than choosing arbitrary units from the world of physical reality (which would make no sense anyway), we instead make only relative measurements. For example, suppose one square is twice the size of another (i.e., it is scaled by a factor of 2).

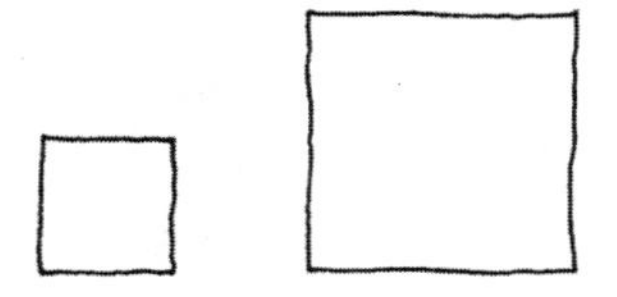

We do not require a measuring unit to say that the larger square is twice as wide; we can compare the lengths directly. Similarly, I need no area unit (e.g., square yards) to know that the larger square occupies four times as much space—I can see immediately how to fill the larger square with four copies of the smaller one. So all of our measurements will be self-contained and relative like this, for the simple reason that they must be. There is no standard length in an imaginary mental universe. Every stick is as good as any other, so we choose none.

On the other hand, it does no harm to choose an ad hoc temporary unit suited to the problem at hand. I am certainly welcome to *call* the side of the small square 1, if I so desire. This would mean that the side of the larger square is exactly 2. Or, if I prefer, I could call the longer side 1 and the smaller $\frac{1}{2}$. It simply doesn't matter, because what we are eventually measuring (in order for it to make sense as a number) is the *proportion* of two geometric quantities, be they lengths, areas, angles, or what have you. Another way I like to think of it is to view our relative measurements as "stretch factors," or *scalars*. When one stick is twice as long as another, we could say that it's a blowup of the smaller one by a factor of 2.

Thus, geometry becomes the fine art of assigning numerical values to geometric configurations. We can compare any two geometric quantities *of the same kind* and express the comparison in numerical terms as a proportion (or stretch factor, if you prefer). It makes no sense to compare angles with areas or length with volume, however. These measurements all behave differently with respect to changes of scale, so their ratios would depend on the units chosen and would thus have no intrinsic meaning. The only proportions to which we can sensibly assign numerical values are the comparisons of area with area, length with length, and so on. The main thing we have to get used to is that there are no culturally agreed-upon units anymore.

Actually, there is a slight exception here with respect to angle. In this case there is in fact a natural unit: one full turn. So we can make a reasonable unit choice for angle measurement, whereas there is no standard imaginary length with which to compare all others. Instead, we compare two lengths directly, without an intervening unit. I myself enjoy the stretch factor interpretation. If one stick is $\frac{3}{8}$ the length of another, then the short stick is simply the longer stick "scaled by a factor of $\frac{3}{8}$." Still another way to look at it is to chop both sticks into smaller pieces:

If we take one-third of the shorter stick as our unit, then the lengths come out to be 3 and 8, in keeping with the 3:8 proportion. In this way, we see that our two sticks are *commensurable*, meaning they can be simultaneously measured with the same unit so that they both come out evenly.

Of course, in the real world of blurry and approximate measurement, all lengths are commensurable, since we can always choose a unit so small (e.g., millimeters) that we are willing to ignore or "round off" any fractional errors or discrepancies. But in the perfect world of geometry, we do not have this luxury. We're not going to round off our measurements because we don't really care what size they are. We're trying to find out what is knowable by the mind of man, not to enter some digits in a spreadsheet.

So we have a serious philosophical question to ask about our idealized geometry: Are any two lengths commensurable?

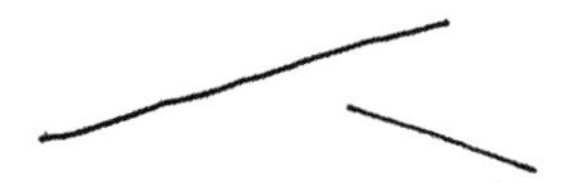

Given two sticks, must there necessarily be a unit (perhaps a very small one) so that both sticks come out perfectly even? Our practical experience with rulers and measuring tapes suggests yes. Every measurement ever made in the real world has been rounded off to some degree of accuracy and can therefore be approximated by a perfect fraction. So we are subconsciously primed to expect our measurements—even those in idealized geometry—to be expressible in this form. In other words, we take it as a given that any proportion (e.g., between two volumes) can always be expressed as a *whole number* proportion—that is to say, as a positive fraction.

This is the historical assumption of commensurability. The ancient philosophers took it as an obvious, unspoken feature of our idealized universe that any two measurements of the same kind must be commensurable and are therefore in whole number proportion. The common unit may be quite small (meaning that the numerator and denominator of the measurement will be quite large), but there will always be one. Commensurability is the foundation of the ancient theory of idealized measurement. It means that our geometric system aligns nicely with our number system, in the sense that our measurements can all be expressed in the language of rational numbers.

Imagine the shock and consternation when this assumption turned out to be false! The fact of the matter is that our idealized world of points and lines is simply too much for $\mathbb{Q}$ to handle. It turns out there are a great many natural geometric quantities that are provably incommensurable, and thus their proportions cannot be expressed in whole number terms. What this means is that we have more lengths in our imaginary world than we have numbers with which to measure them!

This is sometimes referred to as the discovery of irrational numbers, but this is a bit of an oversimplification. It was really more the discovery that *we're screwed*—that all of our most cherished ideas of geometry are built on sand. The ancient Greek philosophers (c. 500 BC) were quite upset and had no idea what to do. Their whole world of shape and measurement was falling apart around them.

The question that ultimately led to the discovery of incommensurables—and thus to the ensuing philosophical crisis—is one of the simplest and most natural mathematical questions one can possibly ask: How long is the diagonal of a square?

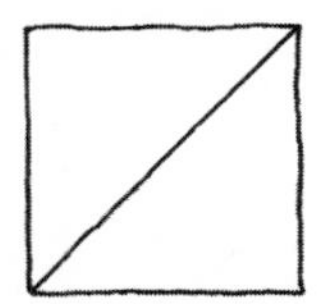

Of course, what we are really asking for is the proportion of diagonal to side. This is a fixed and permanent number, independent of which square you happen to choose.

Let's call this number *d*, for diagonal. We can interpret this in a number of ways, all equivalent: we can think of *d* as the diagonal to side proportion; we can call the side 1 (making it our unit) and then view *d* as the measured length of the diagonal; or, if we prefer, we can regard *d* as the stretch factor that turns the side into the diagonal. However we choose to think of it, it is a certain fixed number and we are wondering what it is.

We could, of course, make some drawings and some approximations: $\frac{3}{2}$ is pretty crude, $\frac{7}{5}$ is better, and so on. Egyptian carpenters typically used $\frac{17}{12}$ as a more precise estimate (e.g., a one-foot square of linoleum tile has a diagonal of about seventeen inches). This could conceivably also be the exact truth for perfect squares as well, though drawings and paper cutouts could never show it. (And doesn't it also seem a bit unlikely that such a beautiful and natural measurement would come out to be some clunky fraction with a 17 in it?)

In order to have exact knowledge, we'll need a mathematical argument. As it happens, the Babylonians came up with a really nice one. We start by taking our square (along with its diagonal) and rotating it around one corner to form a larger square.

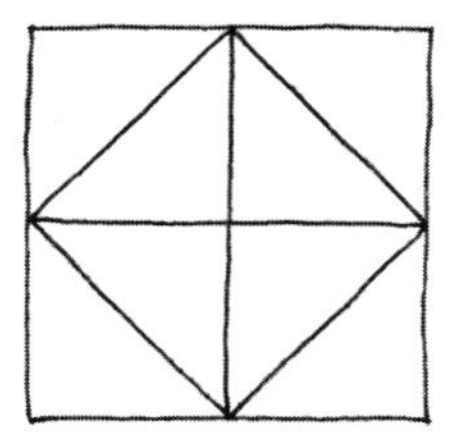

What is so ingenious about this design is the way it forms so many squares. There are the four copies of our original square, the larger (double-size) square they form, and also the diamond shape made by the diagonals, which is also a square.

Why do the diagonals form a square?

Now we can use this elegant mosaic design to make some measurements and some deductions. Let's compare the three sizes of our squares: the small one (our original square), the medium one (formed by the diagonals), and the large one (made of four small ones). Since the side of the medium square is the diagonal of the small square, the blowup factor from small to medium must be d, the number we seek—that is, to turn the small side into the medium side we need to stretch by d. Similarly, to get from the medium side to the large side we would also need to stretch by d, since the large side is the same as the diagonal of the medium square. (Here is an example of a mathematical argument involving almost no symbols or equations; I hope it's understandable!)

Now here's the punch line. In going from the small side to the large side, we need to scale by d twice, so we end up stretching by a factor of $d \times d$. On the other hand, the mosaic design clearly shows that the large side is simply double the small side, meaning that the stretch factor is simply equal to 2. In this way, the ancient scribes were able to determine the precise value of d: it is the number that when multiplied by itself is exactly 2!

So that's a pretty clever little piece of ancient mathematical art, wouldn't you say? Now that we know what our number does, we're halfway to knowing who our number is. We have solved the geometry problem—in the sense that we've converted a geometric measurement into an algebraic description—and now it is time to solve the algebra problem. We have an equation—namely, $d \times d = 2$—so we simply need to determine the explicit value of this unknown number d.

We are now in a very different situation than before, when we were making approximate measurements of ink drawings with rulers. Now we have an actual guaranteed way to test whether a given number is the truth: we simply multiply it by itself and see if it comes out to 2. For instance, it's easy to see that the Egyptian estimate $\frac{17}{12}$ is not exact. This is because $\frac{17}{12} \times \frac{17}{12}$ is equal to $\frac{289}{144}$, which is just a tiny bit larger than 2. If the numerator were 288, then it would be exactly right—but it ain't, so it ain't.

We now have a very amusing arithmetic puzzle: find the numerator and denominator that work perfectly to produce a fraction that when multiplied by itself is exactly 2 and not the slightest drop more or less.

Before we continue this investigation, however, I just want to slip in a few words regarding notation and terminology.

Squares show up a lot in mathematics because they are so simple and pretty. Given a stick, it is very natural to want to build a square with that side.

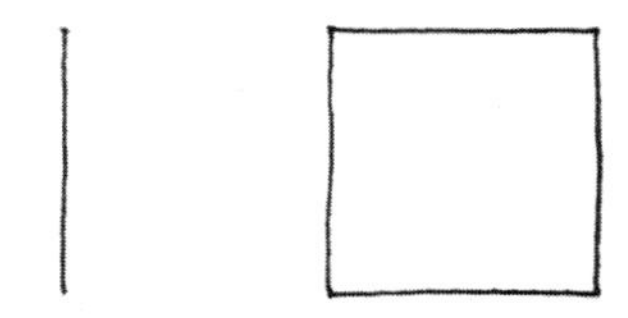

This is what the ancient geometers called *squaring*. Squares are formed from sticks, and in this way the stick is the source, or *root*, of the square. Of course, the length of the stick then determines the measurements of the square. For example, if the stick length is doubled, then the area of the square gets multiplied by 4, as we have seen. If we choose a given stick as our unit of length, and also choose its square as our unit of area, then a stick of length w would have a square whose area is $w \times w$.

Thus, by analogy, the act of multiplying a number by itself has come to be known as squaring as well. For example, the square of 4 is $4 \times 4 = 16$, and we see that four rows of four make a nice square design.

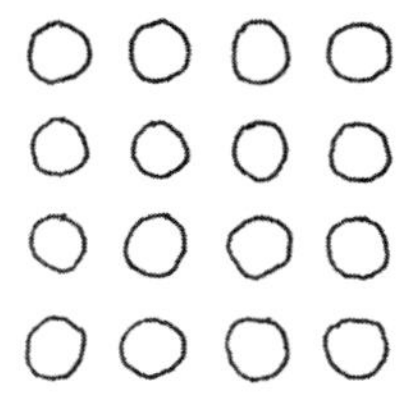

As far as notation goes, we can always write $w \times w$ or ww to denote the square of a number w, but a shorthand notation has also become popular. It often happens when doing algebra that we find ourselves multiplying a certain quantity by itself repeatedly: bbb, $5 \times 5 \times 5 \times 5$, $xyxyxyxyxy$, and so forth. To ease the typographical clutter, the convention is to indicate the number of intended multiplications using a raised counter called an *exponent* (Latin for "on display"). Thus, we write b^3 for bbb, 5^4 in place of $5 \times 5 \times 5 \times 5$, and $(xy)^5$ instead of $xyxyxyxyxy$. There's no real content here; we're just making a convenient abbreviation.

In particular, we have the option of writing w^2 in place of ww. It doesn't really save any paper (it's still two symbols), but it has become quite popular in the past century and a half or so. Squaring is a *unary* operation (meaning it is performed on a single number, like negation), and, like any other arithmetic process, it can be used to help tie and untie algebraic knots. In fact, we already have such a problem. The diagonal of the square comes down to solving the equation $d^2 = 2$. The number d was minding its own business and suddenly it got squared; now it is 2. Who was d before? In other words, as always, we find ourselves wanting to *undo* something.

Following our geometric analogy, we refer to the inverse of squaring as the *square root* operation. The square root of 16 is 4 because 4 is the root, the source, the origin, of the square 4×4. The square root of a number y is simply the number that when multiplied by itself is y.

What is the square root of $\frac{25}{36}$*? How about 0?*

As you may be aware, the notation for the square root of a number w is $\sqrt{w}$. The origin of this symbol is rather amusing, actually. The Latin word for "root" is *radix*, which is also the etymological root of such words as radish, radiation, and radius. One notation popular in the Renaissance was to simply write rw to indicate the square root of w. Gradually, the r became more stylized (so as not to be confused with the letter r) and also included a *vinculum* (overline) to indicate the scope of the operation. Thus, $\sqrt{xy}$ means the square root of the number xy, whereas $\sqrt{x}y$ means the product of the square root of x and the number y.

This means our number d (the diagonal to side proportion of a square) is equal to the square root of 2, so we can write $d = \sqrt{2}$. Of course, this says nothing more or less than $d^2 = 2$, but it does have the psychological attraction of feeling like we are saying who d is, rather than merely what it does.

So our project now is to calculate $\sqrt{2}$ as an exact fraction. This means we're looking for natural numbers a and b so that

$$\frac{a}{b} \times \frac{a}{b} = 2.$$

Why don't we start by going on a hunt? If we were in the jungle looking for a certain species of frog, it would make sense to start collecting frogs and see what we can find.

Suppose we play around with fractions for a while, and we eventually stumble upon the near misses $\frac{7}{5}$, $\frac{10}{7}$, and $\frac{17}{12}$. In the first instance, we find that $\frac{7}{5} \times \frac{7}{5} = \frac{49}{25}$, which is just shy of 2. For a fraction to be equal to 2, the numerator must be exactly twice the denominator. So here we fail, because 49 is not quite equal to 50. I mentioned $\frac{17}{12}$ earlier, and we saw that its square was $\frac{289}{144}$, this time erring on the larger side.

Of course, we can't check every fraction individually because there are infinitely many of them. Instead, we can examine different *types* of fractions and try to narrow our search. Since all natural numbers are either even or odd, we can divide the set of fractions into four types:

$$\frac{\text{odd}}{\text{odd}}, \frac{\text{odd}}{\text{even}}, \frac{\text{even}}{\text{odd}}, \frac{\text{even}}{\text{even}}.$$

In fact, we can even rule out the last type because it is redundant. Since we are looking for a specific positive fraction, we may as well assume that it is written in lowest terms. If the numerator and denominator were both even, we could cut them in half and still have the same fraction, only written with smaller numbers. So let's suppose we do that. Then there are only three types to consider.

If we have a fraction of the form odd/odd (such as $\frac{7}{5}$), then its square will have the same form because the product of two odd numbers is always odd. But no number like this can possibly equal 2 because its top is odd, meaning it can't be double the bottom.

Why is the product of two odd numbers always odd?

The same thing happens with fractions of the form odd/even (like $\frac{17}{12}$). An odd numerator means that when we square our number, we still have an odd numerator. We simply can't get 2 this way.

That means we can narrow our search drastically. The only fractions we ever need to check are the even/odd type. The fraction $\frac{10}{7}$ certainly has this form, and sure enough $\frac{10}{7} \times \frac{10}{7} = \frac{100}{49}$, another near miss—only this time the numerator is at least even, which gives it a fighting chance to be double the denominator. (Of course, it fails here nonetheless.)

In fact, having an even numerator turns out to be too much of a good thing. The square of an even number is not only even but also divisible by 4. (An even-sided square of rocks can be divided in half both horizontally and vertically.) This means that even times even is not just even; it is *twice* an even. Thus, for fractions of the form even/odd, we get

$$\frac{\text{even}}{\text{odd}} \times \frac{\text{even}}{\text{odd}} = \frac{2 \times \text{even}}{\text{odd}}.$$

For this to equal 2 exactly, the top must be twice the bottom. But this would mean that twice an even would have to be twice an odd, which can't happen. So this last type is ruled out as well!

Weary and lost, the steaming jungle rain pouring down upon us, we are finally faced with the truth. Our investigations have led us to only one possible conclusion: the frog we are looking for *does not exist.* There simply is no fraction whose square is exactly equal to 2. You can get close—as close as you like, even—but you will never ever hit it dead on. We've discovered (and proved) a permanent truth about the rational number system: $\mathbb{Q}$ does not contain a square root of 2.

Remember how I felt that $\frac{17}{12}$ could not possibly be the exact value of $\sqrt{2}$ because 17 and 12 just seemed too random? Well, it turns out math feels the same way. There are *no* numerator and denominator good enough to represent a number this awesome.

Geometrically, what this means is that the diagonal and side of a square are incommensurable. If they could be simultaneously measured using the same unit so that they both come out as exact integers, then their proportion (which we know to be $\sqrt{2}$) would be representable as a fraction, which it is not.

It was this discovery that rocked the ancient world. According to legend, the existence of incommensurables was first proved by Hippasus of Metapontum, a disciple of Pythagoras, around 500 BC. The Pythagorean philosophy held that "all is number," meaning the natural numbers (and their relative proportions) are at the root of all things—space and time, the heavens, music, even justice, the soul, and the mind itself. The idea that a geometric measurement could be incapable of being represented as a whole number proportion was nothing short of blasphemy. According to one version of the story, Hippasus was subsequently drowned at sea as a punishment from the gods. (The more lurid version has the Pythagoreans themselves forcibly throwing him overboard.) In any event, it is safe to say that his fellow philosophers were not at all pleased with his startling (and game-changing) discovery.

The upshot is that our beautiful, idealized number system is incompatible with our beautiful, idealized geometry. We have our perfect objects with their perfect measurements all right, but our number system is simply not up to the task of holding the necessary information. We want there to be a square root of 2 because that is exactly how long the diagonal of a square is when measured with its side. However, our number system $\mathbb{Q}$ fails to contain such an entity.

This is not only a geometric disaster, however. Algebraically, we have discovered that our (very natural) squaring activity cannot be inverted. I can square anybody I like, but I can't even unsquare the number 2—not because I'm not smart enough, but because there is no such number in the first place. We now see that $\mathbb{Q}$ is deficient in more ways than one: we can't make the measurements we want to make, and we can't undo the operations we want to undo. That's pretty frustrating.

So it turns out that counting and measuring are rather different after all; their idealized versions do not lead to the same notion of number. Counting gives rise to the rational numbers, whereas ideal measurements concern ratios of geometric quantities. These generally turn out not to be expressible as whole number fractions. So if we want to think of these ratios as numbers (e.g., as stretch factors), then we will have to face the fact that $\mathbb{Q}$ is incomplete: there are more numbers than are dreamt of in our philosophy.

The world handed us our first piece of string in the form of our hands and eyes and conscious minds. We learned to distinguish objects, count and arrange them, make measurements and do arithmetic, and eventually dream up the idea of perfect fractions and the rational number system $\mathbb{Q}$. It certainly is a nice, flexible piece of string, and we can tie a lot of really interesting knots in it, but it is not the only string in town.

And in fact, we find that the system of rational numbers is inadequate for our needs as algebraists. The world of fractions may suffice for counting and bookkeeping, as well as the crude and approximate measurements of the physical world, but it is not sufficient for the purpose of studying algebraic operations and their behavior. Just as we found the positive whole numbers to be lacking, leading to their extension to the rational numbers, we find that the system of rational numbers is similarly flawed, in the sense that it lacks the entities we wish were there.

For a modern mathematician, this is a totally familiar and routine occurrence. A structure of some kind has been constructed and found wanting, so we extend it—happens every day. After all, mathematics is the study of pattern, and patterns practically demand to be continued as far as possible. Mathematical breakthroughs often take the form of a literal breaking through a barrier, allowing a beautiful and interesting pattern to continue into the newly opened realm. To a modern algebraist, extending a number system so that it contains new elements—numbers that can play new roles not assumed by any of the original numbers—is a standard tool of the trade, an almost unconscious, muscle memory sort of maneuver.

But to the ancient philosophers, the discovery that the realm of whole number fractions is in fact flawed (as a model of continuous quantitative measurement) was a complete disaster and a shattering philosophical crisis—and one that took centuries to fully resolve. Once again, it is simply the price of perfection. Not only do we need to be clever and creative—to devise the arguments and explanations necessary for idealized geometry—but now we must also face the fact that our beloved rational number environment is incomplete as well. We need to go back to the drawing board and think more about numbers and what we need them to do.

THE CONTINUUM

Let's return to our number line image. Suppose we build a square with corners at the points 0 and 1.

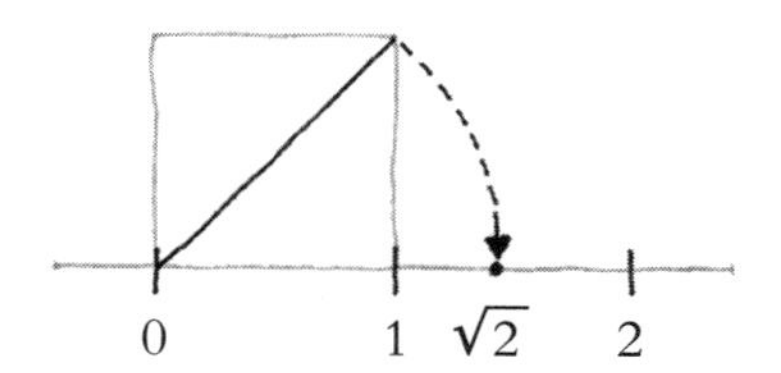

If we imagine the diagonal of this square falling to the ground, the place where it lands would be $\sqrt{2}$. What we now know is that this point on our number line does not correspond to any rational number.

One way to think about it is that our rational number line is not really a line at all; it's more of a Swiss cheese line—full of holes!

Under this interpretation, there simply is no point located where the diagonal would fall.

Perhaps a less disturbing view would be that we do indeed have a line—a complete continuum of points—but it just so happens that a lot of them are unlabeled. There are no missing points, only missing numbers to name them with.

So we're going to need a new kind of number—namely, one capable of representing geometric proportions. In particular, we want entities like $\sqrt{2}$ to exist as numbers. Essentially, our goal is to create a new number system—complete with arithmetic operations like addition and multiplication—whose elements can act as stretch factors for geometric quantities. We want a realm of "geometric numbers" that we can use to make measurements in our perfect imaginary world.

The natural approach would be to simply view numbers as being length ratios from the start, with no reference to units, counting, subdividing, or any of that nonsense. This is a nice idea, but it's also a bit unwieldy. It's not exactly clear how one would go about adding such things, for instance. We can, however, adopt a more concrete approach (if anything pertaining to idealized shapes and measurements can be called concrete) by arbitrarily choosing a certain stick as our length unit.

1

Now every length can be expressed as a single number, its ratio with this chosen unit—in other words, the stretch factor needed to turn our unit into the given length.

One amusing idea is to think of numbers themselves as sticks—that is, instead of our unit stick having *length* 1, we can view it as simply *being* 1. The plan would then be to design a number system consisting of sticks of all possible lengths. The number 1 is identified with our (completely arbitrary) unit, and 0 is simply a point—that is, a stick of no length. Any stick commensurable with our unit will then correspond to a rational number. Two copies of a half unit make a whole unit, so in this way, a half unit behaves exactly like the number $\frac{1}{2}$.

That means we will have two different types of sticks—namely, those that are commensurable with our unit and those that are not—giving rise to two different types of numbers, rational and irrational. An irrational number is simply a geometric proportion (such as the diagonal to side ratio of a square) that cannot be expressed as a whole number fraction. This is the distinction created by the discovery of incommensurables. There was no way of knowing that such quantities existed until this discovery.

Before we investigate the possibility of doing arithmetic with imaginary sticks, I want to point out one bit of comforting news, which is that once we manage to obtain a number system capable of expressing all possible length proportions, we won't need another one for area or volume measurement. This is because *all* geometric proportions can be represented as length proportions.

To see this, suppose we have two regions in the plane.

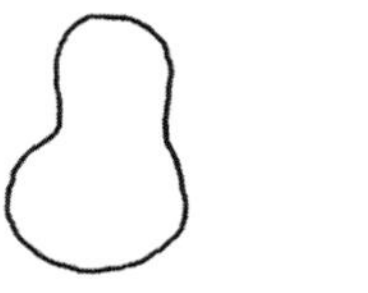

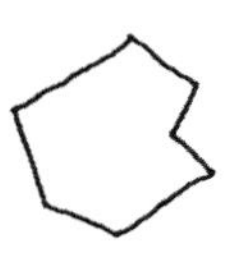

No matter what shapes these may be, they each enclose a certain amount of space, and that same amount can also be enclosed by a rectangle. In fact, given any such region, we can always imagine the specific rectangle of unit height with the same area.

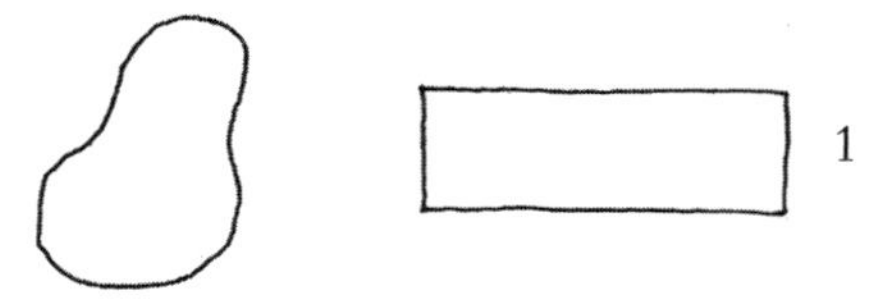

This means the ratio of any two areas can always be expressed in terms of rectangles. Since the rectangles have the same height, their area proportion is then simply the ratio of their lengths.

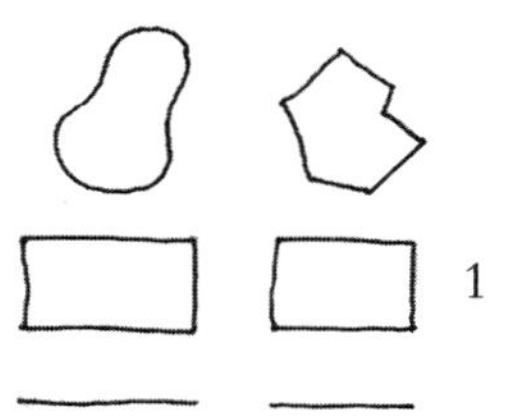

In this way, every geometric proportion can be expressed as a length ratio, so we need only one number system for all of our geometric measurement purposes. (Of course, I say nothing about any *other* purposes we may eventually entertain!)

How can we express the ratio of two volumes as a length proportion? How about two angles?

Now, in order for sticks to play the role of numbers, we will need to have a meaningful (and intuitively satisfying) way to perform our customary arithmetic operations. The first thing we need is a notion of stick addition.

Luckily, this is both obvious and effortless. Sticks, like rows of rocks, naturally combine with each other by *concatenation*—that is, joining together in a chain. We can add any number of sticks by simply lining them up like railroad cars and linking them together. (Notice again how our idealized zero-thickness points cause no trouble at the junctions; we simply identify the end of one stick with the beginning of the next.)

Multiplication is a bit more problematic. It's not clear how we're supposed to multiply two sticks together to obtain another stick. If we look to the world of rocks (i.e., counting), we see that when we multiply two counting numbers—that is, two rows of rocks—we obtain not a row but a *rectangle.*

On the other hand, we can always convert this rectangle into one long row:

Another way to think about it is that our numbers here are rows of rocks, and rows *are* in fact rectangles—trivial rectangles of height one, but rectangles nonetheless. Thus, when we multiply counting numbers we create rectangles and then reorganize them into special rectangles of unit height.

So let's do the same thing in the continuous realm.

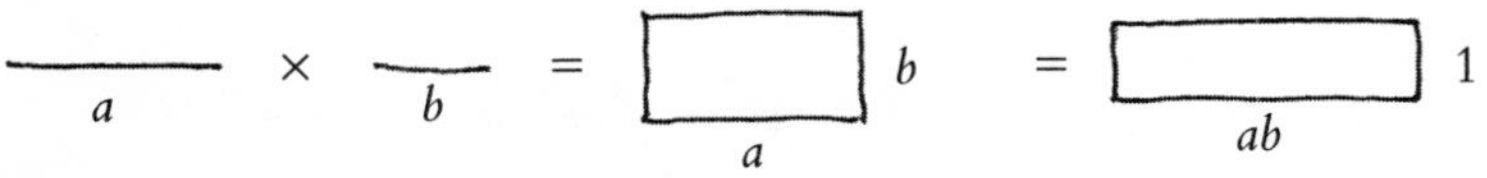

Given any two sticks a and b, we can use them to form an $a \times b$ rectangle. Then we can consider the rectangle of unit height with the same area. The width of this new rectangle is the stick we seek. Now we have a notion of the product of two sticks that agrees with our previous notion of multiplication in $\mathbb{Q}$ but does not rely on counting or commensurability.

Does $2 \times a = a + a$, *as it should?*

Our numbers are now sticks, and we can add and multiply them just as we do whole numbers and fractions. Our previous number environment—the positive fractions—can be viewed as merely a subsystem of the larger world of geometric ratios.

There is (*ahem*) one small problem. Numbers as sticks (or, equivalently, points on a continuum) is all very well and good, but what about negative numbers? Don't we still want all the benefits that come with that notion? In a slightly new way, this is the same old number-as-quantity problem, only now we're dealing with geometric quantities like lengths of sticks. We certainly do want a place where we can freely add, subtract, multiply, and divide in all the usual ways; we just want it to include all of the sticks as well.

One idea is simply to allow ourselves the notion of an "anti-stick" that somehow annihilates the corresponding ordinary stick. We could also take the number line approach and view our sticks as having a forward or backward orientation.

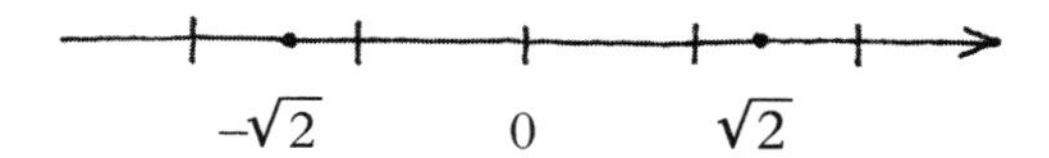

Thus, the number $-\sqrt{2}$ would appear at a distance of $\sqrt{2}$ from 0, but in the opposite direction from $\sqrt{2}$ itself. We can also use our shifting idea from before, if you prefer that metaphor. As before, addition and subtraction then correspond to sliding, and multiplication and division correspond to scaling—stretching (or compressing) the line by a given factor.

What is the geometric effect of multiplying by $-\sqrt{2}$?

However we choose to think about it, we obtain a world of positive and negative sticks that behave in all the ways we have come to expect numbers to behave. We now have a new and improved piece of string with which to tie and untie knots, and it's a much smoother and more flexible string than $\mathbb{Q}$. This new, continuous number environment is known as $\mathbb{R}$, the system of so-called *real numbers*.

This has to be one of the most unfortunate choices of nomenclature ever adopted. The word *real* is a confusing word to use with regard to anything mathematical, let alone an imaginary length. Of course, throughout most of math history no such term existed because no such distinction was needed. We're talking about *all* the numbers on the number line, so we don't need to call them anything. There are rational ones and irrational ones, positive ones and negative ones, but they're all just points on the continuum, and that's all that numbers are.

Until they're not. By the late sixteenth century, algebraists had already discovered the inadequacies of the real number system and were proposing and developing more abstract conceptions of number. This was quite controversial at the time, of course. Whenever anyone suggests something new, there is always backlash from the conservative old guard. So these new, more abstract entities were initially scorned and referred to as "imaginary" numbers (the square root of -1 being an infamous example). The more concrete, traditional numbers were then referred to as "real" by comparison. Eventually, the dreamers won out, as they usually do. The new expanded number system was simply too beautiful, too interesting, and too useful to ignore. But the "real" versus "imaginary" labels stuck, unfortunately. (We will explore these ideas extensively in later chapters, believe me.)

To a modern algebraist, of course, all number environments are equally unreal and abstract, $\mathbb{Q}$ and $\mathbb{R}$ being two particularly beautiful (and historically important) examples. It turns out there are a great many such systems, some with practical utility and others of more theoretical interest. But none of them (especially not $\mathbb{R}$) can walk around claiming to have anything to do with reality. At best, we can say that $\mathbb{Q}$ and $\mathbb{R}$ are figments of our imagination *inspired* by reality. Nevertheless, the name has become ingrained, and so we speak of real numbers as those magnitudes represented on the continuous number line.

One way I like to view the situation is that we have a naked line, and we are using numbers to make a map of it. We choose an arbitrary point on the line as our reference point, labeling it 0; an equally arbitrary direction (to be called "positive"); and finally, an arbitrary choice of unit. This then determines the labels of all the rational points, and every irrational number will live somewhere on this labeled line. The point being that $\mathbb{R}$ is an *extension* of $\mathbb{Q}$ that gives us all the new numbers we wanted without us having to give up any of the nice patterns and intuitive imagery that we enjoy.

There is, however, one glaring problem. Whereas each element of $\mathbb{Q}$ can be referred to easily—that is, by naming an integer numerator and denominator—we have as yet no way to indicate a particular irrational number. Yes, we can point to $\sqrt{2}$ by name, but that's because this particular number does something specific and peculiar—namely, its square is exactly 2. Most irrational numbers don't do anything like this; they just sit there on the line minding their own business, possessing no remarkable or noteworthy properties whatsoever.

Before, we faced the problem of having more measurements than we had numbers. Now we have all the numbers that we need—having basically redefined numbers *as* measurements—but we've run out of *names*. We need to design a nomenclature system for real numbers so that we can refer to each individual, just as we do with rational numbers. It's going to be pretty hard to do arithmetic (let alone tie algebraic knots) with numbers that we can't even name!

Of course, every real number lies between two consecutive integers.

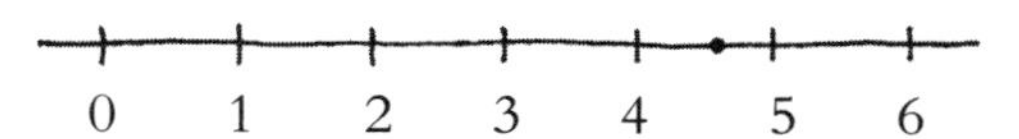

So we can always refer to a number as "4 and a little bit," where that little bit lies between 0 and 1. That means we really only need a naming system for the leftovers in this interval.

Thus, we come to the fundamental question of how to refer to points on a continuum. There are many ways to do this, and they each have their advantages and disadvantages. Ideally, we would like a system of nomenclature that is easy to use and understand and also allows us to do arithmetic quickly and conveniently. But the main thing is to have such a system *at all*. We have a serious philosophical problem here. How do we refer to a perfect point on a perfect line?

One simple way to do this is by repeated halving. We first cut the interval in half and then look to see which side our point is on. If the point is in the left half, we write an *L*; if it's in the right half, we write an *R*.

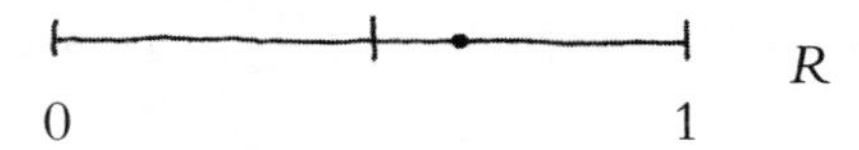

For the next step, we take the half interval in which our point lies and we proceed to cut *it* in half.

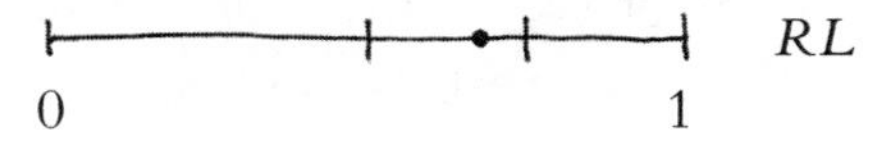

We then record the information as before, using either an *L* or an *R*. As we continue to subdivide, we keep narrowing in on the precise location, squeezing the point into successively smaller intervals.

In this way, every point on the line receives an *LR* code name. The only difficulty is that these names are infinitely long. In order to name a point precisely, we will need to supply *all* of the halving information. If we stop at any particular stage, we will have only described a certain (perhaps very small) interval, not one specific point. Notice that for any two distinct points, there will always be *some* amount of space between them, meaning they will eventually be on opposite sides of some dividing line, and thus their *LR* codes will differ. So we certainly can refer to real numbers in this way; we just have to accept endlessly long names!

Before we start worrying about how we're going to read and write (let alone calculate with) such things, I want to point out a somewhat annoying feature of this naming method. (You may have noticed it yourself while I was describing it.) What do we do if our point lies exactly *on* a dividing line, neither to the left nor the right? We might be happily subdividing away, producing a nice string of *L*s and *R*s, and then all of a sudden: oops! Our point is right smack dab in the middle.

Of course, any such location would correspond to a perfect fraction that already has a name (e.g., $\frac{1}{2}$, $\frac{7}{8}$, $\frac{5}{32}$), so we could just ignore such numbers. On the other hand, it's nice to have a naming system that works for everyone at the same time, such as the numerator and denominator system for $\mathbb{Q}$.

The usual approach is to consider such a point as being on both the left *and* right sides. This has the advantage that these points are now included in our naming system. But it also has the slightly unpleasant feature that such numbers will then have two perfectly valid representations. For example, the number $\frac{1}{2}$, being the exact center of the interval from 0 to 1, would receive both the name *LRRRRR*... (the *R*s continuing indefinitely) as well as *RLLLLL*... (where the *L*s go on forever). This is because if we view $\frac{1}{2}$ as being on the left, then in all future subdivisions it will lie on the extreme right and vice versa. So we just have to get used to the idea that some (very special) numbers will have redundant names in the *LR* system.

Which numbers are these, exactly?

An entirely analogous thing can be done by chopping our intervals into three parts, or seven, or whatever number you choose. In particular, ten is a popular choice, due to the prevailing decimal culture.

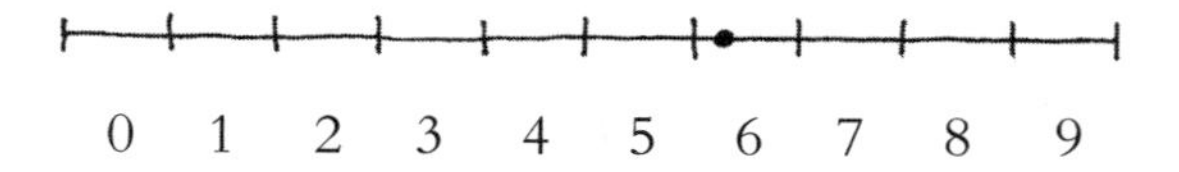

In this version of the scheme, our unit interval is divided into tenths, and each section is labeled with a digit 0 though 9. Now we write a single digit, such as 6, instead of *L* or *R*. Continuing to subdivide into tenths, we produce an endless sequence of decimal digits. Thus, every real number has a decimal representation of the form 4.6322089047... .

Here again we face the redundancy issue when a number lies precisely on one of the decimal division lines. If we view it as being on the left side of the line, then its digit sequence will end in a trail of 9s, whereas if we think of it as being on the right, then it will end in all 0s. For example, the number $\frac{3}{4}$ would have the two distinct decimal encodings 0.749999... and 0.750000... .

So these kinds of naming systems are far from perfect. If we want to use subdivision, then we are not only doomed to infinitely long names, but we must also put up with a certain degree of redundancy. Nobody is happy about it, believe me. I personally detest these kinds of naming systems (e.g., the decimal system) for precisely this reason. As we shall soon see, when dealing with exact real numbers (as opposed to approximations), it turns out that these kinds of infinitely long representations are not of much utility after all. How are we supposed to write such a thing down? Suppose we had two infinite decimal representations and we wanted to add or subtract them. How can we borrow and carry forever? Even if we could manage to make sense of such an infinite process, *convenient* it is not. There is an enormous difference between an extremely long, but nevertheless finite name, and one that in fact goes on *forever.*

So the continuum is no joke. The only means we have by which we can refer to particular points (or numbers, if you prefer) requires infinitely long names that we can never hope to fathom, much less write down. (There are other naming strategies for real numbers to be sure, but they all share this same fundamentally intractable feature.) How can we talk about something if its name takes forever to say?

Of course, there must be some infinitely long names that we *can* talk about, because we can talk about fractions. The number $\frac{1}{3}$, for instance, has the *LR* code *LRLRLRLRLR*… , with the *L*'s and *R*'s alternating forever.

Why does $\frac{1}{3}$ have this alternating code?

The point here is that the sequence has a *pattern*, so we can in fact name the entire sequence by describing the pattern it obeys. In a sense, the pattern itself becomes the name. We can call it $\frac{1}{3}$ (meaning the number that when tripled is 1), or we can call it "*LR* alternating"—either way, we've captured it precisely in a finite number of words, and we won't have to talk for all eternity.

Does every fraction have a repeating LR code?

So the question is whether every real number (i.e., every *LR* sequence) has some sort of pattern to it that can be put into words—finitely many words, that is. We know that every rational number can be so described. The irrational number $\sqrt{2}$ also obeys a simple pattern: its square is 2. The number π, the ratio of circumference to diameter of a circle, is famously known to be irrational (this was proved by Johann Lambert in 1741). Of course, this number obeys a very describable pattern as well—namely, it is the stretch factor that turns the diameter of a circle into its circumference.

If a number obeys a pattern, then it can be named using that pattern. This pattern may be more evident in one representation format than another, however. For example, the *LR* code or decimal digit sequence for $\sqrt{2}$ appears quite random, although it certainly cannot be, given the algebraic property this number enjoys. No matter how complicated or subtle it may be, once there is a pattern then there is a means of description. But is there always a pattern? Does every real number do something that makes it stick out from the herd and thereby be namable?

To clarify matters, let's step away from real numbers and the continuum for a minute and think about something a little simpler. I want to imagine that we are dealing with infinite strings of black and white beads:

●○●●○○○●○●○●●○○●○○○●●●···

Some of these bead strings will have very clear and easily identified patterns—the all black one, the alternating ones, and so on. Others will be much less evident. For example, here is an interesting bead string:

○●●○●○●○○○●○●○○○●○●○○○●○ ···

At first glance, this may seem to be an utterly random sequence of white and black beads. But in fact, the black beads are located at positions 2, 3, 5, 7, 11, 13, 17, 19, and 23—in other words, at *prime* locations. So we can talk about this particular infinitely long bead string even though it never repeats.

Of course, bead strings are pretty much the same as *LR* codes, only we're eliminating the redundancy by simply using black and white beads that have nothing to do with numbers or dividing lines. We are really talking now about information in the abstract and whether an infinite amount of data can be compressed into a finite package. This is what having a pattern *means*, essentially. We want to know whether every infinitely long black and white bead string has a pattern (recognizable or not) or if there might exist truly random bead sequences that simply cannot be captured by language in any way. Look what our idealism has gotten us into!

Of course, no classical algebraist was asking any such questions. When you're busy solving problems and developing new techniques, you're not really worrying too much about language. If your new method allows you to solve your algebra problem and you find the solution to be $\sqrt{1+3\sqrt{7}}$, then you have successfully named your number. The language of addition, subtraction, multiplication, and division—together with the taking of roots—is fairly expressive, and by combining these operations in sequence, a great many numbers can be specified. It's not surprising that you might make the unconscious assumption that *every* point on the number line—that is, every element of $\mathbb{R}$—could, in principle, be expressed in this language. (As it happens, this assumption also turns out to be false.)

So we're talking about a rather modern mathematical issue—what I like to call the *description problem* for infinite collections. This is certainly a detour from our main subject (i.e., classical algebra), but we're taking it anyway because I think it's interesting and I'm at the wheel. The question is whether every bead string has a pattern that can be put into words. Just because we don't see an obvious pattern in the first billion beads doesn't mean there isn't one. So how will we ever know?

I want to show you a beautiful argument that explains why there must exist indescribable bead strings (and therefore indescribable real numbers). Not only is this one of the most breathtakingly simple and elegant works of mathematical art, but it is also one of the most profound. In fact, it would not be too much of a stretch to say that it was this very discovery that initiated the modern era in mathematics. Before I show you this masterpiece, I want to first rephrase and simplify our problem in order to get to the heart of the matter.

Suppose we look at a particular bead string description. It may be short and sweet, or it may be long and convoluted; it really doesn't matter. Whatever it is, it's a finite sequence of letters and punctuation marks that somehow tells us the color of every single bead in the string. In other words, we are trying to associate infinitely long strings of beads with finitely long strings of alphanumeric characters.

One nice feature of finite descriptions (as opposed to infinitely long bead strings) is that it's quite easy to list and organize them. We can start by sorting our descriptions by *length*. We put all descriptions of length 23 (i.e., consisting of 23 total characters) in the same category, and since there are only finitely many descriptions of a given length—there being only finitely many possible characters in each position—we could simply list them all in alphabetical order, say.

Thus, if we wanted to, we could imagine the complete infinite list of all possible bead string descriptions. The list begins with the shortest descriptions (in alphabetical order) and then proceeds through each length category in turn. (The description "all black" would occur in the length 9 section.) In this way, each possible description eventually appears somewhere on this endless list.

Notice that this has nothing to do with the fact that these were *descriptions*—sequences of characters that form meaningful words and sentences describing bead string patterns. That never really came into play. The fact is that *all* finite strings of characters can be listed, whether they are intelligible or not. The complete list of all possible books, poems, and nonsense words is then also included. The point being that all such information is finite in length and therefore can be organized into an infinite list.

Once a collection has been organized into list form, we can then number its members in the order they are listed. Another way to say that an infinite collection is listable is to say that it can be matched with the set of counting numbers 1, 2, 3, etc.

Suppose for a second that we had a *finite* collection—a bag of marbles, say. We can then start making assignments: marble #1, marble #2, and so on, as we remove each marble from the bag. (This is popularly known as counting.) What we are talking about is extending this notion to *infinite* collections such as $\mathbb{N}$, $\mathbb{Z}$, $\mathbb{Q}$, $\mathbb{R}$, and the set of all black and white bead strings.

So let's call an infinite collection *countable* if it can be arranged in a sequential list (or, equivalently, if it can be matched with the counting numbers). The first question is whether this is a distinction worth making. Aren't all infinite collections countable? Just as with marbles, we can keep pulling the next item out of the bag and giving it a number, and neither the bag nor the numbers will ever run out. So what's the problem?

The problem is that our naive intuition about infinite sets is not very reliable—nor should it be. Our brains evolved to survive on the African savannah, not to comprehend infinity. We have no direct experience dealing with infinite collections. It turns out that this "inexhaustible bag" argument is poppycock. Just because we pathetic humans cannot complete the task of listing does not mean that there is (or isn't) a complete infinite list. For a collection to be countable there must exist a matching between it and the counting numbers. Our failure to produce such a matching says more about us than it does about the collection.

One way to show that a given infinite set is countable is to explicitly construct such a matching. Our listing scheme for descriptions is one such example. The set $\mathbb{N}$ is clearly countable, as are the collections of even and odd numbers. It turns out that both $\mathbb{Z}$ and $\mathbb{Q}$ are countable as well.

Can you devise a way to list $\mathbb{Z}$? How about $\mathbb{Q}$?

The question then becomes whether the collection of infinitely long black and white bead strings—a collection I will call X—can or cannot be listed.

It turns out that it cannot. In other words, X is *uncountable*. There simply does not exist a list—even an infinitely long one—that contains every possible bead string. Not only are we humans incapable of producing or describing such a list (or matching, if you prefer), but even the gods are prevented; there is no such list because X is unlistable.

The upshot of this discovery (due to Georg Cantor in 1891) is that there are far more bead strings than there are descriptions, despite the fact that both collections are infinite. This means that not only must there exist indescribable bead strings, but in fact *almost all* bead strings have this property. The same goes for real numbers, plane curves, and many other infinite collections of interest. Almost all numbers and almost all shapes are intrinsically un-talk-about-able. Our language is so limited and weak that it can produce only a countable infinity of descriptions, yet Mathematical Reality presents us with an uncountable infinity of objects to consider.

This means there is a permanent limitation on our ability to have knowledge about mathematical entities. Most of the frogs in this jungle simply cannot be captured by the net of language. Maybe we should have known all along—after all, isn't being namable a thing? Once you say that something can be held in the mind or communicated, doesn't that already make it special and rare? The new outlook we need to adopt is that to be describable *at all* is the real accomplishment for a mathematical object, and to be describable in a particular way (e.g., a real number that happens to be rational) is especially rare.

In fact, the entire mathematics project needs to be reconceived in light of this discovery. Mathematics ceases to be the study of pattern in general and must confine itself to the investigation of *describable* patterns. We, as mathematicians, are operating under a permanent restriction: we cannot talk about what cannot be talked about, and that turns out to be almost everything! In a way, this discovery is for mathematicians what the uncertainty principle is for scientists. In their case, the ability to make measurements of physical reality is impeded by the very presence of the observer and the tools used in the experiment, whereas in our case the obstruction is due to the weakness of our instruments—namely, the finitary nature of language and consciousness.

Viewed in this light, perhaps we should be a bit more appreciative of those things like fractions and symmetrical polygons that can be so easily described. When indescribability lurks around every corner, even the most complicated, blackboard-filling mass of equations should be cherished—whatever information they contain is of that oh-so-rare variety that can be contained at all.

All right, I've blabbed enough. It's time for me to show you Cantor's beautiful argument. How can we know for certain that X, the collection of all infinite black and white bead strings, cannot be listed? Why must every infinite list of bead strings be incomplete?

The idea is almost painfully simple. Given any list of bead strings, we will use the list itself to construct a new bead string not on it. Suppose we had an infinite list of bead strings, say

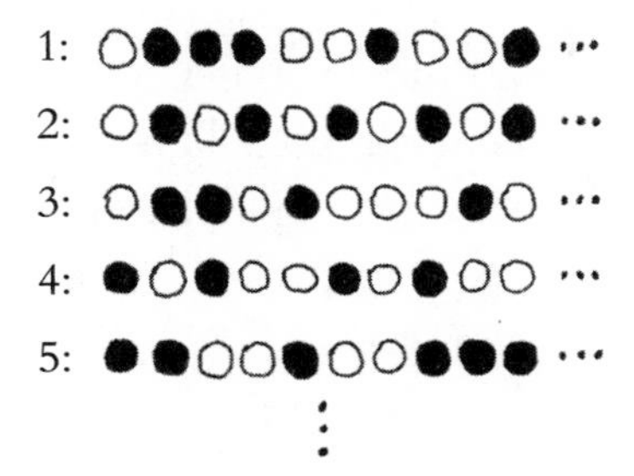

(Of course, being human, we cannot actually exhibit such an infinite list in totality, but we can certainly *imagine* it.) Now we form a new bead string by choosing the first bead of the first string on the list, followed by the second bead of the second string, the third bead of the third string, and so on. In this way, we obtain the "diagonal" bead string

This bead string has the property that it shares at least one bead with every string on the list. That means if we switch colors—exchanging white and black throughout—we obtain a new string that *disagrees* with every string on the list in at least one bead:

More precisely, this new string differs from the first string in the first bead, from the second string in the second bead, from the eighteenth string in the eighteenth bead, and so on. It cannot be the same as the millionth bead string on the list because they disagree in the color of the millionth bead. That means our new string, obtained by swapping colors along the diagonal, cannot appear *anywhere* on the list. It differs (by design!) from every single bead string that the list contains.

This proof, known as *Cantor's diagonal argument*, shows that no matter what list of bead strings you may have, it is always necessarily incomplete. Whether your list is simple and orderly, or utterly random and indescribable, it makes no difference. Whatever list you may have in mind (including one that is incapable of being held in the mind), we can always harness its diagonal to create a bead string that the list does not contain. So our collection X is provably uncountable, as advertised.

That means there are two different types of infinite collections: the countable sets capable of being arranged in an infinite list and the uncountable ones that cannot be so arranged. The number systems $\mathbb{N}$, $\mathbb{Z}$, and $\mathbb{Q}$ are countable; $\mathbb{R}$, however, is not. As we have seen, this is an enormous difference. The complete set of all descriptions of anything is merely countable—infinite, sure, but not all *that* infinite, as it turns out. The real numbers (or the points of a continuum, if you prefer) are a much larger collection. Consequently, most real numbers will never be referred to, for the simple reason that they cannot be.

So we should be very grateful that certain irrational numbers such as $\sqrt{2}$ and π can be talked about at all. The first has an algebraic description (as the number whose square is 2), whereas the second is known to be *transcendental*, meaning that it satisfies no algebraic relationship whatsoever. In other words, the number π doesn't *do* anything particular with respect to addition and multiplication. (This was proved by Ferdinand von Lindemann in 1882.) Of course, most numbers don't. If they did, they would then be describable. So we shouldn't expect our numbers and measurements to be rational, or even algebraic, because these are exceedingly rare qualities for a number to possess. Fortunately, both $\sqrt{2}$ and π are geometrically describable since they are ratios of natural geometric quantities.

Truth be told, Cantor's discovery gives us a smack in the face in more ways than one. Quite apart from the issues pertaining to language, we have also discovered something truly remarkable: not all infinite collections have the same size! Some are countable (that is, matchable with $\mathbb{N}$) and some are not. This opens up a whole new line of investigation, a new river in the jungle to explore and map out. The theory of aggregates—now known as *set theory*—begins as the study of collections and matching. The discovery that not all infinite collections can be matched was the starting point. Not only did Cantor's result force us to question the limitations of language, but it also initiated an entirely new mathematical discipline: the study of collections in the abstract. Moreover, the simple notion of matching finally offered humanity some hope that an understanding of infinity and its properties might actually be attainable.

Of course, Cantor's ideas were not immediately accepted. In fact, the very notion of an infinite collection—a so-called completed infinity—was hotly disputed at the time. Prior to Cantor's work, the fact that the set of even integers can be placed in one-to-one correspondence with the collection of all integers was enough to convince most mathematicians and philosophers that the subject was hopelessly fraught with contradiction. On top of that, Cantor also met with harsh criticism from the leading theologians of the day, who argued that his research was heretical and violated God's exclusive claim to infinity. The hostility of his contemporaries ultimately led to several severe bouts of depression, and Cantor spent the remaining twenty years of his life in and out of hospitals and sanitariums. One supportive journal editor told him that his work had simply come "one hundred years too soon."

As always, what was once seen as radical and subversive is now completely accepted and commonplace. Infinite sets and their counterintuitive properties are now viewed as a vital and essential part of mathematics, the apparent contradictions having been revealed as mere intellectual prejudices. As the German mathematician David Hilbert once remarked, "No one shall eject us from the paradise that Cantor has created for us." Infinite collections are simply too useful (and too interesting) to ignore.

One amusing irony is that although most real numbers cannot be individually described, the collection of *all* of them (namely, $\mathbb{R}$ itself) certainly can. We can talk about $\mathbb{R}$ (or X) because they obey a simple pattern: they are the collections of all possible real numbers (or bead strings). For some reason, I am delighted by the idea that there are collections that can be talked about despite the fact that almost none of their members can be.

I just love how math always manages to get us off our high horse. We thought it was obvious that any two lengths must be commensurable—not! We assumed that every number was algebraically or geometrically describable—not! We thought we were the lords and masters of Mathematical Reality—not even close! We get to play there, and we certainly get to invent and discover and explore, but we most definitely do not own the place. In fact, my favorite feature of mathematics is that it continually blows my mind and teaches me humility. To be utterly and absolutely wrong is one of the greatest gifts one can receive. Luckily, mathematics is extremely generous in this regard!

We have succeeded in creating the perfect number environment for continuous measurement, but it comes at a rather steep price. Not only are most of our numbers indescribable by any means, but we also lack a reasonable naming scheme for those numbers that we *can* talk about. Yes, our numbers can happily engage in all the fun of adding and multiplying, negating and reciprocaling, but we really don't have a convenient way of doing arithmetic in the usual sense. The number $\pi + \sqrt{2}$ is certainly out there, but we simply have no better way to denote it. There isn't anything we can do, as a procedure, the way there is with fractions and whole numbers. Our language is too limited to allow us such an option. We can imagine the *LR* codes or the decimal representations of these two numbers being added (with potentially infinite amounts of carrying), but we can't really perform such an activity ourselves, unless the digit patterns are absurdly simple.

There is one small benefit to this, at least. Since arithmetic with real numbers generally cannot be done, we won't have to do it! If we solve a geometry problem, say, and we find the solution to be $3\pi\sqrt{6} + 17$, then that's simply what it is. (That the number can be said at all is the real news.) Of course, in those rare cases when arithmetic can be done, it is usually worth it to simplify matters. It would be silly to discover that your number is $(\sqrt{5})^2 - 5$ and not recognize it as 0, for instance.

So this is the amusing and intriguing corner we have managed to paint ourselves into. The price of perfection has now been paid. We get to have $\mathbb{R}$, our new, more flexible imaginary string—it even *looks* like a string—but we are forced to accept some fairly humbling truths. In some ways, it feels like the only thing we've really managed to tie into knots is ourselves!

The notion of a continuum is as old as human consciousness, and, in truth, we have *always* imagined numbers as continuous magnitudes. Number has always meant real number; it's just that we assumed they were all rational, or at any rate, describable. We simply didn't understand —until quite recently—what we were really dealing with.

The real numbers constitute the universe for a classical algebraist, the complete collection of all possible numbers—including the controversial negative ones. But the thing about universes is that we tend to get trapped inside them. It is only by stepping outside our own universe that we can really begin to understand it.

ABSTRACTION

What makes algebra—and mathematics in general—so unusual in the realm of human creative endeavor is that it is an entirely abstract mental activity. Mathematics exists only in the mind's eye and is the distillation of our innate human desire for the ideal. Both the rational numbers $\mathbb{Q}$ and the real numbers $\mathbb{R}$ are expressions of that desire. Due to the purely linguistic and nonmaterial nature of the enterprise, we mathematicians are able to more easily alter and adapt our materials. When your only media are ideas, they are much more easily *stretched.*

To a modern algebraist, the value of an environment like $\mathbb{R}$ is that we can freely perform our operations, not that it happens to be inspired by idealized geometry. Similarly, $\mathbb{Q}$ is a nice place to do business because we can add and unadd, not the incidental fact that it arose from counting and arranging rocks. These are two beautiful and fascinating pieces of string, no doubt—but we can also step back and think about what we really want as algebraists.

If we're going to be tying and untying interesting knots, we're going to need to make certain demands on our string—namely, that it can withstand all the twisting and pulling that we require it to undergo. It needs to be flexible enough to handle looping and unlooping, pulling through and pushing out, and so forth. To make and solve interesting and challenging number puzzles, we'll need a number system that is similarly robust. Our basic maneuvers are addition and multiplication, so we require a universe of numbers that can handle both of these operations, as well as their inverses.

One way to think about it is that the string itself is really beside the point. A granny or a clove hitch is the same knot whether it is tied with string, yarn, rope, or even hair. The knot itself is the sequence of moves, not the substrate on which these moves are performed. The only thing we ask of our material is that it not fight us—that it allow us to make the moves we want to make.

So algebra in this sense is not really about numbers at all, only operations. If these operations happen to be addition and multiplication (as in classical algebra), then we need a number system for them to act upon, but it hardly matters what, as long as it includes enough entities that we can freely perform (and unperform) these operations.

The point being, that if knots are to be tied using operations, and we intend to study and develop methods for untying them, then it behooves us to understand our operations as fully as possible. Given that our knots will involve both known and unknown quantities, it is especially important for us to understand the *universal* properties of our operations—that is, the properties and patterns shared by all numbers, not just a specific few. A basic example would be the symmetry of addition:

$$a + b = b + a.$$

This is not a statement about two special numbers, a and b, that just happen to have this amazing property. This is a universal statement valid for *any* numbers a and b whatsoever. In this way, it becomes more of a statement about the behavior of the addition operation than about any particular numbers. In fact, its value is entirely due to its universality. Because it applies to all numbers at all times, it allows me as an algebraist to switch the order of the numbers being added, regardless of whether they are known or unknown to me at the moment. I am free to rewrite $q + 3$ as $3 + q$ if it suits my purposes, just as I can put a strand of rope in my left hand or in my right, if I so choose.

The maneuvers that any artist or craftsman performs are those suited to the medium. Whether it is tying knots, performing ballet, carving wood, or doing algebra, the medium—be it rope, the human body, a piece of wood, or numbers—dictates what can be done to it and how far it can be pushed before the system breaks down. As far as algebra goes, this means exploiting the properties of numbers insofar as they are *valid*. We get to creatively manipulate information so as to untangle it, but the moment we perform a mental or symbolic maneuver that is not justified by the actual behavior of our medium (i.e., numbers), we have essentially cut the rope of reason and broken the leg of logic. (I will spare you the alliterative wood metaphor.)

Since our basic operations (namely, addition and multiplication) are so well behaved, the usual mistake is to unthinkingly divide by zero. This can occur, for instance, when dividing by an unknown quantity that (unbeknownst to us) happens to have been zero all along. Another common source of error is to wish, hope, or demand something of numbers that they simply don't do, perhaps the most infamous example being the desire for addition and squaring to commute—that is, that the square of a sum be equal to the sum of the squares. It certainly sounds nice, and it looks great on the page: $(x + y)^2 = x^2 + y^2$, but it is simply not the case. (Not only is it not true for all numbers, it's barely true for any!)

So, just as woodcarvers don't put too much stress on the wood lest it crack or break, we can't go around asking numbers to obey our whims. On the contrary, we should feel grateful that we have such nicely behaved creatures as we do. The point is that we need to familiarize ourselves with what our numerical entities and operations actually do and also be aware of what they do not.

One thing that makes addition such a nice operation is that we can extend it easily to more than two quantities. If we have collections of sizes 7, 4, 3, and 1, then their total is 15, regardless of how we choose to group or order the individual additions. We could add them serially as $((7 + 4) + 3) + 1$ or as two separate sums, $(7 + 4) + (3 + 1)$; it simply doesn't matter. This is not only convenient mentally, but also has the effect of eliminating the need for parentheses: $7 + 4 + 3 + 1$ is entirely unambiguous. So that's pretty nice.

The simplest instance of this rearrangement property occurs with three numbers:

$$(a + b) + c = a + (b + c).$$

A binary operation with this feature is said to be *associative,* meaning that it gets along well with the neighbors, so to speak. It's not hard to see that by repeatedly employing this associative property, we may regroup a sum in any way that we please.

Aside from this purely notational convenience, associativity actually has a deeper, more important meaning. One way to think about an operation like addition is that it enables us to view our sedate and static numbers in a more active and lively way—to transform our nouns into verbs, as it were. To any number, we can associate the act of adding it. (We talked about this idea before, with shifting dots.) Thus, 3 (the noun) becomes *add* 3 (the verb). Addition of numbers can then be viewed as a compound instruction: $3 + 5$ corresponds to adding 3 and *then* adding 5, which is the same as adding 8. In other words, addition of numbers corresponds to simply following one activity with the other. Essentially, what it means for an operation to be associative is that we can then interpret our elements as actions in a consistent and coherent way.

Luckily for us, multiplication is also both *commutative* (i.e., symmetrical, meaning $ab = ba$ for all numbers a and b) as well as associative: $(ab)c = a(bc)$. This can be easily seen by viewing both products as measuring the volume of an $a \times b \times c$ box.

From an operational, algebraic point of view, these two basic properties—associativity and commutativity—give us the freedom to rearrange our sums and products in whatever way we choose. This also gives us the linguistic and notational freedom to view these operations as nonbinary—meaning they can take not just two but any number of terms as input.

Is the average of two numbers an associative operation?

We now have three ways to view any universal property. We can think of it as a behavior shared by all numbers, or we can instead view it as a property of our operations. Once we have identified such a property, we can then also view it as a *maneuver*—a way of rewriting and reimagining a numerical expression in a way that changes its outward symbolic form without affecting its value or meaning. Anytime we are confronted with a sum $x + y$, we are free to recast it as $y + x$ without fear that we have accidentally altered information in a substantive way.

The universal properties of arithmetic are our guide to which symbolic manipulations we should allow ourselves so as not to jump to unwarranted conclusions. You certainly can rewrite $x + y$ as $y - x$, or xy, or even 0 if you wish, but then you're no longer doing mathematics; you're just playing symbol games that have nothing whatever to do with the actual nature and behavior of numbers. These creatures may be fictional, but they still do what they do and not necessarily what we wish them to do.

So what are the universal truths of arithmetic? What behaviors do all numbers share? The number 6 has the remarkable feature that it is both

$1 + 2 + 3$ as well as $1 \times 2 \times 3$, but this is simply an individual quirk, not a universal behavior shared by all. On the other hand, the fact that $a + 0 = a$, for all numbers a, tells us that the specific number 0 does in fact satisfy a universal property.

As always, we can think of this property of zero in a number of ways. We can view it as the defining feature of this unique and special entity, zero—the thing that when added leaves everybody alone—or we can think of it as an amusing property of the addition operation: the act of adding zero is simply the act of doing nothing. Symbolically, it means that we are free to replace $w + 0$ by w (and vice versa) anytime we like.

Not that this is in any way surprising. The number 0 is supposed to encode the quantity *none*, so of course adding nothing does nothing. It hardly seems worth mentioning. What is there of any interest to say about the act of doing nothing? (Other than being one my favorite pastimes, that is.)

From the practical point of view, as when counting collections, there isn't much to say about an empty pile. The value of zero as a number is not its utility in calculation but rather its theoretical significance in giving meaning to the idea of unaddition. In other words, we require a neutral element to tell us what we mean by the undoing of an activity. If I put on my shoe and then take it off (effectively putting on my anti-shoe), then the result of this "addition" is that I have done nothing. The way we can tell that one action is the inverse of another (e.g., locking and unlocking the door) is that, when combined, they accomplish a grand total of bupkes.

When we enlarged our (reality-inspired) number system to include negative quantities, the reason we conjured up a number like –3 was to represent a new, fictional quantity that was "three less than zero." An alternative view would be that the act of adding –3 is how we reverse the process of adding 3. More simply, all we are really saying about this number –3 is that when added to 3 it makes 0. That is, for each number a, we have its opposite, or *negation*, $-a$. The only demand we make on this number is that $a + (-a) = 0$ so that the acts of adding a and of adding $-a$ are inverse.

So the functional purpose of zero is not just to do nothing but also to provide us with a means of defining opposites. These two universal properties, $a + 0 = a$ and $a + (-a) = 0$, are what allow us to undo the addition of any number simply by adding its opposite. (Associativity also plays a subtle role, if you think about it.)

We have a similar issue with multiplication as well. To undo multiplication by 2, we need to multiply by anti-two, or in this case, $\frac{1}{2}$. The reason this works is that $2 \times \frac{1}{2} = 1$, and 1 is the number that when multiplied by does nothing. So addition and multiplication share the property of having a universal do-nothing, or *identity* element. Zero is our name for the *additive identity* (the thing that leaves everyone alone under addition), whereas 1 is our *multiplicative identity.*

There is almost a corresponding universal statement about multiplicative inverses, but there's that annoying zero nonsense. Multiplication by zero cannot, in fact, be reversed. So instead, our corresponding universal property is that every *nonzero* number a has an opposite (which, for the moment, I will denote by $\bar{a}$) so that $a \times \bar{a} = 1$. (Of course, this number $\bar{a}$ is nothing other than the reciprocal of a.) So if you want to undo multiplication by 2, you can multiply by $\bar{2}$ (that is, ½), and if you want to undo multiplication by ½, you can simply multiply by its inverse, $\overline{½}$, otherwise known as 2.

This is entirely analogous to the additive situation, except for the proviso regarding zero; there is no number that when added causes such destruction. Every addition preserves the distinction between numbers, but multiplication by zero turns everyone into zero, thus losing all information concerning who was who. There's no way to undo something like that. Multiplication by zero is the arithmetical equivalent of dropping an egg on the floor—there's no coming back from that, no way to reverse the process.

Aside from this exception, we're in good shape. Any nonzero rational or real number—positive or negative—has a reciprocal. If your number is $-\frac{2}{3}$, then its multiplicative partner is $-\frac{3}{2}$. Of course, its additive partner (i.e., its negative) would be $\frac{2}{3}$. Each operation has its own sense of inert object (0 or 1), as well as its own notion of inverse (negation or reciprocal). As we've seen already, subtraction and division can then be described (or even defined) in terms of these opposites: $b - a$ is just $b + (-a)$, and $b \div a$ (that is, the thing that when multiplied by a is b) can be rephrased as $b \times \bar{a}$.

So far, we have managed to identify eight universal properties of arithmetic. For any numbers a, b, and c, we have

$$(a + b) + c = a + (b + c) \qquad (ab)c = a(bc)$$
$$a + b = b + a \qquad ab = ba$$
$$a + 0 = a \qquad a \cdot 1 = a$$
$$a + (-a) = 0 \qquad a \cdot \bar{a} = 1 \quad (\text{for } a \neq 0)$$

These are just a few observations we've made about the behavior of our creatures and the way they combine and interact—field notes from the jungle of arithmetic, if you will. Taken together, they say that our two basic operations are quite well behaved: they group nicely and are symmetrical—allowing us to add and multiply in bunches—and they are also invertible so that we can freely do and undo (the zero thing excepted, of course).

Although these are perfectly nice properties and are certainly universal, they are also somewhat trivial, as well as patently obvious. Our universal properties may be true, but they are also rather boring, and the algebraic maneuvers they suggest are not particularly exciting either. "What's that? I get to erase any added zeroes? Whoop-de-do!"

The trouble is that neither addition nor multiplication alone can tell the whole story. The properties that we have listed so far involve only one operation at a time. The real fun—as well as the real complexity—begins when we start combining them. It is the interaction and interrelation of addition and multiplication that give arithmetic its real flavor and provide algebraists with artful and elegant technique.

One simple example of such interaction is the property that adding a number to itself is the same as doubling it: $x + x = 2x$. Of course this is obvious, but it goes to the heart of what we mean by multiplication: making copies. (The word itself comes from the Latin *multiplicatus*, "folded many times.") Originally, what we meant by 5×7 was five copies of seven. If we like, we can picture this as a rectangle consisting of five rows of seven rocks each. We can also view such a product as taking place in the continuous realm, where it would then represent a rectangular area measurement: a bathroom floor measuring 5 by 7 feet would require $5 \times 7 = 35$ square-foot ceramic tiles.

Whether you imagine your rectangles as rows and columns of rocks or as regions to be tiled, either way we are dealing with rectangles, and that is a good thing. Rectangles are extremely reliable and well-behaved objects. One feature that makes rectangles particularly useful and attractive is the way that they stack and subdivide.

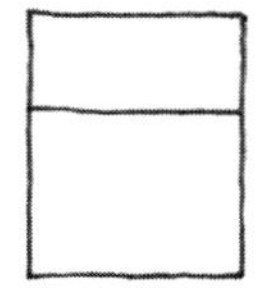

If two rectangles (of the same width) are stacked together, they form a larger rectangle. This is a rare and valuable property. Squares do not stack to form squares, and neither do hexagons or circles. Another way to say it is that rectangles chop up nicely into smaller rectangles.

This feature of rectangles tells us something about the relationship between addition and multiplication: if we have an $a \times b$ rectangle (or a rows of b rocks, if you like) and we join it to an $a \times c$ rectangle, we obtain an $a \times (b + c)$ rectangle.

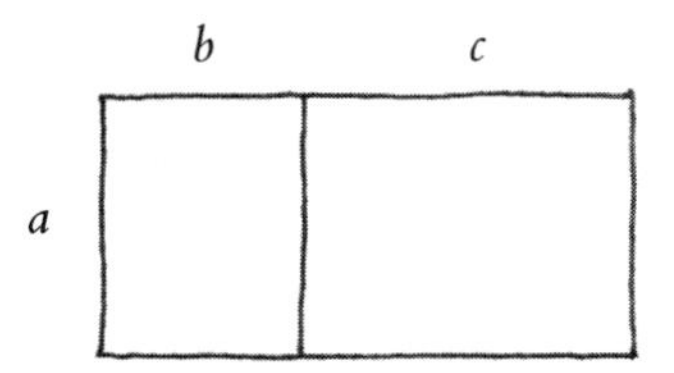

Since the combined area is just the sum of the two parts, we find the universal property

$$a(b + c) = ab + ac.$$

One way to think of this is to imagine two collections of sizes b and c. If we double these collections separately and then combine them, it's the same as putting them together first and then doubling the total. Notice also that if our number a happens to be negative, the pattern still holds.

This simple and elegant feature of our number system is known as the *distributive* property of multiplication, meaning that when a number a is multiplied by a sum $b + c$, that multiplication is distributed through each term of the sum to obtain $ab + ac$. Since multiplication is commutative, this must also mean that $(b + c) \cdot a = ba + ca$, which makes good sense from both the copying and rectangular viewpoints.

The distribution idea can be carried quite a bit further, actually. Rectangles can be chopped up into as many pieces as you like, both horizontally and vertically.

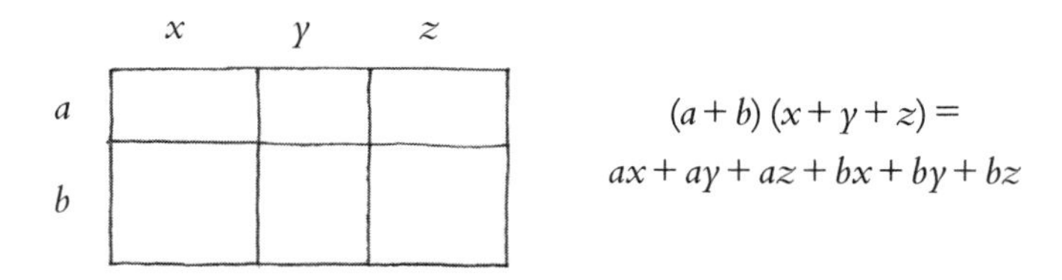

For each way of chopping a rectangle, we obtain a corresponding distribution pattern. The general result is that the product of any two sums can always be rearranged as a sum of individual products—namely, every possible product of terms, one from each sum. Of course, all such patterns can be derived from repeated uses of the original, simplest version: $a(x + y) = ax + by$, just as every chopping of a rectangle can be performed one slice at a time.

What happens if we multiply three sums together?

Although our diagrams suggest that all of the numbers involved are positive (meaning they can be viewed as lengths of sticks or numbers of rocks), the pattern is quite general and holds for any real numbers whatsoever, positive or negative. For example, the product $a(x - y)$ can also be rewritten as $ax - ay$. We can either view this as a separate distribution pattern, or—more elegantly—we can simply view it as the ordinary distribution of a over the sum $x + (-y)$.

Show that for all numbers a and b, we have

$$(a + b)^2 = a^2 + 2ab + b^2.$$

For a working algebraist, distribution means that we have the luxury of being able to take a product of sums and "turn it inside out" so that it takes the form of a sum of separate products. This does not grant us unlimited license to rewrite any expression involving addition and multiplication in any way we choose, however. The distributive property is quite precise: the product of two sums is the sum of the individual products, one factor taken from each sum. If we are dealing with the product $(x + 2)(y + 1)$, we may replace it with $xy + x + 2y + 2$, and if by chance we encounter $pq + 2p$, we are free to call it $p(q + 2)$.

This is the way numbers behave (or if you prefer, the way our operations behave), so we get to perform this maneuver on any numbers whatsoever, regardless of whether they are known to us or not. Numbers exhibit general, as well as personal, idiosyncratic behavior, but it is only the universal patterns that can lead to general methods. If we are going to operate on both known as well as unknown quantities, then we need to make sure that all of our moves are universally valid and independent of the particular values of the quantities involved.

Show that for all numbers a and b, we have

$$(a + b)^3 = a^3 + 3a^2b + 3ab^2 + b^3.$$

What is the corresponding expansion for $(a + b)^4$*?*

A knot, as well as the means of untying it, should be independent of the material being knotted. Whether we are dealing with rope, string, or chain, the pattern of which strand goes over which, and what end gets pulled through what loop—that is, the knot itself, the knot *in the abstract*—is the same. When we perform a distribution, such as multiplying each term of a sum by a certain number, we don't have to stop and wonder whether the numbers can handle it, or whether the system is too "stiff" to allow such a manipulation. The universal truths of arithmetic provide the moves we can *always* make, and this is their great value. Of course, there are way too many of them to possibly list, just as there are infinitely many ways to tangle a piece of string.

On the other hand, if you think about what actually goes into tying a complicated knot, you realize that there aren't really that many basic, fundamental moves. The complexity arises not from the maneuvers themselves, but from the way they are combined. The basic moves of knot tying are almost all entirely trivial and easily reversible—laying the left strand over the right, for instance. The maneuver that gives knots their essential flavor is pulling a strand through a loop. This is the decisive step, the one that is not always so easy to perceive or undo. So yeah, yeah, this over that, around the bend, over and under, but it's when you go *through* that the action really happens in the knot-tying game.

Similarly, our first eight universal patterns of arithmetic—essentially expressing the symmetry of our basic operations—are all rather pedantic and trivial. (The business about zero having no reciprocal being the most interesting aspect, I suppose.) Sure, we get to rearrange the order of our sums if we like, but that's hardly surprising. Also, as we talked about, these properties involve only one operation at a time, with no mixing of addition and multiplication at all.

So it's the distributive property that gives arithmetic (and therefore classical algebra) its characteristic flavor. Without distribution, we would simply have two independent operations spinning their wheels separately. The distributive property binds them together, forcing multiplication to act in the copy-making style. If we like, we could even say that the *reason* that $x + x = 2x$ is that 2 is simply shorthand for $1 + 1$, and distribution tells us that $(1 + 1)x = 1x + 1x = x + x$. In other words, the distributive property is really the distillation of what it means to multiply or to scale a quantity: it means to act appropriately with respect to addition in the sense that scaling a sum should be the same as scaling the individual pieces.

Notice the operational asymmetry of the distribution pattern: $a(b + c) = ab + ac$. Addition and multiplication are being treated quite differently here. Were we to switch the roles of addition and multiplication, replacing all sums by products and vice versa, we would have instead the alternative pattern $a + bc = (a + b)(a + c)$, which is patently false. Addition does not distribute over multiplication; it is multiplication that distributes over addition. So this is the way that our two operations can be fundamentally distinguished. Addition and multiplication may both be symmetrical and associative, as well as invertible, but they behave rather differently when it comes to their interaction with each other.

The distributive property also informs our choice of meaning for multiplication by negative numbers. We may not know what we want -2 copies to mean, but distribution sure does. Whatever $(-2)w$ is, when we add it to $2w$ we'd better get $(2 + (-2))w$, which is the same as $0w$—in other words, 0. So this forces us to choose $(-2)w$ to mean $-(2w)$. Even the obvious fact that $0w = 0$ can be said to be a consequence of the distributive property. We have $0w + 0w = (0 + 0)w = 0w$, and subtracting $0w$ from both sides, we get $0w = 0$.

Which brings us back to our different perspectives on numbers and their properties. The classical view would be that the objects of interest are the real numbers and their behavior. Our list of nine properties is then a set of useful *facts*. The scribal, artisanal perspective would be that these nine properties represent nine universal *moves* that may be performed at any time—distribution being by far the most interesting and important, allowing us to "pull a sum through a loop," so to speak. From this point of view, the numbers themselves are almost irrelevant, serving only as a substrate upon which our operations can act. They constitute the backdrop for the action, not the action itself. Numbers are merely the string, not the clever way in which it is entangled. What are most valued are the universal truths of arithmetic; these are what lead to useful and valuable techniques.

As an example, let's take this puzzle from the Rhind Papyrus:

A quantity, one-half and one-quarter of it added to it, becomes ten.

Expressed in modern notation, this reads

$$w + \tfrac{1}{2}w + \tfrac{1}{4}w = 10.$$

Here I have chosen w to represent our unknown quantity. Now we can employ the distributive property to rewrite this as

$$\left(1 + \tfrac{1}{2} + \tfrac{1}{4}\right)w = 10.$$

I like to call this sort of maneuver a "collection move" because we're taking the various amounts of our unknown quantity and collecting them together as one big multiple. This is an example of viewing a universal property from a technical perspective; the distributive property gives me license to perform collection moves whenever I see fit (as well as many other clever techniques, as we shall see).

If we wish, we can simplify the outward form of this equation (without, of course, affecting its content) by doing some arithmetic: $1 + \frac{1}{2} + \frac{1}{4} = \frac{7}{4}$. Our puzzle then takes the form

$$\frac{7}{4}w = 10.$$

Here we can imagine, if we like, that w was walking down the street minding its own business, and all of a sudden it got multiplied by $\frac{7}{4}$ and is now 10. How can we figure out who it was before this happened? The idea is to simply undo what was done. We started with w, did this thing to it, and now it's 10. So the plan is to start with 10 and then undo the thing, and that will get us back to w itself. To undo multiplication by $\frac{7}{4}$, we simply multiply by its reciprocal, $\frac{4}{7}$. So we find that w must be $\frac{4}{7} \times 10$, otherwise known as $\frac{40}{7}$. (If you like, you can rewrite this as $5\frac{5}{7}$, but it doesn't really do all that much for you.) Just to be sure, let's check that this number really does work: half of $\frac{40}{7}$ is $\frac{20}{7}$, and one-quarter of it is $\frac{10}{7}$. Adding these together, we get $\frac{70}{7}$, which does indeed equal 10.

Of course, no algebraist then or now would care about this particular problem or its solution. What we do care about is the technique, the *way* we untied the knot. In particular, collection moves are a powerful and flexible technique for reducing complexity. The multiterm expression $w + \frac{1}{2}w + \frac{1}{4}w$ is replaced by the sleek and elegant $\frac{7}{4}w$—in effect, reducing the number of occurrences of our unknown from three to one.

The question, then, is what are *all* of the universal properties of arithmetic, and what new and clever maneuvers do they allow us to perform? Surprisingly, it turns out that this is it; there aren't any others. Any universal property of real numbers, expressed as an equation involving only addition and multiplication (along with their inverses) can be deduced solely from our nine basic properties. (We will later see why.)

From the point of view of a classical algebraist, this means that all you've really got at your disposal is distribution and your wits. No further amazing universal truths will be forthcoming, only those built from repeated uses of these nine. For example, the Babylonians noticed that if you multiply a sum $a + b$ by the corresponding difference $a - b$, something very pretty happens:

$$(a + b)(a - b) = aa - ab + ba - bb.$$

The terms *ba* and $-ab$ cancel out, and we are left with a difference of squares. So we discover the universal formula

$$(a + b)(a - b) = a^2 - b^2.$$

Any experienced algebraist has run into this pattern thousands of times and can recognize it instantly. Of course, it's nothing more than a special case of distribution, but it's an interesting special case nonetheless, and one that has a great many uses, as we shall see. So yes, classical algebra essentially comes down to the study of the distributive property (together with its eight friends), but that's more than enough to create plenty of mischief, just as tucking strands through loops is all you need to produce the most complicated snarl imaginable.

From the technical point of view—meaning the creative use of tools to solve problems—the distributive property gives us the ability to rephrase arithmetic information: a product of sums can be rewritten as a sum of products. Aside from the comparatively trivial maneuvers suggested by the other eight properties, this is really the only interesting move we've got. So the bulk of classical algebra consists of increasingly clever and creative uses of distribution to untangle algebraic knots.

Show that for all numbers p and q,
$$(p + q)^2 - (p - q)^2 = 4pq.$$

So far, the main arc of the story is that we have physical reality—actual collections and real-world quantities, along with their approximate comparative measurements—which inspires our idealized and perfected notion of number. The properties of these imaginary quantities (in particular, their behavior under addition and multiplication) suggest improvements—such as the invention and inclusion of negative numbers—that extend the system and make it more symmetrical and flexible. Our numbers and operations satisfy nine basic universal properties, the distributive property being the most noteworthy. From the classical point of view (giving the ancients a little more credit than they deserve regarding negative quantities), we now have an ideal environment in which to formulate (and hopefully solve) algebra puzzles. If the puzzle has its origin in the practical world of reality, we can replace all of the weights of stones and ages of humans by simpler, more pure and elegant numbers. The problem can then be studied and classified, generalized, and solved in the realm of pure mathematics, and then—if need be—the results can be translated back into the grubby language of rocks and people.

What happens, of course, is that you (as a classical or modern algebraist) get really good at manipulating symbolically encoded information, just as a poet or composer does. In our case, it is numbers and operations rather than tones and syllables, and the grammar simply comes down to writing equations that are clear and unambiguous (e.g., using parentheses when necessary). You get really good at moving information around in your head and on paper. Just as a composer can hold a complex harmony in her mind and can hum and dictate each of the separate parts, an experienced algebraist can combine and rearrange abstract numerical information fluently and creatively.

An interesting side effect of symbolic notation systems is that we are establishing a cognitive link between a certain squiggle on paper and a certain object or idea. But the symbol itself also has a look to it that may conjure up entirely different imagery. Music teachers sometimes speak of ants when describing triads to children because the notation (three black dots) suggests it, not because of any intrinsic relationship between chords and insects.

Similarly, when I encounter an equation such as $x^2 = y - x$, my immediate desire is to get rid of that $-x$ on the right-hand side. I want that x over on the left with its friend x^2. So the usual maneuver would be to undo this subtraction with an addition—namely, I want to add x so that it cancels with the $-x$. Of course, since this is an equation—a statement of equality—I can't just go around changing things as I see fit. I need to respect the meaning of sameness. In particular, if I do something to one side of an equation—whatever that action may be—I had better do it to the other side as well or I will have broken the bond of equality between them. So, apart from our nine basic properties of addition and multiplication, we have also this fundamental property of equality: if two things are equal and you perform the same activity on both of them, then the results must also be equal. This is not really a special feature of algebra; it's simply what we mean by the word *equal.*

The point being that I can rewrite my original equation $x^2 = y - x$ in the slightly more attractive form $x^2 + x = y$. This is the result of having added x to both sides of my original equation. However, this is rarely how I think of this particular maneuver. I don't know when this image first came to me—I must have been fairly young—but when I want to move a subtracted quantity from one side of an equation to the other, I always see it (and even sort of feel it) as a monkey (y, in this case) whose prehensile tail (the minus sign) is holding an orange or something (namely, x) and who then throws it to the monkeys on the other side of the river, thus eliminating the orange (and also the tail, apparently) from one side and causing it to reappear on the other.

This is the kind of thing that happens to all artisans and craftsmen. You get to know your tools so well they become your trusted friends. (Though maybe I shouldn't be trusting monkeys with my fruit!) Sailors also employ all sorts of fanciful imagery when learning to tie knots, often involving rabbits and hidey-holes and whatnot.

Anyway, don't be surprised if weird things like that start happening. (I've probably poisoned your mind forever with the monkey-tail meme.) The point is that written symbols are very powerful and suggestive things; they have a sneaky way of taking on a life of their own. (I doubt astrology would have half its appeal if not for its arcane alchemical symbology.) There is a dreamlike quality to the way symbols encode subconscious meanings. Whether we are composing on staff paper or solving an algebra problem, we are engaged in the creation of sense-memory correlations: hand muscles are moved, a symbol is seen, connections—conscious and unconscious—are made with past occurrences of such correlations, mimetic imagery (e.g., monkey tails) is formed, and the symbol acquires additional cognitive baggage well beyond its original intent.

Hopefully, you will not find the cognitive load of our symbols to be too high. Fortunately, we are speaking exclusively of numbers and operations, so there's not a ton of ambiguity. As with any notation system—perhaps especially music notation, with all the black dots looking pretty much identical—one must be careful to say and write precisely what one means. These sorts of symbolic languages tend to be quite brittle and finicky—alter one eighth note here, drop a minus sign there, and the entire structure is ruined; erase a symbol or two, and the information is lost. It is often unclear, even from context, what was intended.

Some forms of communication are more forgiving than others, however. We could even imagine a scale of "information density" for different media. A novel, for instance, would have fairly low information density. We could remove an entire page from *Anna Karenina* and probably still reconstruct what happened; there is plenty of redundancy in the form of dialogue and narrative. A poem has much higher information density; even the omission of a single line does a great deal of damage. A musical composition might be of similar density, perhaps a tiny bit higher depending on the form. Often the key and chord progression, period and style are enough to reconstruct a missing phrase or passage.

Algebraic notation, unfortunately, is at the extreme high end of the information density spectrum: each and every symbol is vital. If you accidentally erase the 3 or the + in the expression $3z + 1$, there may be no contextual clue whatever to tell us what you meant. (I suppose if a left parenthesis were omitted I might be able to make a good guess as to where it belongs, but maybe not!) So we need to be on our toes when we read and write algebraic information; every symbol matters, and there may be no easy way to recover from a transcription error.

When one becomes skilled at something, the tools and media of that craft begin to infiltrate one's dreams and take on various layers of meaning. So an algebraist is going to end up having a very intimate, hands-on relationship with numbers and numerical operations. These "objects" will acquire a new aspect, quite separate from their original quantitative meanings. Numbers and operations begin to take on a functional, behavioral quality: this moves there, that undoes this, flip this over there, cancel those guys—what things *do* starts to matter more than what they are. After all, if they do what I need them to do, who cares what they are?

The approach of the craftsman leads to a utilitarian and artisanal viewpoint that is in its own way quite abstract. What exactly is a chair? Does it matter, so long as I can sit on it? A classical woodworker might object, but I'm sure that a modern furniture designer would happily entertain the functional definition of chair: anything that can be sat upon. That's part of the whole modernism thing, really—the questioning of conventions and the breaking down of meaning into its essential parts.

The modern viewpoint in mathematics is that we are engaged in the study of pattern—pattern in the abstract. These patterns are held by abstract structures such as number systems and geometries, operations and transformations. The aesthetic is simplicity and generality, abstraction and unity.

This viewpoint tends to foster a functional, structuralist philosophy regarding mathematical entities—we don't care what it is, so long as it does what we need it to do. In a sense, what it does *is* what it is. A chair is anything you can sit on, and zero is anything that when added does nothing. But what is addition? Isn't it pushing two (fictitious) sticks or piles of rocks together?

The modern view is that it doesn't matter what numbers are or what addition and multiplication mean. In fact, we prefer that they have no fixed, specific meaning (e.g., a chair requiring four legs) so that we can be as general and as abstract as possible. What we do insist upon—at least in the present context of classical algebra—is that our number system must satisfy the nine properties we have listed. In other words, we intend to mathematically capture the notion of a number system as *any* environment that behaves the same way as our familiar, reality-inspired ones do.

So the plan is to *axiomatize*—that is, to list our demands (also known as *postulates*) and then to assert that anything whatever that meets these demands is worthy of the name "number system" or "chair" or whatever concept it is you are trying to capture. Of course, any axiomatization right away raises the question of whether the generalization may have gone too far and now allows for monstrosities that technically satisfy our demands but utterly fail to jibe with our feelings and instincts about the original intuitive concept. (I'm sure we've all seen chairs that cross the line into stupid and pretentious, regardless of any lumbar support they may offer.) So you'll have to ask yourself whether you feel these nine properties capture and characterize numerical behavior in your opinion.

In the view of the modern algebraist, they do. Let us imagine a putative number environment consisting of a collection of abstract entities, together with two operations that we will call "addition" and "multiplication." These are simply names and may have nothing whatever to do with the usual notions. I will, however, use the same symbols so as to reinforce the analogy. If this system satisfies our nine behavioral postulates, then we call it a *field*. Thus, a field is simply a place where things may add and subtract, multiply and divide, according to all of the usual patterns. An element of a field is called a *number*, whatever it may be.

The point is we don't care what anything is, or what it may possibly connote in human society. We care what it does (and with whom) and what patterns it exhibits. From the modern point of view, classical algebra is essentially the study of fields, whether the ancient algebraists knew it or not.

Of course, no one is saying that numbers (in the traditional sense) aren't amazing and interesting; they certainly are. All we modernists are saying is that by widening our scope to include any possible field environment, we gain a deeper understanding of numbers and their properties. There are specific, particular features of the fields $\mathbb{Q}$ and $\mathbb{R}$, for instance, as well as general behaviors shared by all fields. No matter what particular plant or animal you are interested in studying, you will learn a lot more about it by including its variations and close relatives as well.

So the abstraction game has taken us from rocks in the hand to scratches on paper to abstract entities with symbolically encoded behavior patterns. But isn't that more or less what happened with art, music, and literature as well? The era of the late nineteenth and early twentieth centuries saw the breakdown of the classical viewpoint across the board, including mathematics. Just as the impressionists and cubists were breaking with the objective depiction of reality, mathematicians were also dispensing with the physical universe as a necessary reference point. We study and create imaginary pattern-holding structures determined by axiomatic behavioral demands. So sue us. We try to understand what patterns are created by what other patterns—those pertaining to the physical universe and its mathematical models very much included, of course—and we especially try to make connections between structures that reveal a deeper, even more general class of patterns. Again, the goal is to attain understanding, and it is abstraction and generalization that provide the means of gaining the necessary altitude and perspective.

Not that that's all there is to it. We still have to get in there and solve our problems with whatever tools we have at our disposal (wits and stamina being the most important). We work at the level of generality in which we are comfortable, probing our imaginary structures as best we can, trying to gain intuition from experience. So yes, we are idealistic, pie-in-the-sky dreamers, but we also spend a great deal of time in the jungle, hacking away at the underbrush and sweating like pigs. Our curiosity demands it.

Whether we adopt a classical or modern viewpoint, the result is the same. We have a collection of entities that combine in certain ways according to prescribed behavioral patterns. Using these operations, we can entangle our creatures into algebraic knots, which take the form of equations involving both known and unknown entities. The art is in using these patterns to untangle such knots and thereby deduce the values of these unknowns. If you want to take the position that what we are talking about are actual quantities (i.e., rational or real numbers in the traditional sense), you certainly may. Either way, we are working within a field environment, and the moves and maneuvers are the same whether we think of them abstractly or in concrete terms.

One convenient feature of the axiomatic approach is that we sidestep entirely the philosophical question of what we mean by number. We do not need to wrestle with any thorny, quasi-religious questions concerning the nature of unity and multiplicity or the mind's capacity to grasp infinity. A number is simply an element of a field, and a field is simply a system (of any kind) that satisfies our nine demands.

The status of our real number system $\mathbb{R}$ is now reduced to that of a mere example, or instance, of the larger field concept. The rational number field $\mathbb{Q}$ is another. This is one of the simplest and most interesting fields to be sure, but it is only one of a great many such environments in which algebraic knots can be tied. If we are serious about understanding such knots and how to untie them, then it behooves us to consider *all* of the various possible media—not just string and rope—that can support them. Fields are the natural pattern-holding structures for algebraic information.

So we have opened the barn door fairly wide, allowing any system of entities and operations that satisfies our nine demands to be considered a legitimate number system. What exactly have we allowed in? How large is the class of all possible fields?

Well, it turns out there are a lot of fields out there—an uncountable infinity of them, in fact. That means we could really use a classification system so that we can organize the collection of fields into some semblance of order. It is certainly not my intention to go into great detail here, but it might be amusing to see a few examples of just how wide a generalization we have made. How weird can a number system get and still obey our nine demands?

So far, we have encountered only the fields $\mathbb{Q}$ and $\mathbb{R}$, the number systems arising from counting and measuring. Since every rational number is included in the real number system, we see that $\mathbb{Q}$ is in fact a *subfield* of $\mathbb{R}$—that is, a field living inside a larger field environment.

As it happens, there are a great many subfields of $\mathbb{R}$, all of which are *extension fields* of $\mathbb{Q}$—meaning they include the rational numbers, as well as any other numbers we might decide to throw in. To constitute a field, this new collection will need to form a closed system under our various operations—that is, in addition to whatever numbers we toss in, we will also need to include all of their possible sums, products, reciprocals, and so on.

One such example is the field $\mathbb{Q}(\sqrt{2})$, obtained by adjoining $\sqrt{2}$ to the system of rational numbers. This is not simply $\mathbb{Q}$ with the addition of the single number $\sqrt{2}$ but rather the closed system of all numbers that can be built from $\mathbb{Q}$ and $\sqrt{2}$ by the use of our operations in any sequence. Thus, the field $\mathbb{Q}(\sqrt{2})$ must contain such numbers as $1 + \sqrt{2}$, $\frac{4}{3} + \frac{5}{2}\sqrt{2}$, and $\frac{1}{3-\sqrt{2}}$. Clearly, we are talking about a lot of numbers here. Luckily, the situation doesn't get too out of hand. An extremely convenient feature of this particular extension field is that all of its elements can be written in the same simple format. Every element of $\mathbb{Q}(\sqrt{2})$ has the form $a + b\sqrt{2}$, where a and b are rational numbers.

Why can every element of $\mathbb{Q}(\sqrt{2})$ be written in this way?

In particular, it is not hard to see that $\mathbb{Q}(\sqrt{2})$ is countable. Of course, $\mathbb{Q}(\sqrt{2})$ is still merely a subfield of $\mathbb{R}$, so we really aren't seeing anything particularly new. Is every field just a subfield of the real numbers?

As the Renaissance algebraists discovered, much to their surprise and delight (not to mention mystification and horror), the answer is no. Just as we felt $\mathbb{Q}$ to be lacking—not containing $\sqrt{2}$, for example—it turns out that $\mathbb{R}$ also leaves something to be desired. In particular, the real number system fails to include square roots of negative quantities. There simply is no real number that can play the role of $\sqrt{-7}$, for instance.

So the natural thing to do would be to enlarge and extend $\mathbb{R}$ to include such desirable entities. It turns out to be enough to throw in any one of them, say $\sqrt{-1}$, and that will produce the rest. For instance, we can always write $\sqrt{-7}$ as $\sqrt{7} \cdot \sqrt{-1}$, if we choose. The extension field $\mathbb{R}(\sqrt{-1})$ thus contains not only all of the real numbers, but also $\sqrt{-1}$ and any numbers that can be built from it, such as $\pi + \sqrt{-5}$ and $\sqrt{6} + \sqrt{-163}$. Again we luck out, in the sense that every such number (as we will presently show) can always be put in the standard form $a + b\sqrt{-1}$, where a and b are real numbers.

The extension field $\mathbb{R}(\sqrt{-1})$ is known as $\mathbb{C}$, the field of *complex numbers*—numbers that are the sum of a real number and a so-called imaginary number. This terminology is classical, of course, reflecting a traditional bias against such abstract quantities as $\sqrt{-1}$, but nevertheless we seem to be stuck with it. The trouble with things like notation and nomenclature is that once they catch on and become entrenched, they are very hard to dislodge—even in light of new understanding that makes them obsolete (e.g., the QWERTY keyboard, as a classic example). In any event, $\mathbb{C}$ is a perfectly nice field, regardless of what you choose to call it.

Thus, not only can we view $\mathbb{Q}$ as a subfield of $\mathbb{R}$, but we can also now see $\mathbb{R}$ itself as a subfield of $\mathbb{C}$. The classical algebraists were excited to discover the existence of the complex number field, and for a variety of reasons. First, of course, was the ability to work with new numbers—square roots of negative numbers in particular. In addition, it was soon realized that $\mathbb{C}$ is in fact a vast improvement over $\mathbb{R}$ in that *every* number we could ever hope for—meaning any number describable as the solution to an algebra puzzle framed in the context of $\mathbb{C}$—already exists in $\mathbb{C}$. We say that $\mathbb{C}$ is *algebraically closed*, meaning there is no way to produce by means of algebraic operations and equations any additional numbers. The complex field $\mathbb{C}$ is the holy grail of classical algebra, the perfect setting for all of our (classical) algebraic needs.

There are, of course, some trade-offs. One very beautiful consequence of moving from $\mathbb{R}$ to $\mathbb{C}$ is that we leave the restricted confines of the number line and enter a larger, more expansive *number plane*. The good news is that we can still view our operations as acting geometrically. As we shall see, addition in $\mathbb{C}$ still corresponds to shifting, only now there are two dimensions worth of shifts we can perform. Multiplication turns out to include not only the usual stretching and flipping but also rotation as well.

The bad news is that in moving from line to plane, we sacrifice an important feature: *order*. The number line has a sense of direction, or orientation. We can speak of a number being positive or negative, and we have a clear notion of one number being larger or smaller than another. The real numbers are not only a field, but an *ordered* field. We are used to the sum and product of positive numbers being positive, for instance. There is coherence and consistency to this interplay. The ordering gets along with the algebraic structure, in other words. This is not so for the complex field $\mathbb{C}$. There is simply no way to order the points of the plane so that positive times positive is always positive. Planes aren't lines, and there's not much we can do about that. Order, although we are quite used to it, is not a given: some fields can be ordered appropriately and some cannot.

This means that order is not, in fact, an algebraic issue at all. Algebra is the study of operations, not comparisons. From a modern perspective, we would say that order is part of the spatial, or "topological," aspect of $\mathbb{R}$, not an algebraic feature. In fact, the entire number line image is algebraically irrelevant. It is the fact that $\mathbb{R}$ is a field—a place where we can add, unadd, multiply and unmultiply—that matters to the algebraist. That $\mathbb{R}$ is also an ordered field is nice and is certainly one of the ways in which $\mathbb{R}$ (and its subfields) are special. But it in no way affects our ability to tie and untie algebraic knots.

Here is an example of the wider abstract perspective yielding new insights. Since ℂ is a definite improvement over ℝ (from an algebraic viewpoint) and is also incapable of being ordered, it must be that ordering is irrelevant to the algebra project. In particular, the property of being positive or negative—that is, being greater or less than zero—is not an algebraic notion.

But wait. Wasn't the whole point of moving from ℕ to ℤ that we wanted to include negative quantities? Don't we need a notion of being less than zero? No, we do not. What we require is that every number have a negative—an additive opposite that combines with it to make zero. Thus, the negative of 2 is –2 and the negative of –2 is 2. We need to make a distinction between *having* a negative and *being* negative. Nowhere in our nine demands do we mention anything about anybody being positive or negative. We only require that addition be invertible, hence the demand for *negation* (or multiplication by –1, if you like).

The revelation here is that the number line is a structure with *layers*—different sets of demands being made for different reasons. We are now seeing that the algebraic structure (ℝ being a field) is separate from its geometric structure as an oriented line. Nothing about algebraic entanglement involves this linear image in any way; it's an intuitive picture that we certainly enjoy (and was the motivation for ℝ in the first place), but if we were to pluck each real number from the line and scatter all uncountably infinitely many of them to the four winds, their algebraic behavior would be unaffected. Once the elements and operations of a field are defined, they do what they do—regardless of any mental images you may wish to entertain.

Just as a piece of string can be used for other purposes than tying knots—a game of cat's cradle, for instance—the real number line is the setting for a wide variety of mathematical inquiries, some of which pertain to the algebraic structure, others to the topology or geometry of the situation. The issue of uncountability, for instance, concerns none of these structures—only the raw collection ℝ itself.

I want to be careful not to go off into the clouds too much here. My hope is that even if none of this modernist folderol makes any sense to you, at least I'm getting the motivation and the feeling of it across. Anybody doing anything for long enough will eventually get to a point where they are bored with their familiarity and understanding and will feel the need to push the boundaries of their art. If a stack of colored paper can make it into the Museum of Modern Art, then excuse me if I feel that the investigation of Mathematical Reality by means of deductive argument—math, in other words—is a fine art of the highest order. We push the boundaries, and in math those boundaries tend to lie on the concrete/abstract axis. Breakthroughs almost always entail new ways of seeing, and that tends to mean a *higher* way of seeing—a view from a more abstract perspective, incorporating multiple instances never before put in the same category. When done properly, with adequate attention to the nature of these sorts of epiphanies, mathematics can be a wonderful teacher. As the music of reason, it teaches us how to think; as the art of seeing, it is an end in itself.

Historically, our notion of number has undergone a series of extensions and generalizations from $\mathbb{Q}$ to $\mathbb{R}$ to $\mathbb{C}$ and, finally, to the abstract axiomatic field concept. So far, every field we have considered has been a subfield of the complex numbers. Are there any others? It turns out there are. Not only are there extensions of $\mathbb{C}$—in fact, infinitely many—but there are also field environments that have nothing whatever to do with $\mathbb{C}$ or any of its subfields or extensions.

I will mention only a few of these, just to give you an idea of some of the possibilities. The first example I want to show you is the field $\mathbb{F}_7$. This is the realm of "arithmetic modulo 7," meaning that our elements are the ordinary integers $\mathbb{Z}$ (with the usual notions of addition and multiplication), except that we impose the additional demand that $7 = 0$. The idea comes from the arithmetic of whole number division, where every number has a remainder when divided by 7. The field $\mathbb{F}_7$ is just the ordinary integers endowed with a new notion of equality: we say that two numbers are equal if they have the same remainder on division by 7. Thus, $9 = 2$ and $-3 = 4$, for instance. Stranger still is the fact that in this environment, the number $\frac{1}{2}$ not only exists but is actually equal to 4, since $4 \times 2 = 1$, and that's what it means to be $\frac{1}{2}$. The point is that $\mathbb{F}_7$ is a perfectly good field that satisfies all of our demands, despite having only seven total elements. If you like, these can be represented by the numbers 0, 1, 2, 3, 4, 5, and 6.

Can you work out the complete addition and multiplication tables for this field?

Another example of this sort is the tiny field $\mathbb{F}_2$. In this environment, the only numbers are 0 and 1, and $1 + 1 = 0$. In particular, $1 = -1$, which is a bit weird. One way to think about $\mathbb{F}_2$ is that it behaves like the generic terms *even* and *odd*. Thus, $1 + 1 = 0$ mimics the fact that odd plus odd is even. Notice that in this field, there is no number $\frac{1}{2}$. This is not a defect or a violation of our field postulates but simply reflects the fact that $2 = 0$ in $\mathbb{F}_2$, and 0 never has a reciprocal.

So fields abound, some behaving in more familiar ways, and others not so much. In particular, you may object to environments such as $\mathbb{F}_2$. You may feel that we have gone too far and opened the barn door too wide. Do we really want to do algebra in $\mathbb{F}_2$? What interesting knots can we tie in a string consisting of only two points?

While we're at it, there is the even sillier field $\mathbb{F}_1$, consisting solely of the number 0, where we take $1 = 0$. In other words, we have a number system with only one number—namely, zero—and $0 + 0 = 0$ and $0 \times 0 = 0$. Here I'm going to have to agree with the naysayers. This is a ridiculous place, and there is nothing interesting to say about it. On the other hand, it does technically satisfy each and every one of our demands, albeit for utterly trivial reasons.

The usual decision is to omit $\mathbb{F}_1$ from the pantheon of fields by including in our multiplicative identity postulate the additional proviso that $1 \neq 0$, solely to rule out this absurd example. You, of course, can do as you wish. Maybe your heart aches for poor little $\mathbb{F}_1$, and you want it to be included around the field family hearth. That's fine, only you might find yourself saying, "except for $\mathbb{F}_1$, of course" whenever you are stating and classifying general patterns. It is fair to say that $\mathbb{F}_1$ belongs in its own category, whether or not we wish to call it a field.

A similar situation occurs with prime numbers, where we face the question of whether we wish to consider the number 1 to be prime. It certainly satisfies the condition that it does not factor into a product of smaller numbers (again, for a silly reason). But the main role of primes in arithmetic is to serve as building blocks for natural numbers. Every positive integer is a product of primes, with the exception of the number 1 itself. In other words, the only utility of 1 as a prime would be to build itself.

There is a good analogy here with chemistry. The periodic table contains all of the elements needed to assemble any chemical substance, with one exception: nothingness. The substance consisting of nothing at all is not made of atoms but only of itself. Should we then add it to the periodic table as element 0, for the sake of completeness?

Taking the hard-hearted utilitarian view, the answer is no. No element 0, no 1 being a prime, and no $\mathbb{F}_1$ either. These are aesthetic decisions, and the aesthetic—as always—is simplicity. We are trimming the fat and discarding structures that don't matter to us. We get to do this with chairs as well, drawing the line beyond which we will not sit. Thus, we subtly adjust the barn door, narrowing it slightly when we see that some annoying gnat has managed to get in.

I know I've been talking your ear off, but I do want to mention one more example of abstraction leading to a richer and more sophisticated perspective. Earlier, we were talking about squaring—building a square on a given length. Area considerations then led us to a purely numerical notion of squaring—namely, multiplying a number by itself. Of course, we can do the same thing with volume. Given a stick, we can imagine the *cube* with that stick as its side. Then scaling by a factor w will multiply the volume by $w \times w \times w = w^3$. For this reason, the product of three copies of a number is called its cube. We now have a new unary operation—cubing a number—that then requires an inverse, which, naturally enough, is called the *cube root* operation. Rather than invent yet another symbol, we instead co-opt the existing square root sign and add a little 3 to remind us that it is the cube root version. Thus, the cube root of the number w is written $\sqrt[3]{w}$. In particular, we have $\sqrt[3]{8} = 2$ and $\sqrt[3]{-1} = -1$.

Is there a difference between $(\sqrt[3]{w})^2$ *and* $\sqrt[3]{w^2}$ *?*

You can probably tell where this is going. Since two-dimensional measurement (i.e., area) leads to squaring and three-dimensional volume to cubing, is it then possible to speak somehow of *four-dimensional* objects and measurements in a logically coherent way—to mathematically capture this notion precisely, not just write science-fiction screenplays? Is there such a thing as a hypercube of side w, with hypervolume w^4?

The classical geometers do not seem to have asked this question. Being tied to reality—despite the apparent abstraction of Euclidean geometry—they were incapable of imagining (or even desiring) such a thing. Reality is three-dimensional, and that's simply the way it is. There are points, lines, squares, and cubes, and that's where it stops.

But the thing about patterns is that they don't like to stop; they like to continue—and we like to help! In fact, mathematicians have devised a perfectly sensible means of constructing spaces of arbitrary dimension. We can speak of four-dimensional spheres and seven-dimensional cubes, and we can even prove perfectly sensible theorems about infinite-dimensional spaces. Again, this comes from a conscious decision to break with reality and to worry more about what is pretty than what is practical. Ironically, what modern physics has discovered is that physical reality in fact resembles less the intuitive Euclidean three-dimensional model and more the kind of abstract higher-dimensional spaces that mathematicians invent and explore for their own sake.

So the advent of higher-dimensional spaces opens up whole new vistas of potential discovery. We are free to explore as well as to classify. We might find that a certain spatial arrangement of profound and intense beauty can only exist in twenty-four-dimensional space, for example. The classical Euclidean notion of three-dimensional space then appears quite limited and particular, a special case rather than the be-all and end-all of geometry. If that's not perspective, then I don't know what is.

Notice, however, that we do not require such higher-dimensional imagery to multiply a number by itself a few times. We can speak of "raising a number to the fifth power" (e.g., w^5) without any concept of five-dimensional space or volume. That's one nice thing about algebra—it's finitary and nonspatial. We've got these numbers, and we operate on them; we don't have to imagine or visualize anything, only understand our operations and the patterns they obey.

So let's generalize our notation to include not only raising a number to any power (i.e., exponent) but also the taking of arbitrary roots.

Thus, $3^5 = 243$ and so $\sqrt[5]{243} = 3$. (Of course, a number like $\sqrt[5]{2}$ is hopelessly irrational.) Now our language is a little larger and a little more expressive. Notice that $\sqrt[4]{y}$ is the same as $\sqrt{\sqrt{y}}$, so there is also a certain amount of redundancy, as usual.

The operations of addition, subtraction, multiplication, division, and the taking of roots constitute the classical arsenal of operations. These are the linguistic means by which we will tie (and hopefully untie) our algebraic knots.

ELEMENTARY METHODS

Let us for the moment adopt a classical viewpoint. Our number system is $\mathbb{R}$, the real numbers, and we regard positive numbers as especially real, in the sense that they correspond to geometric measurements. Problems in arithmetic and geometry lead to algebra puzzles involving known and unknown quantities entangled together, with constraints expressed as algebraic equations. Our project is to classify such problems and to devise general methods for their solution.

The simplest problems would be those involving only one unknown quantity subject to only one constraint. This would include what I like to call "walking-down-the-street" problems, where a single unknown quantity undergoes a series of transformations to arrive at a given value. For example, suppose we are faced with the equation

$$\frac{2q-1}{3} = 5.$$

The idea is to view this as a sequence of events: the number q was walking down the street and suddenly she got doubled; after that, she got decreased by 1, and then divided by 3, and now she is 5. Who was she originally?

The nice thing about a sequence of activities is how easy it is to undo: we simply invert each process individually but in reverse order. If I first put on my socks, then my shoes, and finally my galoshes, then in order to reverse the process I need to start by removing the galoshes first, not the socks! Since q gets doubled, then decreased by 1, then divided by 3 to become 5, all we need to do is *start* with 5, then multiply by 3, add 1, and then cut in half to get back to q. In this way we find that $q = [(5 \times 3) + 1] \div 2 = 8$.

As you can see, this sort of algebra problem is a bit on the trivial side. We are taking an unknown quantity and performing a series of (unary) operations on it to arrive at a known destination. To solve such a problem, we simply work backward. It's a bit like twisting a piece of rope instead of tying a knot. Each twist is easily undone, so a series of twists is not really any more difficult.

The key to the walking-down-the-street method is that the unknown is mentioned *only once* in the equation. If instead it were combined with itself in some complicated way, we might not be able to regard the entanglement as a sequence of individually reversible processes. One very convenient (and quite unusual) feature of this method is that it is immune to complexity—it doesn't matter how many processes our unknown undergoes so long as each can be inverted in turn.

On the simpler end of the spectrum would be such pseudo-problems as $3x = 7$. Here x was walking along, got tripled, and is now 7. Clearly, our number x must be $\frac{7}{3}$. (Another way to think about this is that we are simply dividing both sides of our equation by 3.) There is something a trifle ridiculous about this solution, however. Our equation $3x = 7$ says quite plainly that x is the number that when multiplied by 3 is 7. When we "solve" this equation and write $x = \frac{7}{3}$, we are saying exactly the same thing, because that is all that the symbol $\frac{7}{3}$ ever meant.

Another example would be the equation $y^2 = 6$. Our number y got squared, and now it is 6. That means it must be $\sqrt{6}$, meaning the number that when squared is 6. Are we really doing anything at all? Perhaps the silliest example would be the algebra puzzle $z + 4 = 0$. Obviously the solution is $z = -4$, but it is really more by virtue of the meaning of -4 itself than by any clever maneuver of ours. Problems such as these barely qualify as problems at all. They are really just restatements of the definitions of our symbols.

The difference between $3x = 7$ and $x = \frac{7}{3}$ is subtle and psychological. The mathematical content of the two is, of course, identical, but the first equation feels like an implicit description of a number via its behavioral properties, whereas the second comes across as explicit and direct, the behavior now being expressed by the notation itself.

There is a certain humorous quality to maneuvers like these. It's as if the detective has called everyone into the library to announce the identity of the murderer: "I have determined conclusively that it was . . . the person who did it!" We should be no more surprised that $(\sqrt{6})^2 = 6$ than we are that the washing machine actually washes. After all, we built such numbers as $\frac{7}{3}$ and $\sqrt{6}$ for a reason. Despite the lack of real content, however, there is still some small comfort in having an explicit statement of the value of our unknown, especially if we intend to do any arithmetic involving this quantity.

One particularly simple (and frequently occurring) class of problems are those of the form $5w + 3 = 18$. Here we are combining a given multiple of our unknown with another known quantity to make a prescribed total. In fact, the very first problem we talked about—the one involving equally spaced columns in the Egyptian temple—was of this type. (In that case, the relevant equation was $61w + 155 = 420$.) Making like ancient Egyptians, we could get out some papyrus and start listing puzzles of this type, along with their solutions:

$$\begin{array}{ccc} 5w + 3 = 18 & \Rightarrow & w = (18 - 3) \div 5 = 3 \\ 61w + 155 = 420 & \Rightarrow & w = (420 - 155) \div 61 = 4\frac{21}{61} \\ \frac{1}{2}w - \frac{2}{3} = \frac{1}{4} & \Rightarrow & w = (\frac{1}{4} + \frac{2}{3}) \div \frac{1}{2} = 1\frac{5}{6} \end{array}$$

In each case, we have an unknown getting multiplied and then added (in the last example we are adding $-\frac{2}{3}$). So to undo this, we start with the result, then unadd and unmultiply.

We now have a pattern—a class of problems of the same general type that can be solved in a uniform way. The modern approach to communicating this kind of systematic solution is to give symbolic names to the various given quantities as well as the unknown. In this way, we can write the entire pattern in one line:

$$Aw + B = C \Rightarrow w = \frac{C - B}{A}.$$

Here w is our unknown, and the symbols A, B, and C represent known quantities that are given explicit numerical values in any specific instance. In our first example, $5w + 3 = 18$, we are setting $A = 5$, $B = 3$, and $C = 18$ in both problem and solution. The symbols A, B, and C then act as "settable constants"—like dials we can turn to select any particular problem of this type. These are known as *parameters*. They are not unknowns but rather given data that we choose not to specify in order to exhibit the general pattern. Whenever we stumble across a class of problems that can all be handled by the same general method, we then have the option of solving the parameterized version once and for all, as opposed to making a lengthy and redundant list of specific examples.

As another illustration of our unwinding technique, let's look at a somewhat more elaborate example:

$$\sqrt{3(w-2)^2+1} = 5.$$

Here we have only one occurrence of the unknown w, so we should be able to simply unwind all of the twists. The sequence goes: subtract 2, square, triple, add 1, square root. That's how we get from w to 5. So to get back, we start with 5 and undo in reverse:

$$w = \sqrt{\frac{5^2-1}{3}} + 2 = \sqrt{8} + 2.$$

Many algebraists (including myself) prefer to write such a number in the form $2 + \sqrt{8}$ or even $2 + 2\sqrt{2}$ in order to get the square root as far to the right as possible. Of course this is purely stylistic and personal, and does not matter in the least. (Notice that since $(2\sqrt{2})^2 = 2 \cdot \sqrt{2} \cdot 2 \cdot \sqrt{2} = 8$, the number $2\sqrt{2}$ does indeed qualify as $\sqrt{8}$.)

While we're on the subject of square roots, it is worth noting that equations such as $x^2 = 8$ tend to have not just one, but *two* real solutions, both the positive and negative versions. This is because a number and its negative always have the same square, a consequence of our decisions regarding multiplication by negative numbers. We want $(-1) \times a = -a$ for all numbers a—and in fact, the distributive property forces us to accept this even if we didn't want it. In particular, we have $(-1)^2 = 1$, and thus in general (for any number a in any field), we have $(-a)^2 = a^2$.

This means our squaring operation is not as nice as we might wish. Since different numbers go to the same place, our operation actually *loses information*. The symbol $\sqrt{16}$ is consequently somewhat ambiguous. Does it mean 4 or -4? The standard convention (at least when working with real numbers) is to take $\sqrt{w}$ to mean the *positive* square root of w, writing $-\sqrt{w}$ for its negative. This is fine as far as it goes, but in more general contexts (e.g., over the complex field $\mathbb{C}$), we lose the notion of order and positivity, so all we can say in general is that there are two square roots that happen to be negatives of each other, neither being distinguished in any particular way.

I feel it is best to regard $\sqrt{w}$ as meaning any and all square roots of w—that is, there may be several (usually two) distinct meanings of $\sqrt{w}$, and we need to be aware of that. The number 0 is an amusing exception, in that $\sqrt{0}$ has only one possible meaning—namely, 0. (This is simply due to the fact that $-0 = 0$, so in this case there is no ambiguity.)

The other unpleasant feature of the square root operation on the real numbers is that it can fail to have any meaning at all. Since the square of a real number is always positive (or at worst, zero), it means that equations such as $x^2 = -3$ can have no solution. There is simply no real number that plays the role of $\sqrt{-3}$.

The classical view would be that positive numbers have two square roots (one positive and the other negative), zero has one square root (namely, itself), and negative numbers have none. Symbols such as $\sqrt{-3}$ are meaningless, just as $\frac{1}{0}$ is meaningless: the alleged number supposedly being described does not in fact exist. The blame lies with the squaring operation. Since not every number is a square, the inverse operation (i.e., square root) is not always meaningful.

The modern view is that it's not the squaring operation that's causing the trouble; it's the system of real numbers itself that is the problem. We're the ones trapped in the quantitative, geometric mindset of the number line. So the fault lies not in our squares but in ourselves. It's not that $\sqrt{-3}$ is meaningless—in fact, it is a perfectly sensible symbol for any number whose square is -3. It just so happens that $\mathbb{R}$ fails to contain any such entity. This number may exist quite happily in other fields, however. The field $\mathbb{C}$ certainly contains a number whose square is -3 (in fact, two such numbers), so $\sqrt{-3}$ is perfectly meaningful, as long as we accept the usual ambiguity. These two square roots of -3 are, of course, negatives of each other (meaning they add up to zero), but neither one can be said to be positive or negative, because these words have no meaning in $\mathbb{C}$. If our number environment happens to be $\mathbb{F}_7$, then we also have two square roots of -3, these being the numbers 2 and 5. This is because $2^2 = 4$ and $5^2 = 25 = 4$, and in this weird realm $4 = -3$. As usual, these two square roots are negatives of each other, since $2 + 5 = 0$.

Historically, of course, it took several millennia for the blame to be shifted from the squaring operation to the real numbers. If negative quantities were already suspect and considered second-class citizens (not being lengths of sticks and all), how much more impossible would it be for an ancient scribe to conceive of—let alone accept—quantities that cannot even be situated on the number line at all! The modern philosophy is not that $x^2 = -3$ has no solutions but rather that it happens to have no *real* solutions. Once you pop out of your own universe and start comparing it with other universes, your original viewpoint can't help but appear provincial and naive. So we'll adopt a classical attitude for now, meaning we will confine ourselves to $\mathbb{R}$, but always with an eye toward the larger, more abstract view.

The upshot is that any walking-down-the-street problem is easily untangled, the only slight hitch (so far) being the ambiguity involved in taking square roots.

Are there any ambiguities concerning cube roots?

We can thus regard any algebra problem that involves only one unknown mentioned only once as essentially solved. But what if our unknown is entangled with itself multiple times?

We have already encountered several problems of this kind, involving various multiples of the unknown combined together. As yet another example of this ancient theme, we have this puzzle from the Rhind Papyrus:

A quantity, two-thirds, one-half, and one-seventh
of it added to it becomes thirty-three.

Translating this into modern algebraic notation, we arrive at the equation

$$x + \tfrac{2}{3}x + \tfrac{1}{2}x + \tfrac{1}{7}\,x = 33.$$

Here the unknown x appears four times, so there is no way to view it as x walking along. (I suppose we could concoct some fanciful tale about x replicating itself and so forth, but it will not be a sequence of unary activities.)

On the other hand, we have already found ways of dealing with this sort of scenario. Since we have a sum of multiples of x, we can use a collection move (i.e., distribution) to put them all under the same roof:

$$x + \tfrac{2}{3}x + \tfrac{1}{2}x + \tfrac{1}{7}x = (1 + \tfrac{2}{3} + \tfrac{1}{2} + \tfrac{1}{7})x.$$

Our equation then reads $(1 + \frac{2}{3} + \frac{1}{2} + \frac{1}{7})x = 33$, and now we have only one occurrence of our unknown. That means we *can* view our problem as being of the street variety. All we need to do now is to divide by the quantity in parentheses. We don't even have to do any arithmetic if we don't want; we can simply write

$$x = \frac{33}{1 + \frac{2}{3} + \frac{1}{2} + \frac{1}{7}},$$

and be done with it. If you are curious about exactly which rational number this happens to be, then you are welcome to do some arithmetic and rewrite it as a simple fraction. (In this case, it happens to be $\frac{1386}{97}$.)

So a collection move is a useful tool for reducing complexity. Whenever we have a sum of various multiples of an unknown quantity, we can always rewrite it as a single multiple, reducing the number of occurrences of the unknown and thereby lowering the complexity of the problem. As a somewhat more convoluted example, let's consider this puzzle from ancient Babylon:

I found a stone but did not weigh it.
After I weighed out six times its weight, added two gín,
and added one-third of one-seventh of this weight multiplied by
twenty-four, I weighed it. The weight was one ma-na.
What was the original weight?
(A weight of one ma-na equals sixty gín.)

Here again we have only one unknown, the original weight of the stone (measured in *gín*). Let's call it w for weight. The translation in this case is a bit of a problem in itself, but the way I interpret it would lead to the equation

$$(6w + 2) + \tfrac{1}{3} \cdot \tfrac{1}{7}(6w + 2) \cdot 24 = 60.$$

Again we see the enormous advantage of the modern format. The paragraph (which is itself a potentially flawed translation of the original cuneiform) is confusing and ambiguous to say the least. The algebraic equation, on the other hand, is crystal clear, despite its apparent complexity. Our unknown makes two separate appearances, however, so we'll need to do some work before we can simply unravel it.

Here we have the opportunity to do something both clever and convenient. The puzzle mentions the combined weight $6w + 2$ twice. This means we not only have two occurrences of our unknown, but both of these are also trapped inside the same (parenthesized) expression. One useful technique in such cases is *abbreviation*. At the moment, w is an unknown quantity, meaning that $6w + 2$ is equally unknown. Suppose we simply give this quantity a name, say q. With this shorthand (and a dash of arithmetic), our equation now takes the simpler form

$$q + \tfrac{8}{7}q = 60.$$

Next, we can perform a collection move to rewrite this as $(1 + \frac{8}{7})q = 60$, so that $q = 60 \div \frac{15}{7} = 28$. Of course, we haven't quite solved the original puzzle, since we need to find w, not q. But q was just our temporary name for $6w + 2$, so what we are really saying is $6w + 2 = 28$. Finally, we can simply unwind this to get $w = \frac{26}{6}$. So the stone weighed $4\frac{1}{3}$ *gín*, it turns out.

Now, you may have taken one look at this ridiculous, contrived piece of Babylonian busywork and turned up your nose in disgust—I certainly couldn't blame you. As a problem in mathematics, it is uninteresting and ugly and of no value whatsoever. As an exercise for an apprentice scribe, on the other hand, it is quite good. Nobody cares about the problem or its solution; what we care about are the *methods*—collection, abbreviation, unwinding—that the problem inspires. It's like a still-life exercise in a drawing class. Nobody needs to see another amateur bowl of fruit, but each individual student needs hundreds of hours of practice, nonetheless. We are learning to tie and untie algebraic knots, and the best way to develop techniques and get them into our fingers is to practice a lot. A practice problem need not be elegant or interesting (although it's always nice when it is), only useful as a tool for playing and learning.

The abbreviation technique is never necessary, only typographically convenient. Most experienced algebraists would be quite comfortable working with a simple expression like $6w + 2$ without needing to give it a name. We can simply rewrite $(6w + 2) + \frac{8}{7}(6w + 2)$ as $\frac{15}{7}(6w + 2)$ straightaway, so that our equation reads $\frac{15}{7}(6w + 2) = 60$ from the start. Now we can unwind directly to get

$$w = \left(\tfrac{7}{15} \cdot 60 - 2\right) \div 6.$$

The rest is merely arithmetic. So feel free to use abbreviation if you like, or not if you don't. All I'm saying is that it's an option, and it can be very useful for getting a simpler and clearer picture of what's going on. We need all the freedom and flexibility we can get, in fact. Each new puzzle is an opportunity to be clever and creative, and the more methods and maneuvers we have, the more likely something is going to work.

Let's look at a more extreme example. Suppose we have an unknown y satisfying the absurdly convoluted constraint given by the equation

$$3(y + 4) + 2(1 - y) = 3(2y + 1) - 3y.$$

Here we have four appearances of y, and we would like to reduce that to one. Perhaps the first order of business would be to expand all of the various products using a "distribution move." Thus, for instance, we could replace the product $3(y + 4)$ with the somewhat simpler and more convenient $3y + 12$. This has the advantage of unlinking our unknown y from the known quantity 4. Another way to think of it is that distribution acts as a means for y to break out of the jail of parentheses. Now $3y$ is on the loose and able to combine with other such multiples. Expanding each of the terms leads us to

$$3y + 12 + 2 - 2y = 6y + 3 - 3y.$$

Now it is time to rearrange all of these pieces so that the various multiples of y are all together, as well as the known quantities. We have knowns and unknowns on both sides, making it difficult to perceive what is really being said.

Remember the monkeys and their tails? This is exactly the situation that calls for such a maneuver. Do you see that obnoxious $-3y$ at the end there? One nice way to get rid of it is to simply add $3y$ to both sides. This has the visual effect of moving the subtracted $3y$ from the right side to the left—only now it is *added*. (The monkey $6y + 3$ threw it over using its minus sign tail.) Similarly, by subtracting $6y$ from both sides we can eliminate the first term on the right, making it reappear on the left, only *negated*.

The justification for these moves may be that we are adding the same thing to two equal quantities, but the visual and kinesthetic feeling is that we are swinging chunks of information from side to side, so I tend to view such maneuvers as "rearrangement moves." The important thing to remember is that when you move a term from one side to the other, its sign changes, so if it was added on the left, it will then be subtracted on the right.

Swinging all of our y terms to the left side and the known quantities to the right (and praying that we didn't make a sign error), we get

$$3y - 2y - 6y + 3y = 3 - 12 - 2.$$

Now we can collect and do some arithmetic to obtain the (much simpler) equation $-2y = -11$, and we find our solution to be $y = \frac{11}{2}$. So the creative use of distribution, rearrangement, and collection allows us to reduce the number of explicit mentions of our unknown, vastly simplifying and clarifying what is being asked of our mystery number.

My favorite number z has the property that

$$2(z + 1) + 3(z + 2) + 4(z + 3) = 5(z + 4).$$

What is my favorite number?

Naturally, things get more difficult—and therefore more interesting—when we are dealing with more than one unknown quantity. In this case, we will usually require more than one equation. The more mysteries there are, the more clues we need. Let's start with this classic dialogue puzzle from the *Greek Anthology*:

> *A: "Give me ten minae, and I become three times as much as you."*
> *B: "And if you give me the same, I am five times as much as you."*

Here we have two speakers, both providing information about the quantities (of *minae*) they possess. The first challenge is to understand precisely what is being said. (This is one of the main reasons such verbal descriptions have gone out of fashion; they tend to be ambiguous, overly poetic, and hard to decipher.) Let's use f and s to denote the amounts held by the first and second speakers, respectively. (We could, of course, use x and y, p and q, or whatever names we like, but we may as well benefit from the mnemonics.)

My take on the clues is that the speakers are comparing the amounts they *would* have, were they to exchange ten *minae* with each other. Thus, the first speaker's statement (at least under my interpretation) would translate to the algebraic equation

$$f + 10 = 3(s - 10).$$

The idea here is that if the second speaker gave ten *minae* to the first, then f would increase by 10 and s would decrease by the same amount. After this transaction, we are told that the first speaker has triple the amount of the second. Similarly, the second clue would tell us that

$$s + 10 = 5(f - 10).$$

So here we have a system consisting of two equations involving two unknowns. How are we going to untangle such an incestuous pair of relationships?

We can start by recasting both equations in the simplest possible form. For the first, we can distribute and rearrange to obtain $f = 3s - 40$. The second equation becomes $s = 5f - 60$. We still have two unknowns, and they are still intertwined in two equations, but at least these equations are as simple as they can be. Thus, translated and simplified, our puzzle becomes finding the numbers f and s that satisfy the simultaneous demands

$$f = 3s - 40,$$
$$s = 5f - 60.$$

Notice that both of these equations are simply telling us what f and s are, albeit in terms of each other. This inspires a very natural (and extremely effective) technique known as *substitution*. Since the first equation is being so generous as to tell us what f is equal to (namely, $3s - 40$), we can then replace any and all occurrences of f with this expression. Sure, we don't exactly know (yet) what this number happens to be, but that doesn't matter. The nature of equality is that whenever two quantities are equal, then either can be used in place of the other whenever we feel like it. In this way, our symbolic expressions essentially act as pronouns, standing in for certain particular numbers whose values are still a mystery.

The first equation authorizes us to replace the unknown f in the second equation by the (equally unknown) expression $3s - 40$. Our second equation now reads

$$s = 5(3s - 40) - 60.$$

Before we proceed to untangle this, let's take a second to observe the effect of our substitution maneuver. We started with two equations in two unknowns, and now we have something a bit different. Our first equation, $f = 3s - 40$, has become less of a clue and more of a prescription: it literally tells us what f is, as soon as we know s. So the problem has been reduced to one equation in only one unknown (namely, s), which we can rewrite in the distributed form $s = 15s - 200 - 60$. (If we prefer, we could just as easily go the other way, using the second equation to substitute for s in the first equation.)

The point here is that we have lowered the complexity by essentially eliminating f from the discussion. We are not yet ready to unravel our equation, since the unknown s still occurs twice, but this is easily remedied by collecting everything together to get $260 = 14s$. Thus, we find that $s = \frac{130}{7}$. (This is a rather strange amount of *minae* to possess, but so be it.)

Now that we know the value of s as an explicit fraction, we can return to our first equation to obtain

$$\begin{aligned} f &= 3s - 40 \\ &= 3 \cdot \tfrac{130}{7} - 40 = \tfrac{110}{7}. \end{aligned}$$

So both f and s turn out to be certain numbers of sevenths, 110 and 130, respectively. (I'm not crazy about the awkward fractions, but if the solutions were whole numbers, it might be too easy to guess or to use trial and error.)

This puzzle involved a fair amount of computation, both with unknown expressions as well as explicit arithmetic. This makes it fairly likely that we may have made a mistake of some kind. Whenever I find myself moving a lot of information around in my head, I start to get a little nervous. It's just so absurdly easy to lose a term or to make a sign error. As any algebraist will tell you—and one is telling you right now—the better you get at these mental gymnastics, the more you attempt, and the more likely it is that you will fall from the balance beam of truth. I simply do not trust myself to perform such computations flawlessly, having seen far too many of my careless errors lead to hours of wasted effort. So I routinely check both my process as well as my results. Did I move everything around properly, according to the actual behavior that numbers exhibit (i.e., our nine field postulates), or was I the victim of my own wishful thinking? Did I perhaps make an arithmetic mistake? What about my alleged solution: Does it actually work?

Let's try it. We're claiming that the first speaker has $\frac{110}{7}$ and the second has $\frac{130}{7}$. After the first exchange, these numbers become $\frac{180}{7}$ and $\frac{60}{7}$, which are indeed in 3:1 proportion. And if we go the other way, we get $\frac{40}{7}$ and $\frac{200}{7}$, the second number being five times as large as the first. So our solution actually works, and we have successfully solved a 2500-year-old puzzle!

Let's try another example. Suppose I take a certain known quantity, say 100, and I break it into two pieces whose identities I keep secret. If

I give you some information about these pieces—for instance, that they differ by 13—then you have an elegant little puzzle on your hands:

$$a + b = 100,$$
$$a - b = 13.$$

Here I have named the larger number a and the smaller one b. This is what I like to call a "broken bones" puzzle. In this case, the number 100 is the bone, broken into parts a and b, and we want to mend this break by determining the two pieces from the given information. (I have no idea if this is the actual origin of the romantic translation of *al-jabr*, but I like to think so, anyway.)

At the moment, neither of these equations has the form of a prescription telling us how one unknown explicitly depends on the other, but we can easily make it so. The second equation, for instance, can be rephrased as $a = b + 13$, telling us the value of a the moment we have determined b.

Substituting this into the first equation, we find that b must satisfy the condition $(b + 13) + b = 100$. Once again, we have successfully eliminated a, reducing the problem to one unknown and one equation. Now we can collect and unravel to get $b = \frac{87}{2}$. The prescription $a = b + 13$ then tells us $a = \frac{113}{2}$. My suspicious, paranoid, and insecure alter ego then demands that we make sure: $\frac{113}{2} + \frac{87}{2} = \frac{200}{2}$, which equals 100—so that's good. We also have $\frac{113}{2} - \frac{87}{2} = \frac{26}{2}$, which is 13—so it all works out. The broken bones are $56\frac{1}{2}$ and $43\frac{1}{2}$, in other words.

Substitution is a marvelous technique when we are fortunate enough to be able to apply it. Essentially, what is required is that at least one of our equations can be written so that one unknown is expressed solely in terms of the others. If, for instance, we are faced with five unknowns, and one of them, say w, is equal to some combination of the others, then we can regard that equation as merely a prescription for w, eliminating w itself from every other equation. In this way, we reduce both the number of (remaining) unknowns as well as the number of equations. The trouble, of course, is when none of our equations are amenable to this plan, meaning that we have no way to isolate any of the unknowns so that it is alone on one side of the equal sign.

Luckily, there are alternative methods for reducing complexity. Let's take our same broken bones problem and solve it a different way:

$$a + b = 100,$$
$$a - b = 13.$$

Here is the original puzzle, prior to any rearrangements or substitutions. We have a system of two unknowns and two equations, and we would like to reduce this to one equation involving only one unknown.

The substitution method relies on the principle of equality—that if two things are equal, then either may be used in place of the other. Another pretty maneuver that makes use of this principle is what I like to call a "combination move." The idea is that if we have any two equations, say □ = △ and ☆ = ○, then we can confidently assert that □ + ☆ = △ + ○. If you like, you can think of this as a simultaneous double substitution, in that we are replacing both □ and ☆ with their alternate names, △ and ○. In any event, the visual appearance is that we are simply adding the equations together, forming a new equation from both.

In the present instance, this would mean adding the left and right sides of our two equations separately, to obtain

$$(a + b) + (a - b) = 113.$$

Although we cannot fault the logical validity of this step, we can certainly question its value. What good does this (admittedly correct) inference do for us, exactly? We started with two equations each involving both a and b, and now we have a third equation in the same two unknowns. How does that help?

In general, it doesn't. Given any set of equations you can always combine them in all sorts of ways that are valid—multiply two of them together, add one to five copies of the other, whatever—but that doesn't mean the resulting equation will be in any way helpful or less complicated than the original clues. The art is in choosing a particular combination that helps to simplify matters.

As it happens, our above combination actually does help. In the expression on the left side, $(a + b) + (a - b)$, the two occurrences of a

reinforce each other to form $2a$, whereas the b and $-b$ cancel each other out. Our equation reduces to $2a = 113$, so we get $a = \frac{113}{2}$, as before. It is now a simple matter to replace a with this value and to determine (from either of our equations) the value of b. Or, if you prefer, we can instead *subtract* the second equation from the first (another combination move) to get $2b = 87$, and thus obtain $b = \frac{87}{2}$ directly. In the case of this particular puzzle, we see that substitution and combination are equally effective, so it's basically a matter of taste. (I prefer the combination approach myself because it involves slightly less arithmetic, but that's just me.)

Can you solve the general broken bones puzzle

$$x + y = A,$$
$$x - y = B,$$

with arbitrary parameters A and B?

We have now compiled a fairly versatile set of techniques for untangling algebraic information: rearrangement and collection, substitution and combination, distribution, and, of course, walking down the street. We can now turn our attention to the larger classification project. What are the different types of algebra puzzles, and how are they solved? Are there problems for which none of our methods work? Just how bad can things get?

For the moment, let's stick with the simplest type of constraints, those involving only sums of unknowns and their multiples, such as $x + 2y - 5z = 11$. These are known as *linear* equations. If we view our unknowns x, y, and z as sticks, then we are doing nothing more complicated than scaling them and linking them together to form one long stick, as opposed to multiplying two or more unknowns together (e.g., xy or z^3), which would correspond to making rectangles and cubes.

Suppose we have a problem involving several unknown quantities related by a set of linear constraints. Can we develop a general methodology for this simplest family of algebra puzzles? We can start by ridding ourselves of any confusing verbal descriptions involving weights of stones and ages of Diophantus (thus sidestepping the translation issue entirely) and simply taking our class of problems to be systems of linear equations (written in the modern style) from the outset.

Here is a typical example of a puzzle of this type:

$$3x - 2y = 1,$$
$$2x + 3y = 4.$$

No need for any irrelevant backstory; the equations themselves tell the tale. True, it's a pretty stripped-down, no-nonsense narrative, but that is the mathematical aesthetic at work. Simple is beautiful, and anything unnecessary is just excess baggage to be unloaded and dispensed with. This puzzle is pure and is speaking the language that numbers speak. Since we are the foreigners here, we should do as the Romans do. I can understand if this feels like a loss of romance and connection to the human saga, but the pursuit of simple beauty is *also* a romantic endeavor, and the mathematical aesthetic—austere as it may first appear—is itself part of humanity's search for meaning and beauty.

So let's be craftsmen and see if we can use our tools creatively—and if possible, with style—to solve our problem. One approach would be to try substitution to eliminate one of our unknowns. This is feasible, but it will require some rephrasing. For example, we could take our first equation, $3x - 2y = 1$, and reorganize it as $3x = 2y + 1$, and then divide both sides by 3 to get $x = \frac{1}{3}(2y + 1)$. This is the great value of linear constraints: they *always* allow us to write one unknown in terms of the others, simply by rescaling and reorganizing.

Now that we have a prescription for x, we can substitute into the second equation, $2x + 3y = 4$, to obtain

$$\tfrac{2}{3}(2y + 1) + 3y = 4.$$

We have successfully eliminated x and now have an equation in only one unknown—unfortunately, still mentioned twice. So our next step should be to distribute and collect so that we can walk down the street with only one mention of y. After distributing the $\frac{2}{3}$, our equation becomes $\frac{4}{3}y + \frac{2}{3} + 3y = 4$. Collecting and rearranging terms, we get $\frac{13}{3}y = \frac{10}{3}$, and thus $y = \frac{10}{13}$. Returning to our prescription $x = \frac{1}{3}(2y + 1)$, we find that $x = \frac{11}{13}$. (I'm leaving out a fair amount of annoying arithmetic here, so as not to distract from the methodology.)

Of course, I do not entirely trust myself, so let's check our solution just to be sure. We have $3x - 2y = 3 \cdot \frac{11}{13} - 2 \cdot \frac{10}{13}$, which makes $\frac{13}{13} = 1$, and $2x + 3y = 2 \cdot \frac{11}{13} + 3 \cdot \frac{10}{13}$, which is $\frac{52}{13} = 4$, so it does indeed work out. As always, neither this (reasonably pretty) puzzle nor its (slightly unpleasant) solution matters in the least. The issue is our method and whether it is universally applicable, or whether it might fail in certain circumstances. Can we always use substitution to eliminate in this way?

For problems of linear type, the answer is yes. We can always rewrite any linear constraint so that it expresses one unknown in terms of the others and then substitute to reduce the complexity. The number of unknowns will decrease by one, as will the number of equations, one of them having become a mere prescription for the substituted unknown.

Thus, if we begin with four unknowns, we can use this method to get down to three, then two, then one. Finally, we walk down the street to freedom—that is, unless something goes wrong. What can go wrong is that we run out of equations too fast. Imagine that we are eliminating our way along, and we get ourselves down to two unknowns but only one equation. What kind of puzzle is it to say that $p + q = 14$ but provide no additional information? A lame one, that's for sure. Indeed, we can choose any number we like for p, and then q will be whatever it needs to be (namely, $14 - p$). So this so-called puzzle not only has (uncountably) infinitely many solutions, but also is essentially saying, "pick any value of p you like." From the number detective point of view, everyone is still a suspect. So that's no fun. Such a puzzle (that is, one where the number of independent clues is smaller than the number of mysteries) is said to be *underdetermined.*

One common way this can occur is if our equations happen to be redundant. Consider the system

$$a + b = 3,$$
$$2a + 2b = 6.$$

The second equation is simply a rescaled version of the first, which means it is providing no new information. "The murderer was wearing a red scarf, as well as a scarf that was red"—thanks a lot! Essentially, this means that we have no second equation, and our unknowns are free to take on infinitely many values.

But suppose we were somehow unaware of this redundancy, and we proceeded to use our substitution method. The first equation could be recast as $b = 3 - a$, and this could then be substituted into the second equation to obtain $2a + 2(3 - a) = 6$. After expanding and rearranging, we would then get the exciting news that $0 = 0$. This is usually a signal that we lack enough information and the puzzle is underdetermined.

Even more disturbing is when our clues turn out to be *contradictory*. Suppose I tell you that I've got numbers a and b such that

$$a - 3b = 1,$$
$$3a - 2 = 9b.$$

Two numbers, two clues. Let's see what happens when we substitute $a = 3b + 1$ into the second equation: $3(3b + 1) - 2 = 9b$. When this is distributed and collected, we arrive at the confusing equation $1 = 0$. Far from being redundant, this assertion is patently *false* (unless, of course, you are one of those free spirits who enjoys working in the trivial field $\mathbb{F}_1$).

What's happening is this: whenever we slap down a bunch of equations involving a set of unknown quantities, we are making a tacit assumption. We are assuming (and with no real justification) that there are indeed such numbers satisfying the given constraints. But this may simply not be the case. Some constraints are never satisfied, and many puzzles have no solution at all, despite the fact that the equations appear to be perfectly sensible assertions about mystery quantities. Just as witnesses can lie to detectives, equations can be inconsistent with each other as well. So it is important to go into any algebra puzzle with the understanding that what we are really doing is prefacing our entire problem with a healthy dose of supposition: *Assuming* there are numbers that do this, what can we then conclude? When one concludes that $1 = 0$, then it is clear what has happened. Our assumption that there was a solution has led us to a contradiction, meaning our assumption must have been wrong. There never was a solution in the first place, and all of our subsequent reasoning—however logical and correct it may have been—rested on an initial fallacy.

Does the system

$$3x + 6y = 14,$$
$$8 - 4y = 2x$$

have any solutions?

One of the easiest ways for a puzzle to contradict itself is when the number of constraints exceeds the number of unknowns. In this case, we say that the puzzle is *overdetermined*. It doesn't necessarily mean that it has no solutions—some of the equations might be redundant, for instance—but it does put more stress on the unknowns than they can usually handle.

Generally speaking, three unknowns require at least three independent (i.e., nonredundant) constraints, and it is dangerous to impose more. Of course, we will encounter all sorts of strange scenarios that defy this principle, but the general rule of thumb is that we need as many independent clues as we have mysteries. Two unknowns want two equations, and seventeen unknowns will require seventeen independent constraints—unless you are happy with redundancy, wildly infinite and untethered sets of solutions, or wholesale contradiction.

In any event, we now have a workable generic strategy for any number of linear equations in any number of unknowns. We choose one unknown to eliminate at each stage, lowering the complexity each time until we either run out of equations (underdetermined), contradict ourselves (no solutions), or arrive at a single, unique solution. This gives us both a working methodology as well as a classification system for this very common (and extensive) family of algebra problems.

Of course, the combination approach is equally applicable in these situations and is often preferable to substitution, especially when the number of unknowns gets large. Let's return to our previous puzzle:

$$3x - 2y = 1,$$
$$2x + 3y = 4.$$

Rather than rewriting one equation and then using substitution, we can instead look for a clever way to combine the two equations so that we eliminate one of our unknowns. If it were $2y$ in the second equation instead of $3y$, we could simply add the two equations together and eliminate y in one fell swoop. As it is, however, this move doesn't do us much good. We would merely get another equation (i.e., $5x + y = 5$) involving both unknowns.

It's not that this is bad, or wrong, or false in any way. If the original equations have a solution, then this equation must also be valid. So it's perfectly correct, just not all that helpful. You can go ahead and produce thousands of such truths by combining our equations in all sorts of different ways, but that is a scattershot approach of obtaining information; we want something more strategic than that. We want a combination that actually gets us somewhere. Instead of making random combinations and hoping they eliminate, we need to design combinations so that they eliminate.

So here is the clever maneuver. (Boy, it must have been awesome to be the first human being to come up with this beautiful and ingenious trick!) The problem is that the amounts of y in the two equations don't agree, so when we add or subtract the equations, they don't cancel as we want them to. The idea is to rescale both equations (without in any way affecting their content, of course) so that the numbers *do* agree. In this particular case, we can scale the first equation by 3 and the second by 2 to get

$$9x - 6y = 3,$$
$$4x + 6y = 8.$$

This is the exact same pair of constraints, only rephrased (by design) to create a coincidence: the amounts of y are now synchronized and ready to cancel. Adding the two equations, we then get $13x = 11$, and we're as good as done. The rest is arithmetic (and also checking arithmetic, if you are a worrywart like me).

I really like this method. It is reminiscent of finding a common denominator when adding fractions. I especially like it when you can pull off a double elimination by combining equations in an artful way. For example, there is the metal crown problem I mentioned earlier:

Make for me a crown of sixty minae, mixing gold and brass and with them tin and much-wrought iron. Let the gold and bronze together form two-thirds, the gold and tin together three-fourths, and the gold and iron three-fifths.

Translated, this puzzle leads to a system of four linear equations in four unknowns:

$$\begin{aligned} g + b + t + i &= 60, \\ g + b \phantom{{}+ t + i} &= 40, \\ g \phantom{{}+ b} + t \phantom{{}+ i} &= 45, \\ g \phantom{{}+ b + t} + i &= 36. \end{aligned}$$

I have suggestively written these equations so that the unknowns are aligned in columns; this makes it easy to spot coincidences and agreements.

The first thing I notice is that the unknown g figures in every equation. In fact, except for the first, each equation involves only one other unknown in turn. This means I can do something totally awesome (and exceedingly rare). The combination I have in mind is to add all three of the lower equations and subtract the top one. This gives me the combined equation

$$(g + b) + (g + t) + (g + i) - (g + b + t + i) = 40 + 45 + 36 - 60.$$

When the smoke clears, we are left with $2g = 61$. This one move eliminated three unknowns! Of course, this was due to the extremely contrived nature of the puzzle. The author clearly designed it with this maneuver in mind. In a sense, he was making an algebraic joke, and we were lucky enough to get it—even after more than two millennia. Is that intellectual culture, or what?

The punch line, of course, is what happens now that we know the value of g to be $\frac{61}{2} = 30\frac{1}{2}$. Each of the lower equations then hands us the corresponding unknown on a silver platter:

$$\begin{aligned} 30\tfrac{1}{2} + b &= 40, \\ 30\tfrac{1}{2} + t &= 45, \\ 30\tfrac{1}{2} + i &= 36. \end{aligned}$$

So without really doing much work at all, we find the solution to be $g = 30\frac{1}{2}$; $b = 9\frac{1}{2}$; $t = 14\frac{1}{2}$; $i = 5\frac{1}{2}$. (In this case, checking our solution is practically redundant.)

Now that we have a general method, we no longer need to confine ourselves to specific, individual instances; we can solve entire families of puzzles at a single blow. The most important distinguishing feature of a linear system is the number of unknowns—what I like to call the *dimension* of the system. The generic one-dimensional puzzle would then take the form $Ax = B$, where A and B are parameters. As we saw before, this is essentially trivial. Assuming that A is nonzero, we simply rescale to get $x = B/A$.

What if A is 0?

In the two-dimensional case, we find ourselves faced with *six* generic parameters:

$$Ax + By = P,$$
$$Cx + Dy = Q.$$

The goal is to solve this system so that x and y are expressed explicitly in terms of the parameters A, B, C, D, P, and Q. If we can do this, we will have solved every two-dimensional linear system at the same time. What we are looking for is a *formula* for x and y that will allow us to simply plug in the values of our parameters and obtain the solution, without us having to do any actual work. Or perhaps I should say that the work is in solving the parameterized version once and for all. The only real difficulty here is getting used to the abstraction. (I am purposely using capital letters for parameters and small letters for unknowns to help us remember the status of our various symbols.)

Of course, not knowing what these supposedly "known" quantities are, we will have to argue in the abstract. The method calls for us to rescale our equations so that we get an agreement that we can use to eliminate an unknown. Let's say we choose to eliminate y from the discussion. The easiest way is to rescale the first equation by D and the second by B to obtain

$$ADx + BDy = PD,$$
$$BCx + BDy = BQ.$$

Now we can subtract these equations to get

$$(AD - BC)x = (PD - BQ).$$

The point again being that these parameters represent known quantities, so the expressions in parentheses would be concrete values that could be computed in any specific instance. (Notice that by arguing abstractly like this we spare ourselves from having to do any *actual* arithmetic.)

Now that we have successfully eliminated y, we can easily rescale this equation to obtain x. All we need do is divide both sides by the number $AD - BC$. This gives us

$$x = \frac{PD - BQ}{AD - BC}.$$

Rather than substitute this ungainly and convoluted expression into our equations, it's actually a bit simpler to just obtain y directly, by eliminating x in the same way. Thus, we scale our first equation by C and the second by A to get

$$ACx + BCy = PC,$$
$$ACx + ADy = AQ.$$

By subtracting the first equation from the second, we obtain the eliminated equation $(AD - BC)y = (AQ - PC)$, and hence

$$y = \frac{AQ - PC}{AD - BC}.$$

And there we have it: all two-dimensional linear systems solved simultaneously. Now, if faced with a specific example, such as

$$3x - 5y = 11,$$
$$4x + 2y = 8,$$

we would set our dials to $A = 3$; $B = -5$; $C = 4$; $D = 2$; $P = 11$; $Q = 8$, and simply plug them into our formulas:

$$x = \frac{(11)(2) - (-5)(8)}{(3)(2) - (-5)(4)} = \frac{62}{26},$$

$$y = \frac{(3)(8) - (11)(4)}{(3)(2) - (-5)(4)} = -\frac{20}{26}.$$

The important thing here is not so much the formulas themselves, but rather the fact of their existence. Our method is so general that this sort of thing is possible. Now we can place all such puzzles in the same category—linear, two-dimensional—and know that all such puzzles can be solved in a uniform way. What an explicit formula is saying is that there is a pattern to the way such problems can be solved, and *that* is the headline. The details are the details, but in this case the pattern happens to be quite pretty:

$$x = \frac{PD - BQ}{AD - BC},$$

$$y = \frac{AQ - PC}{AD - BC}.$$

Notice that in both of these formulas we have similar pieces showing up in the numerators and denominators. In particular, both share the denominator $AD - BC$. (This was the scaling factor we divided by at the end to obtain x and y themselves.)

Oh, but wait. What if this number should happen to be zero? That would invalidate our division step. So we'd better investigate this possibility. The first question is whether it can even occur, and if so, what does it tell us? Suppose we have our system

$$Ax + By = P,$$
$$Cx + Dy = Q,$$

and it just so happens that $AD - BC = 0$. This would mean that $AD = BC$, which can certainly happen.

Let's try to engineer such a scenario and see what it looks like. For instance, we can choose $A = 3$, $B = 1$, $C = 6$, and $D = 2$. Then our system takes the form

$$3x + y = P,$$
$$6x + 2y = Q.$$

Now we can see what is going on. The left side of the second equation is just a scaled version of the first. If we multiply the first equation by 2, we get

$$6x + 2y = 2P,$$
$$6x + 2y = Q.$$

So either $2P = Q$, in which case the equations are redundant (and the system is underdetermined), or else $2P$ is not equal to Q, and we have a contradiction. Either way, it means that the puzzle is flawed, and the two equations are not really independent.

Show that $AD - BC$ is zero precisely when the two equations are either redundant or contradictory.

Thus, the number $AD - BC$, known as the *determinant* of the system, acts as a detector of whether a puzzle of this type is degenerate in one way or another. When the determinant is nonzero, our above formulas provide the unique solution.

The same approach can be applied to linear systems of higher dimension, only the number of steps in the procedure will be significantly greater, as will the complexity of the resulting formulae. (There is a corresponding determinant in each dimension as well.) Nowadays, such problems are routinely given to machines to solve (both numerically and symbolically), so modern algebraists have moved on to greener pastures and greater challenges. For both practical and theoretical purposes, we can regard all such problems in linear algebra as essentially solved.

Solve the system:

$$a + b + c = 7,$$
$$2a + 5b + c = 8,$$
$$3b + 4c = 9.$$

Can the parameterized three-dimensional system

$$\begin{aligned} y + z &= A, \\ z + x &= B, \\ x + y &= C, \end{aligned}$$

ever fail to have solutions?

QUADRATIC METHODS

Perhaps the most delightful aspect of elementary algebra and geometry is the subtle and surprising relationship between the two—the amusing and intricate conversation in which they are constantly engaged. Shape configurations lead naturally to numerical relationships among their measurements, and conversely, purely algebraic statements can often be viewed geometrically. As we have already seen, the distributive property essentially reflects the chopping and stacking behavior of rectangles, and the diagonal measurement of the square satisfies the algebraic equation $d^2 = 2$.

Of course, not all questions in geometry necessarily lead to algebraic equations. The measurement of curves and surfaces, for instance, tends to give rise to transcendental constraints beyond the power of algebra to describe. Nevertheless, geometry has always been (and continues to be) a source of great inspiration to algebraists. Many ancient algebra puzzles arise naturally from architecture and engineering, and the idealized versions then become problems in pure geometry.

As a classic example, let's try to measure the diagonal of a regular pentagon.

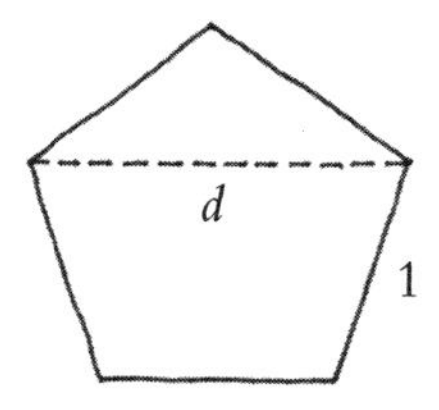

As usual, we'll take the side of the pentagon as our measuring unit and write d for the diagonal length. Since our pentagon is *regular* (meaning entirely symmetrical), we can see that the value of d is completely forced by the geometrical constraints and is therefore beyond our control. So now we have a mystery, and we will need an ingenious argument.

There are a number of clever ways to get at this measurement, and I want to show you one of my all-time favorites. (One reason I like this argument so much is that I was lucky enough to discover it for myself. You can't help being proud of your own offspring!)

The first thing to notice is that because of the symmetry, the diagonal must be parallel to the side opposite. That means when we glue two pentagons together something remarkable occurs:

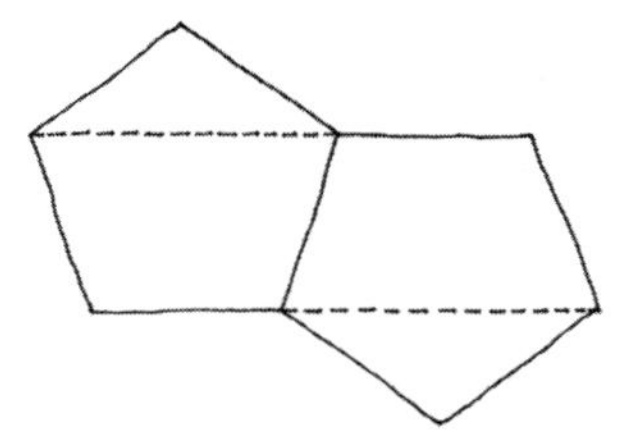

The second pentagon is simply a rotated version of the first—that is, our original pentagon turned upside down. This means the diagonal of each one lines up perfectly with the side of the other! It is this beautiful coincidence that allows us to get at the diagonal measurement relatively easily.

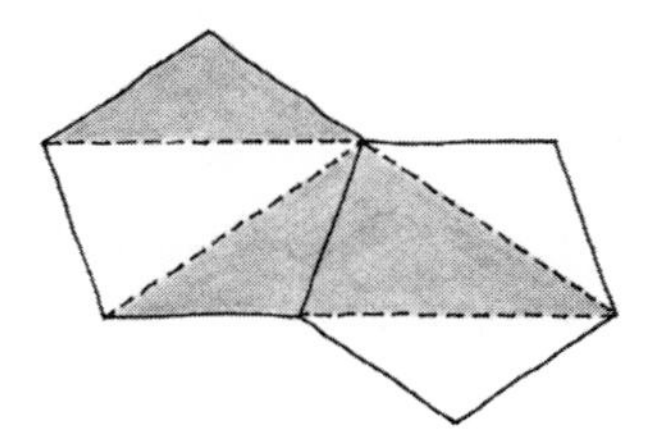

Taking advantage of these alignments, we notice that the two shaded triangles must be *similar*—that is, they have the same shape, only different sizes. This is because each side of the small triangle is parallel to the corresponding side of the larger one, so the two triangles must have the same angles. That means the large triangle is a blowup of the small triangle.

Our unit is the side of the pentagon, so the short sides of the small triangle have length 1. Since these blow up to become sides of length d, the blowup factor must be d itself. (This may ring a few bells, if you recall our argument for the diagonal of a square.) The long side d then also scales by d, and this must equal the long side of the larger triangle, which is clearly $d + 1$. So we find that the diagonal-to-side proportion of a regular pentagon satisfies the algebraic equation

$$d^2 = d + 1.$$

(For an alternative derivation of this result, see *Measurement*.)

We have thus discovered an intriguing connection between a purely geometric description—the diagonal measurement of a symmetrical pentagon—and a purely algebraic one: the number whose square is one more than itself. This number, whoever it is, must be very proud of itself. Like its friend $\sqrt{2}$, it gets to brag that it not only enjoys a simple and elegant algebraic property, but also has a natural geometric interpretation as well.

For the diagonal of the square, our corresponding equation was $d^2 = 2$. Here we discovered the alarming fact that we had no language for this number, so we were forced to add square root as a new verb and then name it $\sqrt{2}$. Now we have a somewhat more complicated and self-referential equation, but we can still ask the same questions. Can d be expressed in the language of fractions, or is it irrational? (Always bet on irrational; there's just so many more of them!) If d is irrational, can it at least be expressed in the language of roots, or must we again cop out and simply make up a name? For instance, we could always rewrite our equation in the form $d^2 - d = 1$, and then give a name to the process on the left. Let's say that to "snork" a number is to subtract it from its square. The snork of 3 would then be $3^2 - 3 = 6$, and the snork of 1 would be 0. Since our number d has the property that its snork is 1, we could just call it the "snork root of 1" and be done with it.

What is the snork root of 12*?*

Algebra is a linguistic enterprise at heart. We are naming things—both implicitly and explicitly—and our questions always come down to what language is needed in order to say what we want to say. The diagonal of the square forced us to adopt an expanded language. Will the diagonal of the pentagon do the same, or is the square root operation enough of a linguistic tool to detangle this equation?

Let's first notice that none of our customary methods are of any avail here. We can rearrange our equation $d^2 = d + 1$ in a multitude of ways, such as

$$d^2 - d = 1, \quad d = \sqrt{d+1}, \quad d = 1 + \tfrac{1}{d},$$

and so on, but in each case we are still faced with two mentions of our unknown. There doesn't seem to be an easy way to eliminate anything. This means we need an entirely new technique.

Algebraic constraints involving squares of unknown quantities are known as *quadratic* equations (from the Latin *quadratus*, meaning "square"). They typically arise in geometry, especially when area measurements are involved. The product of two different unknowns is also considered quadratic information, pertaining as it does to the area of an unknown rectangle. In the old days, the determination of area was known as *quadrature*—the goal being to construct a square of the same area as a given region. Thus, problems in quadrature lead to quadratic equations, and these typically involve products of two unknowns, squares in particular.

This turns out to be rather unfortunate nomenclature, actually. The word *quadratic* clearly pertains to the number 4 (a square having four sides), yet area is a two-dimensional measurement leading to products of *two* unknowns. To an algebraist, the word *quadratic* always means two, as opposed to *cubic* (three unknowns multiplied together). The trouble is that two-dimensional squares happen to have four sides, and now we're stuck with the name. (It reminds me of the time I told my teacher how neat it was that my birthday is October 8, the eighth day of the eighth month. Then she told me the awful news.)

The ancient algebraists were clearly intrigued by quadratic equations. Here is a typical Babylonian example:

I have added together seven times the side of my square
and eleven times the area: six and one-quarter.

Discarding the irrelevant geometric language, the puzzle asks for a number s with the property that

$$7s + 11s^2 = \tfrac{25}{4}.$$

Again we have a certain (slightly messy) combination of an unknown with its square, and we need a way to deal with this sort of entanglement.

Fortunately, the Babylonians discovered a simple and versatile method for handling this very situation. The idea, not surprisingly, comes from the measurement of rectangular areas. Suppose we have a rectangle of a certain width and height, say 6 by 10.

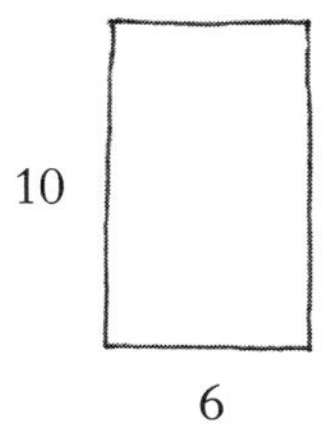

What the Babylonians noticed is that if we chop off the right amount from the top and glue it onto the side, we can form a perfect square—except for a missing square notch.

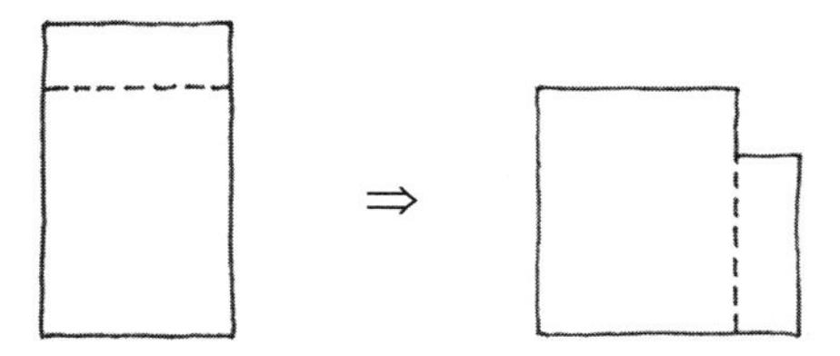

To get this to work, we need the new (larger) width to match the new (smaller) height, meaning we need to "average out" the height and width—that is, we need to reduce the height and increase the width so that they agree at a happy medium. In our case, that means cutting off 2 from the height and adding it to the width, so that we make an 8×8 square with a 2×2 notch.

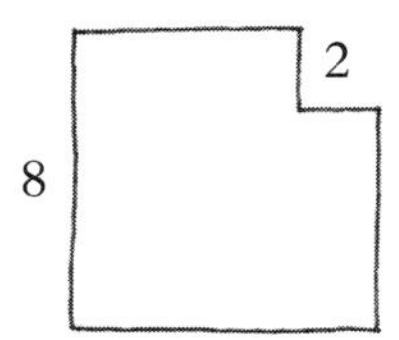

Since chopping and rearranging pieces does not affect area, the upshot is that the product of any two numbers can always be rewritten as a difference of two squares: $6 \times 10 = 8^2 - 2^2$. In fact, we can be a bit more precise. Given any two numbers, their *average* is the number halfway between them. This is the same as half their sum, in fact.

Why is the average of two numbers equal to half their sum?

Because the average lies right in the middle, the distance from it to each of the original numbers is the same—what I like to call the *spread*. Thus, for the numbers 6 and 10, the average is 8 and the spread is 2. Now we can state the Babylonian discovery in complete detail: *the product of any two numbers is equal to the square of their average minus the square of the spread.*

Although this pattern was first observed in the crude physical world of approximate rectangular areas (e.g., farmland), it is also an exact statement in geometry and is thus true for any pair of positive real numbers. As a matter of fact, it works for any numbers in any field and follows directly from the distributive property—again, no surprise since it concerns rectangles. (Actually, there is one modern technical proviso. To define the average, we need to be able to divide by 2. So the result only holds for fields where $2 \neq 0$, which, fortunately, is most of them.)

Let's try it out a few times, just to get used to it:

$$4 \times 12 = 8^2 - 4^2 \qquad (48 = 64 - 16)$$
$$7 \times 13 = 10^2 - 3^2 \qquad (91 = 100 - 9)$$
$$2 \times 3 = (\tfrac{5}{2})^2 - (\tfrac{1}{2})^2 \qquad (6 = \tfrac{25}{4} - \tfrac{1}{4})$$

Another way to think about this pattern is that if two numbers have an average of a and a spread of s, then the numbers themselves must be $a + s$ and $a - s$. So the Babylonian rectangle maneuver is really saying that $(a + s)(a - s) = a^2 - s^2$, which is easily seen by distributing the product on the left. (We saw this "difference of squares formula" before, you may recall.) That means there's nothing really new here, just a special instance of distribution with an amusing rectangle interpretation. So how does this help?

The value of the method becomes clear once unknown quantities are involved. Suppose we have a mystery rectangle with sides w and $w + 4$. The average of these two numbers is $w + 2$, and the spread is 2. That means we can rewrite the product as a difference of squares:

$$w(w + 4) = (w + 2)^2 - 2^2.$$

Notice that the unknown w appears in the average but not in the spread. This means that although we have two occurrences of w on the left, there is only one on the right. The Babylonian technique becomes an elimination tool!

To illustrate the use of this marvelous ancient instrument, let's imagine that we have an algebra puzzle such as

$$w^2 + 4w = 7.$$

Since w appears in both terms on the left, we can "factor out" w (using distribution) to get

$$w(w + 4) = 7.$$

Of course, we still have two mentions of w, but now we can employ the Babylonian technique to rewrite the left side, giving us

$$(w + 2)^2 - 2^2 = 7.$$

Magically, we are back on the street again, and our problem is easily solved: $w = \sqrt{7 + 2^2} - 2 = -2 + \sqrt{11}$. What a clever little species we are sometimes!

Show that this value of w does indeed satisfy $w^2 + 4w = 7$.

Now let's return to the diagonal of the pentagon. Our equation is $d^2 = d + 1$. The first thing to do is to rearrange this into a product of the right sort. Putting all of the unknown quantities on the left, we get $d^2 - d = 1$, and distribution then tells us that

$$d(d - 1) = 1.$$

So there's a mildly annoying preparatory stage in which we must recast our quadratic information as a product of terms, each involving the unknown. (We will soon eliminate this step in favor of a simpler, more streamlined approach.)

Now we are ready to take a trip to Babylon. The average of d and $d-1$ is $d-\frac{1}{2}$, and the spread is $\frac{1}{2}$. Our new and improved equation becomes

$$\left(d-\tfrac{1}{2}\right)^2 - \left(\tfrac{1}{2}\right)^2 = 1.$$

We can now unravel this to obtain $d = \frac{1}{2} + \sqrt{\frac{5}{4}}$. Since $\sqrt{\frac{5}{4}}$ is the same as $\frac{1}{2}\sqrt{5}$, this number can also be written in the attractive form

$$d = \frac{1+\sqrt{5}}{2}.$$

Of course, the numerical value of this (provably irrational) number is irrelevant. (If you're curious, it's approximately 1.618.) What matters is that it is sayable in the language of whole number arithmetic with square roots. It's also pretty cool that the number 5 makes an appearance, being the number of sides of a pentagon. (This is not at all typical; the diagonal of a regular heptagon has nothing to do with $\sqrt{7}$, for instance.)

The Babylonian technique allows us to untangle quadratic equations using nothing more elaborate than square roots. We do not need to incorporate *snork* into our vocabulary, nor any other extension of our language. We required the square root operation in order to solve the simplest quadratic equations (e.g., $x^2 = 13$), and luckily (thanks to some clever Babylonian scribes), this is all the language we need to solve *any* such equation.

While we've got the pentagon sitting here, I want to point out an interesting feature of quadratic equations arising from geometry. From a purely algebraic point of view, such equations typically have two solutions, corresponding to the two possible meanings of the square root. In the case of the diagonal of the pentagon, we have $d = \frac{1}{2}(1+\sqrt{5})$, but in the context of the real numbers $\mathbb{R}$, there are two such entities. If we adopt the notational convention that the symbol $\sqrt{5}$ denotes the *positive* square root of 5, then we have $d = \frac{1}{2}(1+\sqrt{5}) \approx 1.618$, the alternate solution then being $d = \frac{1}{2}(1-\sqrt{5}) \approx -0.618$. It is not hard to show that this number is in fact the negative reciprocal of d—that is, the product of the two solutions is exactly -1.

Why is the alternate solution equal to $-\frac{1}{d}$?

Of course, this is not a solution to the original geometry problem, since a negative length has no meaning—or does it? Is there some way to view this weird negative number as being an alternate geometric solution?

In this case, there actually is an amusing interpretation. We wanted the diagonal of a symmetrical pentagon, but did we really specify exactly what we meant by that? Suppose that a naive (or possibly malicious) fellow mathematician were to interpret it to mean a symmetrical five-pointed star?

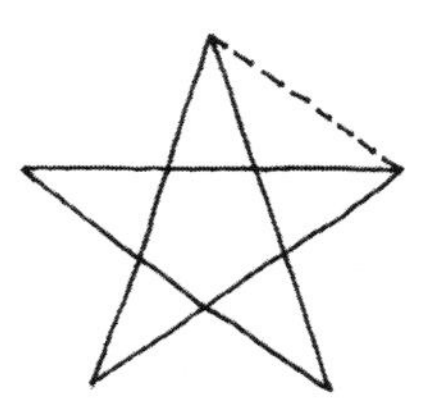

This shape has five straight sides of unit length, making a closed figure. Only now the "diagonal" would be the short distance between consecutive points of the star. The diagonal to side proportion of the star is therefore the same as the side to diagonal proportion of the pentagon. This is clearly equal to $\frac{1}{d}$, which (apart from the minus sign) is the exact value of our alternate solution! The side and diagonal relationship has been inverted and turned inside out, which is what the negative reciprocal is trying to tell us. Our equation thus incorporates both solutions and both interpretations because our argument cannot tell the difference.

This happens frequently, as a matter of fact. A simple and elegant problem in geometry leads to a quadratic equation with two algebraic solutions. Typically, one of these is the intended solution to the original problem (with the conventional interpretation). The other is considered spurious or not meaningful geometrically. But I always like to see if I can concoct a way to give it meaning, and usually this requires going through the looking glass in some way or other. This also comes from being a modernist. I'm listening to these jungle creatures chatter in the night, and I want to understand what they are trying to say.

From a purely algebraic standpoint, we are now in great shape. Our new method gives us a guaranteed elimination procedure, allowing us to convert two appearances of our unknown into one. Putting rectangles and notched squares aside, we can think about this pattern more abstractly. Suppose x is an unknown and $\square$ is a given quantity. (I am thus using $\square$ as a parameter.) If we add these two numbers and then square the sum, we get

$$(x + \square)^2 = x^2 + 2\square x + \square^2.$$

Here we see the crucial feature: one mention of x on the left, two mentions on the right. Reorganizing this a bit, we can write

$$x^2 + 2\square x = (x + \square)^2 - \square^2.$$

What this is saying is that anything that looks like the left side (square plus multiple of the unknown) can always be rephrased so that the unknown appears only once. Notice that $\square$ is not quite the multiple of x on the left side, but rather *half* that value. So if we are dealing with $x^2 + 8x$, for example, then $\square$ takes the value 4 and we would write $(x + 4)^2 - 4^2$. The point being that we don't really need all the fancy Babylonian geometric trappings; we only need to remember to cut a certain number in half.

Here's the way I like to operate. (This process often goes by the name "completing the square.") Let's take as our example the quadratic equation

$$q^2 + 3q = 19.$$

The goal is to rewrite this so that q is mentioned only once. I don't want to do a lot of work here, mind you; I want to make my arithmetic life easier. Since I know I'm going to be constructing a certain square, I first write down a kind of template for myself:

$$(q + \quad)^2 = 19 +$$

From the general pattern we just observed, I know that what needs to go in the parentheses is not 3, but half of 3. So I write in the $\frac{3}{2}$:

$$\left(q + \tfrac{3}{2}\right)^2 = 19 +$$

Now I see that the square on the left will produce not only a q^2 as well as a $3q$, but also the constant $(\frac{3}{2})^2 = \frac{9}{4}$. So the left side of this equation will not quite match that of the original but is instead increased by this amount. That means I need to increase the right side as well:

$$\left(q + \tfrac{3}{2}\right)^2 = 19 + \tfrac{9}{4}.$$

In this way, I obtain the desired equation in one go.

In the old days, of course, negative numbers did not quite exist, and the freewheeling maneuvers that we now routinely allow ourselves were held in suspicion. In fact, quadratic equations were even placed in different categories depending on which terms—all positive, of course—appeared on which side of the equation. There were equations of the form "square and things equal number" (e.g., $x^2 + 3x = 5$) as well as "square and number equal things" (e.g., $x^2 + 5 = 3x$). This sort of classification has the effect of making the subject seem more complicated than it really is.

Fortunately, there is a certain rigidity and structural integrity to algebraic equations and relationships. The patterns that govern numerical behavior (e.g., distribution) are not confined to positive numbers or rational numbers or whatever pet numbers you may hold dear. They are perfectly general, and therefore so is our method of completing the square. There was never any need to classify quadratic equations in this way; it is an unnecessary artifact caused by a fear and mistrust of abstraction. The point is that our nine properties do not stop to ask questions about whether anybody is positive or negative, so neither should we.

One popular format is to put all the terms on the left, positive or negative, leaving nothing at all on the right side. In this way, all quadratic equations then take the same form: $Ax^2 + Bx + C = 0$, where A, B, and C are known quantities. If we like, we can apply our method to this general (parameterized) equation and solve all such problems simultaneously.

Before we tackle this project, let's practice our new method by solving our Babylonian problem from before:

$$7s + 11s^2 = \tfrac{25}{4}.$$

To get this equation into the form we need, let's switch the terms and divide by 11 so that the square term is unmolested:

$$s^2 + \frac{7}{11}s = \frac{25}{44}.$$

Now we can complete the square by cutting $\frac{7}{11}$ in half and adjusting the right side appropriately:

$$\left(s + \frac{7}{22}\right)^2 = \frac{25}{44} + \left(\frac{7}{22}\right)^2.$$

The numbers are starting to get a little ugly, but the whole point is that the method doesn't care—positive or negative, rational or irrational, pretty or ugly, describable or indescribable, it simply doesn't matter. This is the great value of universal properties like distribution—they are true of all numbers, whether or not you have a prejudice against them.

It is now a simple matter to untangle our equation to get

$$s = \sqrt{\frac{25}{44} + \left(\frac{7}{22}\right)^2} - \frac{7}{22}.$$

The funny thing about this ghastly mess of a number is that it happens to be exactly equal to $\frac{1}{2}$! In other words, the creator of this puzzle designed it to have this simple solution, despite its outward complexity.

Can you verify that $s = \frac{1}{2}$?
What is the alternate (negative) solution?

Had we been clever, of course, we might have simply noticed that the number $\frac{1}{2}$ solves our problem, because $7 \cdot \frac{1}{2} + 11 \cdot (\frac{1}{2})^2$ equals $\frac{7}{2} + \frac{11}{4}$, and this is the same as $\frac{25}{4}$. On the other hand, it's nice to know that we have a systematic method that works whether we are clever or not.

As a final example, let's look at yet another ancient Babylonian algebra puzzle:

The igibum exceeded the igum by seven.

Here the *igum* is the unknown quantity itself, and the *igibum* is its reciprocal, only scaled by a factor of 60 (don't even ask). In modern notation, writing p for the unknown, our equation becomes

$$\frac{60}{p} = p + 7.$$

Maybe you are one of those people who plays with numbers a lot, and you immediately notice that $p = 5$ happens to work: 60 divided by 5 is 12, and that is indeed 7 more than 5. That kind of insight can save a lot of time. You may even have found the alternate solution ($p = -12$) by inspection as well.

As a test of our method, let's see what completing the square would look like here. First, we need to reformulate our equation a little. Multiplying both sides by p, we get

$$p^2 + 7p = 60.$$

Completing the square, we obtain the equation

$$\left(p + \tfrac{7}{2}\right)^2 = 60 + \left(\tfrac{7}{2}\right)^2,$$

leading to the solution $p = \sqrt{60 + \frac{49}{4}} - \frac{7}{2}$.

Generally speaking, numbers like this tend to be irrational, unless there is an amazing coincidence of some kind. Here, that coincidence is the fact that $60 + \frac{49}{4}$ happens to be the same as $(\frac{17}{2})^2$. That means the two possible values of the square root are $\frac{17}{2}$ and $-\frac{17}{2}$, leading to the two solutions $\frac{17}{2} - \frac{7}{2} = 5$ and $-\frac{17}{2} - \frac{7}{2} = -12$. Of course, this second solution would not have been allowed (or even imagined) by the ancients.

While we're on the subject of coincidences, I want to mention an alternate approach to quadratic equations that is sometimes quicker, though it requires a bit more intuition and experience. Putting all the terms on the left, our equation becomes

$$p^2 + 7p - 60 = 0.$$

The idea is to then use distribution to rewrite the left side as a product of the form $(p + \square)(p + \triangle)$ for some clever choice of $\square$ and $\triangle$. When expanded, this product takes the form

$$\begin{aligned}(p + \square)(p + \triangle) &= p^2 + \square p + \triangle p + \square\triangle \\ &= p^2 + (\square + \triangle)p + \square\triangle.\end{aligned}$$

It is now a simple matter of reverse engineering: we try to find $\square$ and $\triangle$ so that their sum is 7 and their product is -60. Suppose we insightfully recognize that -5 and 12 do the trick. Then our equation $p^2 + 7p - 60 = 0$ can be rewritten as

$$(p - 5)(p + 12) = 0.$$

(If we were to multiply this out and collect terms, we would simply recover our original equation.)

Of course, this doesn't help us eliminate; the unknown is still mentioned twice. On the other hand, the form of our new equation is very suggestive: the product of two numbers equals zero. How can that happen? What it means is that one or the other (or possibly both) numbers must be zero. In fact, this is a property of all fields: the product of two nonzero numbers can never be zero.

Why is this true?

Thus, the equation $(p - 5)(p + 12) = 0$ leads us to two different possible conclusions: either $p - 5 = 0$ or $p + 12 = 0$. These clearly yield our two solutions $p = 5$ or $p = -12$.

This factorization method essentially reduces a quadratic equation to two linear ones, performing the elimination in a subtle and artful manner. It's a great technique when you can pull it off. The trouble is that it's not always so easy to just "see" the two numbers whose sum and product are required to match the given data.

Can you rewrite $w^2 + 3w - 10$ as a product?

Now let's see what the solution to the general quadratic equation looks like. We have

$$Ax^2 + Bx + C = 0.$$

We want to solve this equation by expressing x in terms of the parameters A, B, and C. The first thing to do is to isolate x^2 by dividing both sides by A. (We do not have to worry about the possibility of A being zero, since in that case our equation would no longer be quadratic.) Putting the unknown terms on the left and the constant on the right, we get

$$x^2 + \frac{B}{A}x = -\frac{C}{A}.$$

Now we can complete the square to obtain

$$\left(x + \frac{B}{2A}\right)^2 = -\frac{C}{A} + \left(\frac{B}{2A}\right)^2.$$

(It is this step that cannot be performed in fields where $2 = 0$, for obvious reasons.)

Thus, we find our general solution to be

$$x = -\frac{B}{2A} + \sqrt{\left(\frac{B}{2A}\right)^2 - \frac{C}{A}}.$$

If we like, we can do a little tidying up to obtain the somewhat more compact and attractive formula

$$x = \frac{-B + \sqrt{B^2 - 4AC}}{2A}.$$

As usual, there are really two solutions here, arising from the ambiguity of the square root operation.

So we see that the square root language is all we need to solve any and all quadratic equations in one unknown, and we even have a general formula showing us the precise shape of our solution in algebraic terms.

The most interesting feature of this expression (known as the *quadratic formula*) is the number appearing under the square root sign: $B^2 - 4AC$.

This is the quantity that determines whether or not a quadratic equation has a solution in a given field. When this number—the so-called *discriminant* of the equation—is not a square, then it will have no square root, and thus our equation will have no solution in that field. If the discriminant is a square, however, then it will have two square roots, leading to two distinct solutions. The borderline case is when the discriminant is equal to zero. In this event, there is only one square root, and the equation will then have a unique solution.

The discriminant $B^2 - 4AC$ therefore acts as a detector, telling us how many solutions to expect. In the case of the real numbers, we already know which numbers are squares: the positive ones. Thus, if A, B, and C are real numbers, the quadratic equation

$$Ax^2 + Bx + C = 0$$

will have no solution if $B^2 - 4AC$ is negative, a unique solution if it is zero, and two distinct real solutions if it is positive.

For other fields, it will depend on which numbers are squares. In $\mathbb{Q}$ this is a somewhat complicated issue depending on the prime factors occurring in the numerator and denominator (e.g., $\frac{4}{9}$ is a square in $\mathbb{Q}$ but $\frac{5}{9}$ is not). For finite fields like $\mathbb{F}_7$, it's simply a matter of squaring every number and seeing which values show up. (The nonzero squares in $\mathbb{F}_7$ are 1, 2, and 4, as it happens.) The complex field $\mathbb{C}$ has the beautiful property that *every* number is a square, so the discriminant is only interesting when it is zero, detecting one solution instead of the customary two. For fields where $2 = 0$ (such as $\mathbb{F}_2$), our quadratic formula is simply invalid, and we have no discriminant to speak of.

Show that the equation $x^2 + x + 1 = 0$ has two solutions in $\mathbb{F}_7$ and no solutions in $\mathbb{R}$ or $\mathbb{F}_2$.

The simplest quadratic equation with no real solution is $x^2 = -1$. The Old World attitude was to regard such an equation as ludicrous. "What, a square of side x with negative area? You must be mad for even proposing such a thing!" The discriminant then performs the vital task of determining whether the alleged constraint is even meaningful. When the discriminant is negative, it simply means that a false statement is being made—there are no numbers that engage in such a nonsensical dance.

From a modern perspective, a negative discriminant no longer means there are no solutions, only that the solutions are no longer real. There are perfectly good complex numbers that satisfy the given equation; they just happen not to live in $\mathbb{R}$. In other words, they correspond to points in the number plane that do not lie on the number line. Even if your field of interest is exclusively $\mathbb{R}$, it is often useful to employ the expanded field $\mathbb{C}$ so that all of your potential solutions can be talked about, real or not.

Generally speaking, if one is working with a quadratic equation over an arbitrary field $\mathbb{F}$, it usually pays to extend the field to include all of the various square roots that you may require. In particular, if the discriminant of the equation is not a square in $\mathbb{F}$, then we can always enlarge our field to include its square root. Now the solutions live in the expanded field but not in the original. Rather than throwing our hands up in despair and saying that the equation has no solutions, we are instead pulling back to a wider perspective and seeing that in fact it has a perfectly good pair of solutions, only they inhabit a larger numerical realm than the one we began with.

In this way, we can see algebraic equations both as puzzles as well as engines for the creation of new numbers. If an equation has a solution in a given field, then we certainly want to know that, but if it does not, then we can read it as a description of an entity that we do not yet possess, and we can look for (or construct) extension fields where such a number does exist. Thus, the modern study of algebraic equations has essentially become the study of fields and field extensions in the abstract.

What the quadratic formula is saying is that in order to explicitly state the solutions to a quadratic equation, at worst all we require is a square root. If such a number already lives in our original field, then great; we've solved our problem. Otherwise, we can extend our field to include it. The solutions may not be the numbers we were hoping them to be, but at least they have a place to live and are sayable in a reasonably simple and uniform way.

We should probably stop for a moment and be thankful that there even is such a thing as a quadratic formula. It could have been that the general quadratic equation breaks up into several cases, each requiring different methods and involving a mass of complicated charts and tables and whatnot. Instead, we have a single uniform pattern. If, for whatever reason, you want to know the solutions to $5x^2 - 8x + 2 = 0$, then you can simply write them down:

$$x = \frac{1}{10}\left(8 + \sqrt{8^2 - 4 \cdot 2 \cdot 5}\right) = \frac{8 + \sqrt{24}}{10}.$$

This can also be written $\frac{4+\sqrt{6}}{5}$, if you prefer smaller numbers.

All I'm saying is that we should be happy. A lot of times people get upset and discouraged by mathematical formulae and equations because they are offended by the seeming complexity. But true complexity is having no description at all. We should be grateful for every pattern and every formula that we've got. These are the distillations of order and harmony in Mathematical Reality. Being sayable at all makes them precious, even if they require a dozen pages of equations and diagrams.

Speaking of equations and diagrams, I want to say a few more things about the ubiquity of quadratic equations in elementary geometry. We saw before that the measurement of the diagonals of both the square and regular pentagon led to quadratic relationships, and this is in fact true of a great many of the simpler shape configurations. For instance, the relationship between the sides and diagonal of a rectangle is also quadratic. This is the famous Pythagorean theorem: for any rectangle, the square of its diagonal is equal to the sum of the squares of its sides.

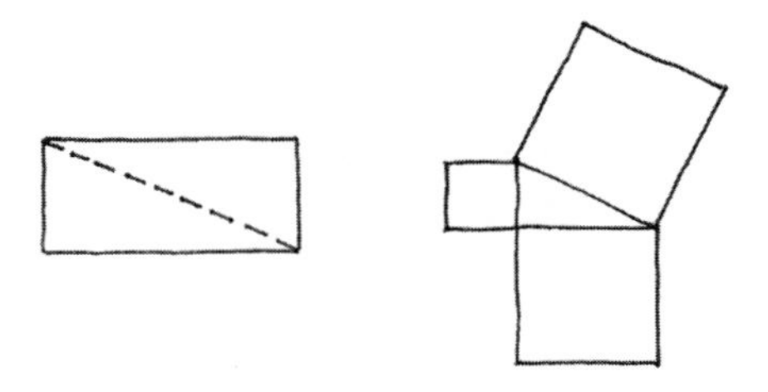

This means that if we cut a rectangle diagonally and then build squares on each side of the resulting triangle, the areas of the squares will add up: the large square contains exactly as much space as the two smaller squares combined. This is one of the most surprising and beautiful patterns ever discovered, and there are literally dozens of ingenious ways to prove it. (President James Garfield even came up with one, in fact.)

My favorite explanation is due to Albert Einstein. It is often referred to as the Three Little Pigs argument. Like many proofs of the Pythagorean relation, it relies on a certain special feature of right triangles—those triangles that include a right angle (i.e., a quarter turn, or 90 degrees). Since the angles of a triangle always make a half turn, the other two angles of a right triangle must then total a right angle together.

Now suppose we place our right triangle on the chopping block, long side down. Chopping vertically through the tip, we form two smaller right triangles:

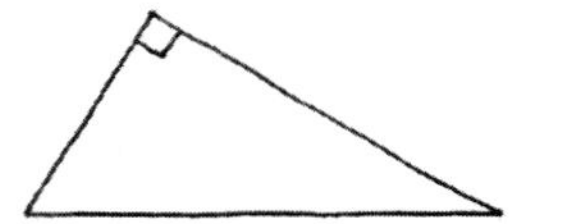

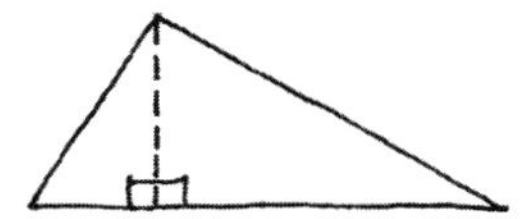

The beautiful pattern here is that the two smaller triangles turn out to have the exact same shape as the original! Each one is a scale model of the triangle we started with, only rotated and flipped. Maybe a simpler way to say it is that the three triangles all have the same angles. That means they must have the same shape.

Why do all three triangles have the same angles?

Now let's build squares on each side, as before.

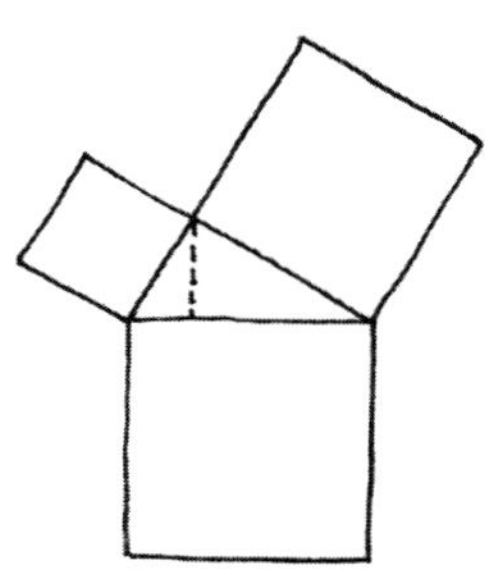

We want to understand why the large square must have the same area as the two smaller squares put together. The ingenious idea is to break our diagram into three "houses," each consisting of a square room and a triangular roof:

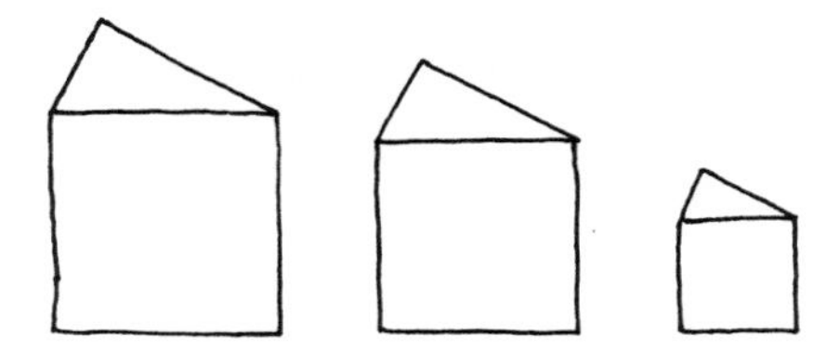

Each house comes from one of the squares in the diagram, together with its own portion of the right triangle. The roof of the large house is the entire triangle, and those of the smaller houses are the subdivided parts. Since the three triangles all have the same shape (as do all squares), the three houses are the same shape as well, only at different scales. But here's the thing: the two smaller triangles *obviously* combine to make the larger triangle, so the squares must also combine in the same way, since each square room is in the same proportion with its triangular roof. This is one of those arguments that is so breathtakingly simple and free from technical details that it may appear *too* simple, like some sort of magic trick. It is certainly a very minimalist piece of work, but well worth taking the time to understand and appreciate. (For a more traditional, no-nonsense approach, see *Measurement*.)

What makes this result so interesting is not that the sides and diagonal of a rectangle are interrelated; of course they are. If you lock down the sides of a rectangle, then its diagonal is forced. So there must be *some* sort of mutual dependence among these measurements. What is not so obvious is that this relationship is algebraic, let alone quadratic. We now know why this is so, but it is still a great gift—consider how the diagonal might have depended on the sides if we *hadn't* been so fortunate!

And what pattern would we have wanted anyway, if we could have our druthers? We certainly couldn't expect the diagonal to simply be the sum of the two sides because clearly it is always shorter to take the diagonal path than to go around the block. So if the diagonal and sides do not themselves relate this simply, at least their *squares* do—it's really about as simple a relationship as we have any right to expect. Plus, it happens to be the truth.

To illustrate the mathematical utility of this discovery, let's use it to measure a symmetrical triangle (also known as *equilateral*, Latin for "same-sided"). Specifically, let's see if we can figure out how much of a square it occupies.

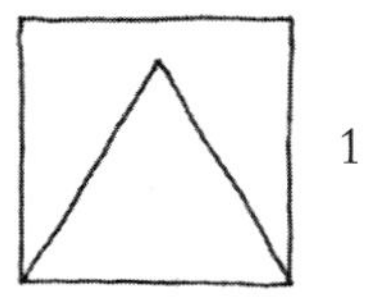

Since all the edges in this diagram have the same length (which we may as well take as our unit), we can see that when the left and right sides of the square fall inward to form the sides of the triangle, the tip must then be lower than the top of the square. This means that the square is not quite the triangle's box (i.e., carrying case). That would instead be the rectangle with the same height h as the triangle.

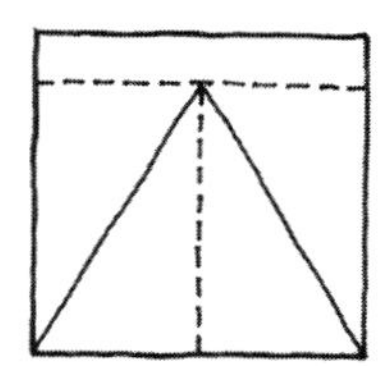

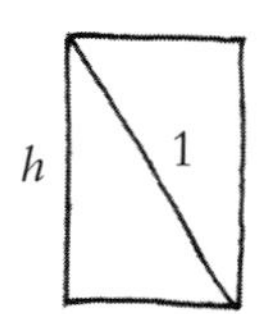

So it all comes down to measuring this height. The key is to notice the rectangle formed by h and half the width of the box. Its diagonal has length 1 because our unit is the side of the square, which is also the side of the triangle. The Pythagorean relation tells us that h must satisfy the quadratic equation $h^2 + (\frac{1}{2})^2 = 1^2$. Thankfully, there is not much untangling to do, and we find that $h = \sqrt{\frac{3}{4}}$ (or $\frac{1}{2}\sqrt{3}$, if you prefer). The height of an equilateral triangle is $\sqrt{\frac{3}{4}}$ times its side. This number is irrational, of course, but obviously describable with square roots.

Now we can return to the original question regarding area. The triangle takes up half its box, and the box is $1 \times h$. This means the area of the triangle is $\frac{1}{2}\sqrt{\frac{3}{4}}$, approximately 0.433. (I've always been rather fond of this particular number because it goes 1, 2, 3, 4 as you write it.) For the practical minded, we discover that an equilateral triangle takes up a little over 43 percent of a square.

Here we have a classic example of how quadratic equations arise in geometry. Simple shapes arranged in simple ways have a tendency to contain symmetries and alignments that produce rectangles, and the length and area measurements of these lead to quadratic entanglements.

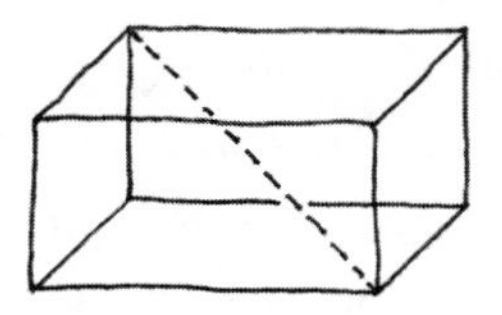

How does the diagonal of a three-dimensional box depend on its sides?

I want to show you a few more amusing examples, but before I do, I need to say a few words about mathematical relationships like the one expressed by the Pythagorean theorem. A lot of people are so freaked out by symbols and abstraction that they end up memorizing things like this as a senseless mantra: "*a* squared plus *b* squared equals *c* squared," or "negative *B* plus or minus the square root of *B* squared minus four *AC*, all over two *A*"—and so on. This is no good. This is not real understanding; this is not having an honest conversation with yourself or your subject. You can memorize formulas all you like, but it won't create a feeling, a touch, or a sensitivity to pattern. The Pythagorean relation is not $a^2 + b^2 = c^2$; that's just alphabet soup. The Pythagorean relation says that for any right triangle, the *long side* squared is the sum of the two short sides squared. And this makes complete sense because the big one should be the sum of the smaller ones, whatever they may be called. Also, suppose *c* happened to be the name of one of the short sides? Then the alphabet mantra would only cause confusion. Similarly, the quadratic formula is about the parameters and the way they are treated, not what they happen to be named.

At some point, in order to really play a musical instrument, the symbols on the page need to mean something to your fingers, without the intermediary of names. You need to perceive a run up the minor scale or a sequence of similar phrases moving with the chord changes. It can't be Every Good Boy Does Fine for each note. The Pythagorean relation and the quadratic formula need to be *felt*, not just notated. The distributive property is not a code; it's a behavior. Arithmetic is not symbols and rules; it's patterns and choices. Mathematics, though an entirely mental art form, is surprisingly tactile and visual—sensory by proxy, if you will. Not only is math all in the mind's eye, but we also find ourselves needing to employ the mind's *hands*.

This is especially true in the realm of geometry, where so much depends on a feeling of physical *pressure*—the ways that shapes constrain themselves and each other in order to do what they need to do. When an object is required to be symmetrical or to contact another object in a prescribed way, this puts tremendous pressure on its measurements, and this pressure needs to be felt and understood if we are to have any hope of solving our problems. A geometer is really a translator, converting shape information into numerical patterns and vice versa.

Take, for instance, this amusing configuration of circles:

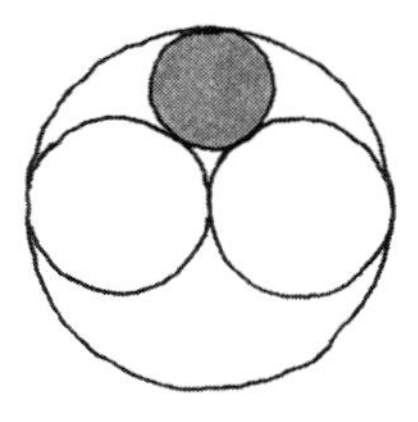

Here is a fine example of a *sangaku*, a Japanese temple geometry problem. This is a wonderful mathematical and cultural tradition dating from the late seventeenth century, in which Buddhist monks would design elegant and challenging geometry problems and display them in the shrines for contemplation. (It is gratifying to know that at least *one* human culture has recognized mathematics as a path to spiritual enlightenment.)

The nice thing about *sangaku* problems is they do not require a lot of explanatory words. We clearly have a large circle with two half-size circles meeting in the center. The mystery is the shaded circle. How big is it? This is the problem facing the geometer (or Buddhist adept). As always, the goal is to convert the geometric constraints on our shapes into algebraic constraints on our measurements.

Let's start by trying to feel all of the various pressures this little circle is under. Suppose I were to say, "Place a small circle somewhere in the space above the two half-size ones." Well, we can certainly do that, but it doesn't put much pressure on the new circle; it can be lots of different sizes. If we demand that the small circle contact both of the middle circles, that would limit it more. But it can still grow and shrink smoothly. When the small circle grows to the extent that it contacts the outer circle, that's when it gets locked in. So we can imagine the shaded circle wanting to be as large as it can be, but the three other circles are pushing on it, keeping it confined and fixed.

Here we can feel three important "pressure points," as I like to call them: the contact points between the small circle and the three larger circles. These points simply have to figure in anyone's argument because these are the points that determine the circle we are trying to measure. Each contact point is therefore a sine qua non (Latin for "without which, not"). We simply cannot determine the size of this circle without taking these points into account. So here's a bit of geometric strategy: locate and identify pressure points.

The centers of circles form another important class of pressure points. Without the center coming into play, how are we to inform our argument that the shape is circular? We will surely get a different solution if one of the circles is allowed to be an ellipse or a random blob. I'm not saying that the *only* property possessed by a circle is having a center, but that's certainly its defining feature. So I always try to keep track of centers, as well as any other points of special symmetry.

Let's mark all the pressure points in this diagram and see if we notice anything.

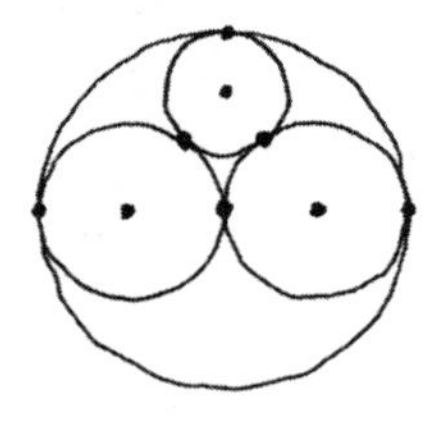

There are nine in all—four centers and five points of contact. The most striking pattern is that the five lowest pressure points appear to line up horizontally. (This is fairly clear from the symmetry and doesn't involve the mystery circle.) There is a more subtle alignment, however, involving the center of the mystery circle, the center of one of the middle circles, and the point of contact between them. This is a routine occurrence, actually. Whenever two circles touch (whether externally or internally), the point of contact is always in line with both centers.

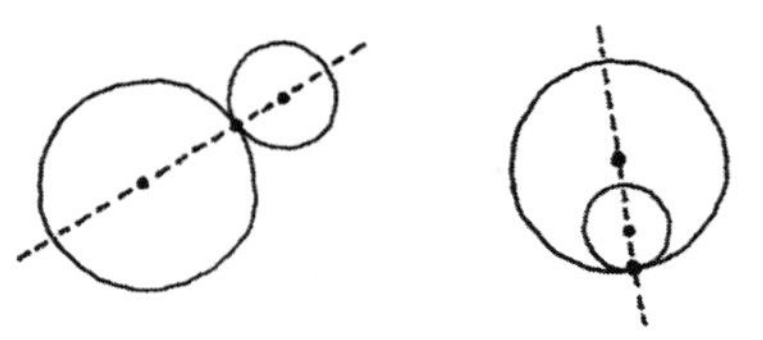

Why do the centers and contact points always line up?

Now we can see all the various ways that our pressure points are aligned, and this data can then be extracted as the "skeleton" of our problem. All the information is now contained in this system of points and lines.

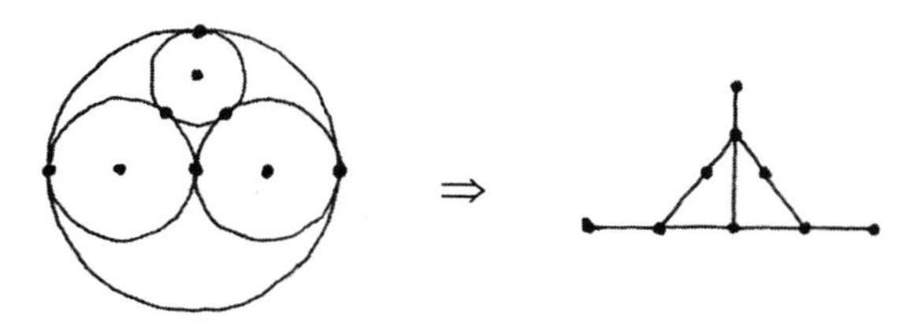

In a way, this process is a lot like converting ancient algebra puzzles written in poem or story form into simpler systems of algebraic equations. The narrative ingredients fall to the wayside, just as the original circle shapes in this problem have disappeared, leaving only their bare bones to tell the tale. The truth is that most of the points on these circles aren't really doing much work for us, so we simply get rid of them.

Now it's time to do some measuring. For simplicity, let's measure our circles using their radii. After all, this is pretty much the defining measurement of a circle—its diameter, circumference, and area being somewhat secondary and derivative. We can take the radius of the outer circle as our unit so that the middle circles each have radius $\frac{1}{2}$. Suppose we let r denote our mystery radius. Transferring this information to our skeleton, we arrive at the following diagram (I'm ignoring the left half, due to symmetry).

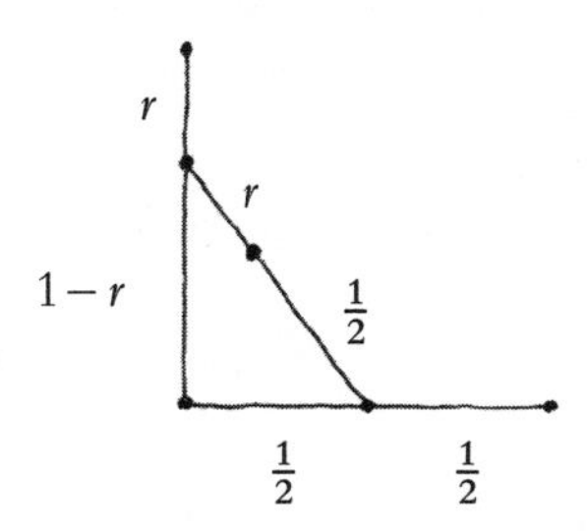

The only clever bit here is to notice that the vertical line, being a radius of the large circle, must have length 1, making the height of the small right triangle $1 - r$. Since its width is $\frac{1}{2}$ and the diagonal length is $r + \frac{1}{2}$, the Pythagorean relation tells us that r must satisfy the equation

$$\left(r + \tfrac{1}{2}\right)^2 = (1 - r)^2 + \left(\tfrac{1}{2}\right)^2.$$

We have successfully translated the geometric pressure into an algebraic constraint. At this point, our work as geometers is done, and we must now become algebraists. (Of course, what we must really be are mathematicians, sensitive to *all* forms of pattern and beauty.) The first order of business is to expand the quantities in parentheses:

$$r^2 + r + \tfrac{1}{4} = 1 - 2r + r^2 + \tfrac{1}{4}.$$

After collecting terms, this equation reduces to simply $3r = 1$, and we find that $r = \frac{1}{3}$. The small circle is a $\frac{1}{3}$-scale version of the large circle. How pretty!

One amusing consequence of this discovery is that the small right triangle in our diagram then has sides $\frac{1}{2}$, $\frac{2}{3}$, and $\frac{5}{6}$. These are in the same proportion as 3, 4, and 5, so the centers of the small, medium, and large circles form a 3:4:5 right triangle! Once again we see the way that simple and elegant geometrical configurations lead to simple and elegant numerical relationships, as if the same underlying pattern is trying to express itself through both media.

Notice that despite the use of the Pythagorean relation and all of the consequent squaring, our equation ended up being linear; the r^2 terms on both sides fortuitously cancelled. This is what I like to call a "faux quadratic"—to all appearances the equation involves squares of unknowns, but when the dust settles, none remain.

As a final example, here is another *sangaku*—this one designed by one of my students:

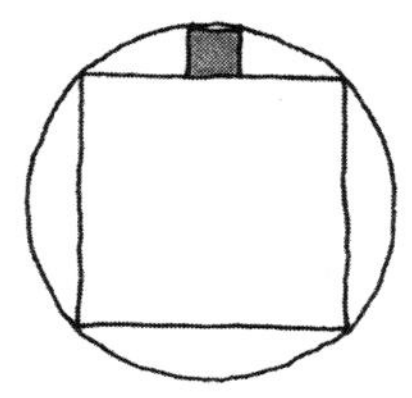

Here we have a square inscribed in a circle, with a smaller square tucked in the space between. The pressure on the little square comes from contact with the outer circle and the upper side of the big square. Let's take the side of the big square as our unit and write s for the side of the small square. (Thus, we expect s to be something small, like $\frac{1}{4}$ or so.) We need to find a way to convert these pressures into algebraic constraints on our unknown.

Identifying the various pressure points, I am immediately struck by the desire to symmetrize:

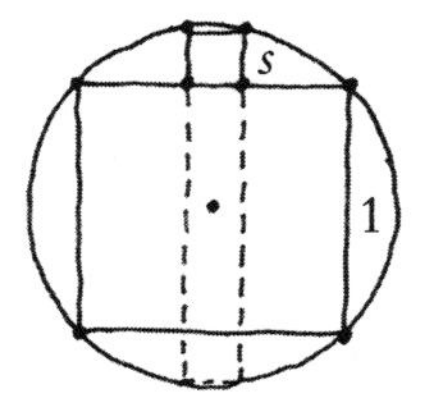

Now I can see two different rectangles inscribed in the same circle. The first is a square of side 1, and the second is a skinnier one of width s and height $2s + 1$. The thing is, whenever *any* rectangle sits in a circle, its diagonal must be a diameter.

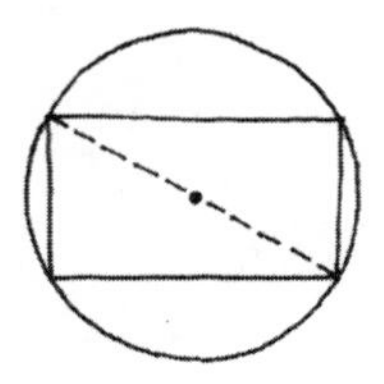

Why is the diagonal of an inscribed rectangle always a diameter?

This means the diagonal of our skinny rectangle must be the same as that of our unit square, which we know to be $\sqrt{2}$. So our rectangle has sides s and $2s + 1$, and its diagonal is $\sqrt{2}$. The Pythagorean relation then tells us that

$$s^2 + (2s + 1)^2 = 2.$$

This is the algebraic equation corresponding to the geometric configuration "little square in between square and circle." After expanding and collecting, we arrive at the quadratic equation

$$5s^2 + 4s = 1.$$

Now we can proceed to solve this equation in whatever manner we see fit. If we prefer the hands-on approach, we can begin by rescaling to isolate the square term:

$$s^2 + \tfrac{4}{5}s = \tfrac{1}{5}.$$

Completing the square as usual, we get

$$\left(s + \tfrac{2}{5}\right)^2 = \tfrac{1}{5} + \left(\tfrac{2}{5}\right)^2.$$

Thus, we find $s = -\frac{2}{5} + \sqrt{\frac{1}{5} + \frac{4}{25}}$. As luck would have it, the expression under the square root turns out to be $\frac{9}{25}$, which is the square of $\frac{3}{5}$. That means that our number s is in fact equal to $-\frac{2}{5} + \frac{3}{5} = \frac{1}{5}$. The small square is exactly one-fifth the size of the big square. How charming!

Of course, we could also use the quadratic formula. In this case, our equation reads $5s^2 + 4s - 1 = 0$, so we immediately get $s = \frac{1}{10}(-4 + \sqrt{36})$. Since $\sqrt{36}$ is either 6 or -6, we find the two solutions to be $s = \frac{1}{5}$ and $s = -1$. Clearly, the first is the intended side length. Does the second solution have any geometric meaning?

Our demand was for a square that touched both the side of the big square as well as the circle. We decided to let s denote the side of this little square. Another way to say it is that s represents the height *above* the big square needed to hit the top of the little square. This suggests that $s = -1$ may refer to a square of side 1, only *below* the line this time. In other words, our equation is trying to tell us that the big square *itself* is a solution to our problem. How funny! Notice that its lower corners do indeed touch the circle, and it certainly also rests on its own upper side, only hanging down instead. So sometimes our equations know more than we do and include alternate solutions that still meet our stated demands, if not exactly our intentions.

Incidentally, had we been extra intuitive and observant, we might have noticed that our equation $5s^2 + 4s - 1 = 0$ can actually be written in factored form:

$$(5s - 1)(s + 1) = 0.$$

From this, we can immediately deduce that either $5s - 1 = 0$ (in which case $s = \frac{1}{5}$), or else $s + 1 = 0$, providing the alternate solution $s = -1$. Factorization can be a quick and fun alternative to completing the square.

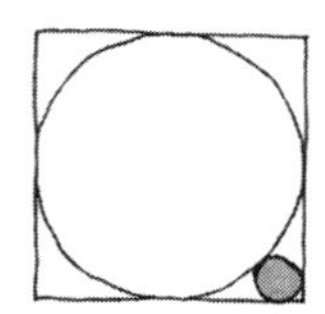
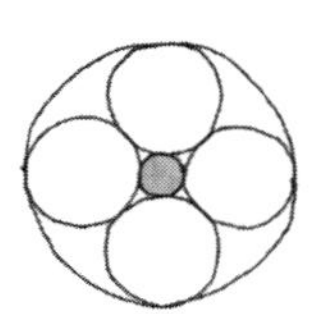
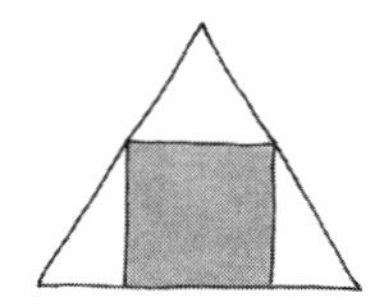

Can you solve these sangaku?
Design your own elegant sangaku puzzle.

FORM AND CONTENT

Having assembled a small library of basic techniques (most of which boil down to distribution in one form or another), maybe it's time for us to pull back and think about the larger picture. What does an algebra problem look like in general? How complicated can such puzzles get?

The number of unknowns certainly plays a role, as does the number of equations, but the main thing that determines the difficulty of an algebraic knot is the extent of the entanglement itself—that is, the complexity of the equations. One way to measure this would be to look at the total number of operations involved and how they are arranged hierarchically.

For example, suppose we are faced with the equation

$$\sqrt{x^2 + 2} + \frac{1}{\sqrt{x^2 + 2}} = x + 1.$$

Here we have only one unknown, but it is being made to perform a rather complicated dance involving square roots and reciprocals. It's not going to be easy to extract x from underneath all that rubble. At the moment, we can see that x is being squared, added, square-rooted, and reciprocal-ed. That's four layers of hierarchical nesting. Naturally, the way to rescue x will involve undoing these operations, but we have to be a little careful with the order in which we perform these activities so that we don't make matters even worse.

Maybe the first thing we should do is get rid of the reciprocal by multiplying both sides of the equation by $\sqrt{x^2 + 2}$. This is known as "clearing denominators." Notice that in this particular case, we also benefit from the elimination of the square roots on the left side. We thus obtain the new equation

$$(x^2 + 2) + 1 = (x + 1)\sqrt{x^2 + 2}.$$

This is a marked improvement over the original. Though we haven't affected the *content* of our equation (i.e., the information it provides about our unknown quantity), we have radically altered its *form*—that is, the manner in which the information is being conveyed. Of course, simplicity is in the eye of the beholder, but most algebraists would find our new equation vastly preferable, merely because there is no division present.

We still have a square root, however, and that means our unknown is still trapped inside it. Naturally, the way to extricate it would be to square everything. Combining the terms on the left and then squaring both sides, we obtain the equation

$$(x^2 + 3)^2 = (x + 1)^2(x^2 + 2).$$

Now there are no roots or reciprocals left, and the only things going on are addition and multiplication. In fact, we can use distribution to simplify even further. Expanding the products completely, we get

$$\begin{aligned} x^4 + 6x^2 + 9 &= (x^2 + 2x + 1)(x^2 + 2) \\ &= x^4 + 2x^2 + 2x^3 + 4x + x^2 + 2, \end{aligned}$$

and after collecting terms, this becomes

$$2x^3 + 4x = 3x^2 + 7.$$

This is certainly a vast improvement over our original equation. I'm not saying that we can immediately see how to *solve* such an equation, mind you. I'm only saying that if I have to be creative and think hard about a certain entanglement, I'd prefer to have it presented in its simplest possible form. Here, we have at least managed to reduce our problem to an equation involving only sums of products of unknowns, so that's something.

In fact, it turns out this can always be done. No matter how snarled and tangled up our equations may be, involving however many nested additions and subtractions, multiplications and divisions, square, cube, and higher roots, we can always process them in this way—clearing denominators, raising to powers, and distributing products until we are left with nothing but a sum of terms, each of which is a simple product of knowns and unknowns.

Let's try a somewhat more convoluted example. Suppose we have a system of two unknowns, say α and β (these are alpha and beta, the first two letters of the Greek alphabet—and, incidentally, the source of the word *alphabet* itself). Let's say that one of our equations takes the form

$$\frac{\sqrt{\alpha + \beta}}{1 + \sqrt{\alpha}} = \sqrt{\beta + 1}.$$

Nobody even wants to *look* at this mess, let alone think about it. We need to give it a makeover, so to speak. (If this equation were to show up at a formal algebra dinner party in its present state, it would only embarrass itself.)

The first thing to do is to clear denominators by multiplying both sides by $1 + \sqrt{\alpha}$. This gives us

$$\sqrt{\alpha + \beta} = (1 + \sqrt{\alpha})\sqrt{\beta + 1}.$$

This is still a godawful mess, but at least it fits on one line. Since there are square roots all over the place, it makes sense to square both sides. This will certainly help on the left, but we have to be careful on the right when we square the sum. The result is

$$\alpha + \beta = (\beta + 1)(1 + 2\sqrt{\alpha} + \alpha).$$

This is starting to look like Greek alphabet soup, but at least the square roots are almost all gone. Here, it pays to isolate the remaining square root term by expanding the right side, taking care not to introduce any additional roots. We have

$$\begin{aligned}\alpha + \beta &= (\beta + 1)(1 + 2\sqrt{\alpha} + \alpha)\\ &= (\beta + 1)(1 + \alpha) + (\beta + 1)2\sqrt{\alpha}\\ &= (\alpha\beta + \alpha + \beta + 1) + 2(\beta + 1)\sqrt{\alpha}.\end{aligned}$$

After a bit of rearrangement, our equation now reads

$$-2(\beta + 1)\sqrt{\alpha} = \alpha\beta + 1.$$

That's certainly a lot nicer—we got lucky there with the $\alpha + \beta$ cancelling. Now we can square both sides to get

$$4(\beta + 1)^2\alpha = (\alpha\beta + 1)^2.$$

All of the roots and denominators are gone, and what remains is a combination of additions and multiplications only. These are still a bit nested and convoluted, so the usual idea would be to multiply everything out and collect terms. When the smoke clears, the final result is

$$4\alpha\beta^2 + 6\alpha\beta + 4\alpha = \alpha^2\beta^2 + 1.$$

This is what the original equation was trying to say all along, only now it is expressed in the primal language of our field operations (i.e., addition and multiplication), as opposed to the more indirect language of inverse operations. Every division can be rephrased in terms of multiplication, and every square or cube root can be removed by squaring or cubing (and the same goes for any higher roots as well). You might think that the next step would be to remove any subtractions using addition, but this is unnecessary. Since subtraction is merely adding the negative and negation is the same as multiplying by -1 (which every field is guaranteed to contain), we can achieve any subtraction using addition and multiplication alone.

In fact, it's simplest to put *all* the terms on one side, regardless of whether they are added or subtracted. Thus, with everything on the left, our equation becomes

$$\alpha^2\beta^2 - 4\alpha\beta^2 - 6\alpha\beta - 4\alpha + 1 = 0.$$

Such an equation is said to be in *polynomial* form (Greek for "many terms"). Any algebraic equation, no matter how absurdly tangled and convoluted, can always be rephrased in this way—as a sum of terms, each of which is a product of known and unknown quantities, equaling zero.

That means we can fit all our problems into the same general framework. We have a field in the background (e.g., the real numbers $\mathbb{R}$) providing us with our numerical constants, and a set of unknown quantities constrained by various algebraic equations, each expressed in polynomial form. In this way, classical algebra becomes the study of polynomial equations.

So what makes one algebra puzzle more difficult than another? What is it about a polynomial equation that makes it hard to untangle? One simple measure of complexity would be the number of unknowns being multiplied together. Suppose we select a given term, say $-4\alpha\beta^2$. This is then referred to as the $\alpha\beta^2$-*term*, with -4 being its *coefficient*—the multiplier telling us how much that term is counting toward the total sum. (If a term is absent, then it has a coefficient of zero.) This particular term involves the product of three unknowns: $\alpha\beta^2 = \alpha \cdot \beta \cdot \beta$. So we would say it has *degree* 3 (as in "degree of difficulty"). Similarly, the first term, $\alpha^2\beta^2$, would have degree 4. The -4α term has degree 1, and it is customary to say that the constant term 1 has degree 0, since it involves no unknowns at all.

Each term then has its own degree, and we can take the degree of difficulty of a polynomial to be the maximum degree of all of its terms. In particular, our equation has degree 4 and would be classified as a fourth degree (or *quartic*) equation in two unknowns. Equations of degree 1 are also called *linear*; degree 2, *quadratic*; degree 3, *cubic*. People also say *quintic* for degree 5, but after that it starts to get a bit silly. (If you're dealing with a polynomial equation of degree 18, you have bigger problems than what to name it!)

The dimension of a polynomial system (i.e., the number of unknowns) along with the degrees of the individual equations provide us with both a rough measure of complexity as well as a classification system for algebra puzzles. For example, any problem that reduces to the polynomial system

$$\begin{aligned} p + q + r - 12 &= 0, \\ 3p^2 + pq - 2r + 1 &= 0, \\ p^3 + qr + 3pqr - 5 &= 0, \end{aligned}$$

could then be placed in the "dimension 3, degrees 1, 2, and 3" category. (Actually, as we shall soon see, this method of classification turns out to be a bit naive.)

The important thing is that we never have to put up with any denominators or roots, and all our problems can now be placed on the same footing. Before we address the question of how to solve such systems, I want to mention a somewhat simpler approach to polynomial reduction that is often preferable to seat-of-the-pants manipulation. This is our old friend *abbreviation.*

Let's go back to our first example:

$$\sqrt{x^2+2}+\frac{1}{\sqrt{x^2+2}}=x+1.$$

The ugliness here is caused by the square roots and the reciprocal. So let's get rid of them in the laziest possible way—by simply giving them names. We can write y for $\sqrt{x^2+2}$, and let z stand for $\frac{1}{y}$. Our equation now reads $y+z=x+1$, which looks much nicer. Of course, all we've really done is sweep the complexity under the rug. On the other hand, it is almost effortless to put our information into polynomial form. We now have *three* unknowns satisfying the polynomial constraints:

$$\begin{aligned} y+z-x-1&=0,\\ x^2+2-y^2&=0,\\ yz-1&=0. \end{aligned}$$

The benefits of this approach are its generality and simplicity. Anyone can abbreviate at any time without restriction—and with very little demand on one's creativity or technical skill.

The downside, of course, is that we are increasing the dimension. In practice, this can make the abbreviation technique inadvisable, despite its relative ease and simplicity. For theoretical purposes, however, it can be quite convenient. For one thing, it clearly shows that any system of algebraic equations can always be put into polynomial form, at the cost of possibly increasing the number of unknowns.

There is a small technical point here that I hesitate to mention. When we operate on equations in order to simplify them or express them in a particular format, we are (as always) making the assumption that such numbers exist—that our equations are true statements concerning a set of (currently unknown) numbers. We are then deducing consequences from this assumption, in the form of ostensibly simpler statements.

However, depending on the actions we take, this deductive process may not be reversible—that is, we might deduce new equations that are not entirely equivalent to the original ones. The culprit, as usual, is the possibility of multiplication by zero. Whenever we clear denominators, we are multiplying by an expression containing unknowns. This means we have no guarantee that we aren't multiplying both sides by zero—essentially losing information by performing an irreversible maneuver. As an example of this phenomenon, consider the equation

$$1 + \frac{1}{w-1} = \frac{w^2}{w-1}.$$

Naturally, the first thing to do would be to clear denominators by multiplying both sides by $w - 1$. This leads to the much simpler (quadratic) equation $w^2 - w = 0$, with solutions $w = 0$ and $w = 1$. Going back to our original equation, we see that 0 is a perfectly fine solution, whereas 1 is a bit problematic: when $w = 1$ neither side of the equation makes any sense. Do we want to say that 1 is a solution because nonsense equals nonsense?

The point being that when we clear denominators, we occasionally introduce spurious solutions that are perfectly legitimate as far as the polynomial version of the equation goes but do not in fact satisfy the original constraint (in the sense that we encounter a division by zero). In practice, all this really means is that we have to check our solutions, but since we do this anyway as a matter of routine, it's really no big deal. In any event, I'd much rather work with polynomials and have an extraneous solution or two than have to deal with roots and ratios.

The other thing I want to point out is that our ability to manipulate equations into polynomial form is a direct consequence of our nine field postulates. Any further universal property—let's say, one enjoyed by any pair of real numbers a and b—could then be expressed in the form

$$(\text{polynomial combination of } a \text{ and } b) = 0.$$

We would then have a certain polynomial equation that is satisfied by *every* possible value of a and b, which is absurd—unless the left side cancels completely. This shows that our list of universal algebraic properties is complete, as advertised.

Let's assume from now on that all of our algebra puzzles are expressed as a system of polynomial equations. As we have seen, this is no real loss of generality. Our problems can now be organized by dimension and degree, and we can begin to develop systematic techniques for their solution. As always, our goal is *elimination*—finding ways to reduce the complexity by removing an unknown from the conversation.

If we are exceedingly fortunate, we may be able to massage one of our equations so that it directly expresses one unknown in terms of the others (thus allowing for substitution), but this is usually too much to hope for. If one of our equations happens to be linear (i.e., degree 1), then we can always do this, but in general our equations will have higher degree, and each unknown will be mentioned multiple times.

Since linear equations always make life easier, and since we already know how to deal with quadratic equations in one unknown, the next simplest case would be two quadratic equations in two unknowns. One such example would be the system

$$\begin{aligned} 2x^2 + xy + 3y^2 - 1 &= 0, \\ x^2 + y^2 + x - 3y &= 0. \end{aligned}$$

(Such systems arise quite frequently in geometry, as we shall see.) Notice that both unknowns x and y feature twice in each equation, so there is no simple way to employ substitution.

Instead, we will need to elaborate on our method of combination. The idea, as usual, is to find clever ways to scale our equations so that we create alignments and agreements that we can use to cancel terms. Let's say we decide to eliminate x in order to obtain a single equation involving y only. The natural move would be to double the second equation and subtract it from the first to obtain

$$xy + y^2 - 2x + 6y - 1 = 0.$$

This eliminates the x^2-term. Now we only have xy and $-2x$ remaining. In fact, we can even collect these two multiples of x together and write our equation in the compact form

$$(y-2)x + (y^2 + 6y - 1) = 0.$$

At this point, we have several options. If we like, we can forego polynomial form and simply write

$$x = -\frac{y^2 + 6y - 1}{y - 2},$$

allowing for substitution. This is what I would call a quick and dirty solution—we immediately obtain an equation in y alone, but it's a fairly grimy business, with unpleasant denominators all over the place that will need to be cleared.

A somewhat prettier (though more lengthy) process would be to continue seeking combinations that eliminate and simplify, while remaining at all times safely inside the confines of polynomial form. We have the two equations

$$\begin{aligned} (y-2)x + (y^2 + 6y - 1) &= 0, \\ x^2 + x + y^2 - 3y &= 0. \end{aligned}$$

We can align the first terms of these equations by scaling the upper equation by x and the lower one by $y - 2$. (The only new twist is that we are scaling by unknown quantities, but that's perfectly fine; numbers don't care if we know them or not.) The newly scaled versions become

$$\begin{aligned} (y-2)x^2 + (y^2 + 6y - 1)x &= 0, \\ (y-2)x^2 + (y-2)x + (y-2)(y^2 - 3y) &= 0. \end{aligned}$$

Subtracting these (and expanding the products) gives us

$$(y^2 + 5y + 1)x - (y^3 - 5y^2 + 6y) = 0.$$

This looks quite a lot like our previous equation,

$$(y - 2)x + (y^2 + 6y - 1) = 0.$$

In both cases, we have multiples of x that involve y. (In a sense, we are choosing to view these as *linear* equations in x with unknown coefficients involving y.) Once again, we can scale the first equation by $y - 2$ and the second by $y^2 + 5y + 1$ to align the terms, and then subtract to obtain the equation

$$(y^2 + 6y - 1)(y^2 + 5y + 1) + (y - 2)\,(y^3 - 5y^2 + 6y) = 0.$$

We have successfully eliminated x, all the while remaining in the polynomial setting. Expanding and collecting, we finally get

$$2y^4 + 4y^3 + 46y^2 - 11y - 1 = 0.$$

Our problem has been distilled down to its essence: an equation of degree 4 in one unknown. Assuming we can solve such an equation, we would then obtain a certain set of explicit y values that work. We could then substitute each of these into either of our original equations to obtain an equation in x alone. Or, if you prefer, we could go through the same elimination process again, only this time eliminating y. Either way, our problem reduces to solving a polynomial equation in only one unknown.

So this is our strategy: stay within the polynomial world and find clever combinations that reduce complexity, eliminating one unknown at a time until we get down to an equation involving only a single unknown quantity. Obviously, we will not always be successful because not every algebra puzzle has a solution. But the hope would be that if a problem is well designed and noncontradictory, then such an elimination procedure could always be successfully performed.

Notice that we began with two equations of degree 2 and we ended with a single equation of degree 4. This is typical. The price of elimination is usually a dramatic increase in degree. As a simple example, consider the system of equations

$$u - t^3 = 0,$$
$$u^2 - t = 0.$$

The first equation tells us that $u = t^3$, so we can substitute this into the second equation to obtain $t^6 - t = 0$, an equation of degree 6. In general, a decent rule of thumb is that the degree of the resulting equation in one unknown is equal to the *product* of the degrees of the original equations. This is not strictly true in all cases, but it is a good general guide to the level of complexity of a polynomial system.

While we're on the subject of elimination strategy, it's worth noting the advantages of expressing our equations in the form

blah blah blah = 0.

Our methods involve scaling equations and combining them via addition or subtraction. The nice thing about putting everything on the left is that we then don't have to pay attention to the right side at all—it's always zero! Scaling and combining don't affect that, so there's less to worry about. In fact, if we like, we can dispense with the right-hand side altogether.

What we will be doing in actual practice is combining and manipulating the left sides of our equations only, meaning that the objects in our hands (so to speak) are not really equations but *polynomial expressions*—that is, polynomial combinations of known and unknown quantities. We are multiplying these by other polynomial expressions and adding such things together. The "equals zero" part goes without saying—which means we can literally go without saying it and simply replace our equations with a list of polynomial expressions ready for processing.

The more we play around with complex structures like molecules or polynomial expressions, the more they start to feel like individual entities. We ourselves are made of billions of constituent cellular subsystems. I think it is fairly clear that our so-called minds are composed of multiple submodules and competing personas as well. (One of mine is now trying to figure out how to extricate myself from this digression, in fact.) My point is that individual building blocks—such as atoms and unknowns—can be put together in complex ways to form larger structures, and the more we engage with and manipulate such structures, the more they become atomic, in the sense that we hold them in our minds as single objects.

So the idea is to construct a new class of entities to hold the information about how our unknowns are being combined and entangled. These objects will be called *polynomials.* To distinguish these new creatures from polynomial equations involving a particular choice of unknowns, we will require a new notational device to represent the idea of a "generic unknown." This is not a number whose value we seek, nor a parameter that will be set at some point. What we need is a set of placeholder symbols with no value whatsoever. Such things are referred to as *indeterminates.* I'll denote them using capital Roman letters near the end of the alphabet. For example, if one algebraist is working with the equation $3w^3 - 2w + 5 = 0$, and another is looking at $3p^3 - 2p + 5 = 0$, then they are both concerned with the abstract polynomial $3X^3 - 2X + 5$.

One natural way to think of such an object is as a process: the polynomial $3X^3 - 2X + 5$ encodes the procedure whereby we take a number, triple its cube, subtract twice it, and add five. We're not saying what that is supposed to equal, and we're not saying that there even is a number that we are acting on; we're just naming the process itself. A polynomial is a bit like a washing machine with no laundry in it yet—we can still talk about what it does and how it works, even though no clothes are being washed.

Thus, our previous example (involving alpha and beta) pertains to the abstract polynomial

$$X^2Y^2 - 4XY^2 - 6XY - 4X + 1$$

in two indeterminates. The puzzle may be to find actual numerical values that we can insert in place of X and Y so that we obtain zero, but the polynomial itself doesn't care. It is an abstract entity happily enjoying its abstract existence, whether you choose to evaluate it at specific inputs or not.

Given any field $\mathbb{F}$, and any set of indeterminates—say X, Y, and Z—we can imagine the space of all possible polynomials in X, Y, and Z, with coefficients in $\mathbb{F}$. This would constitute a library of sorts, containing all of the ways that three numbers can be polynomially entangled. This large collection is written $\mathbb{F}[X, Y, Z]$ (i.e., the name of the field, together with the indeterminates in brackets). For example, the collection of all polynomials in one indeterminate with rational coefficients would be denoted $\mathbb{Q}[X]$. The reason I am bothering to give names to such environments is because these are the places in which we are now doing business.

We have already seen that any algebra puzzle can always be expressed as a list of polynomial equations, but now we are simplifying even further by getting rid of both the equations as well as the unknowns. An algebra puzzle is simply a list of abstract polynomials. The goal, of course, is to find the sets of inputs that cause these polynomials to simultaneously *vanish*—that is, equal zero.

For instance, rather than saying that we have unknowns g and h satisfying the system of equations

$$3g + 2h - 5 = 0,$$
$$g^2 - 4gh + 3 = 0,$$

we could instead view the situation as being that we have the two polynomials $3X + 2Y - 5$ and $X^2 - 4XY + 3$, and we want to find their *simultaneous zeros*—the pairs of numbers we can substitute for X and Y to make both polynomial expressions vanish.

Obviously, we aren't really doing anything here but changing the psychology. We're shifting the emphasis from numbers to polynomials because these are the objects that we need to manipulate. Rather than working with unknown numbers, we are instead dealing with known polynomials.

In fact, we should really think of polynomials as a *generalization* of numbers. We can add and subtract polynomials just as we can numbers, and we can also multiply them using distribution in the usual way. Thus, for example, we have

$$(X + 2Y) + (X^2 + Y) = X^2 + X + 3Y,$$
$$(X + 2Y) \cdot (X^2 + Y) = X^3 + XY + 2X^2Y + 2Y^2.$$

In this way, the polynomial environment $\mathbb{R}[X, Y]$ becomes a sort of number system in its own right. Since constants are included as polynomials (of degree 0), we see that $\mathbb{R}[X, Y]$ is in fact an extension of $\mathbb{R}$, obtained by throwing in the abstract indeterminates X and Y, and allowing them to freely add and multiply. Thus, the system $\mathbb{R}[X, Y]$ satisfies almost all of our demands for a field: the addition and multiplication behave in all the usual associative, commutative, and distributive ways, and the constant polynomials 0 and 1 still act as additive and multiplicative identity elements. Since the negation of a polynomial can be achieved by multiplication by -1 (thus negating all of its coefficients), there is no problem there, either.

The trouble is with reciprocals and division. The whole point of polynomial form was to get rid of denominators and other such complexities. So $\mathbb{R}[X, Y]$ is not a field because the reciprocal of a polynomial is not usually a polynomial—an upside-down washing machine will not wash. Thus, we have a system of polynomial entities that is very reminiscent of $\mathbb{Z}$, the realm of integers. An environment of this kind—where we satisfy all nine demands *except* for the existence of reciprocals—is called an *integral domain* (or simply *domain*, for short). Basically, this means we are in a place that behaves like the ordinary integers, in that we can add, subtract, and multiply in all the usual ways.

Actually, to be precise, there is one additional demand that we need to make in order that our domains behave properly. Though the domain of integers does not possess reciprocals, it does behave reasonably with respect to multiplication. In particular, although multiplication by zero always destroys information, at least no *other* number has this disastrous effect. If the product of two numbers is zero, then one of them has to be zero. This feature is shared by integers and polynomials, so we build it into the definition of integral domain. Thus, a domain can be viewed as a sort of defective field, satisfying the softer demand that products of nonzero elements be nonzero, as opposed to the stricter requirement of having actual reciprocal elements.

Which integers have reciprocals?
Which polynomials do?

Our elimination procedure can now be seen as taking place entirely inside such a polynomial domain. We are given a set of polynomials, and we allow ourselves the freedom to multiply them by whatever polynomials we like and to add and subtract the results, generating further polynomials until we (hopefully) eliminate our way down to only one remaining indeterminate. In this way, we can view classical algebra as a kind of strategy game, where the pieces are polynomials.

However you want to think of it, polynomials are out there adding and multiplying with each other, and we can choose to investigate or not. For the practical purpose of solving algebra problems, we will certainly find ourselves doing a fair amount of arithmetic with these beasts, and the more familiar we are with their behavior, the more creative we can be as algebraists.

In particular, we will find ourselves wanting to talk about a generic polynomial—without necessarily having to specify its coefficients or even the number of indeterminates. As usual, we require a new level of abstraction and a notational system to match. The typical solution is to use *indexing*. This is where instead of choosing names like a, b, c, which depend on a limited alphabet, we select a single symbol and attach a numerical label: a_1, a_2, a_3. In this way, we can easily extend to four, or five, or any number of names. In fact, we don't need to commit to this number either. Suppose we had n entities we wished to name. Then we could call them $a_1, a_2, a_3, \ldots, a_n$ (The ellipsis is necessary here, since we don't know the value of n.)

With this device in hand, we can now refer to generic objects more easily. For instance, an unspecified polynomial of degree n in one indeterminate X can be written

$$a_0 + a_1X + a_2X^2 + a_3X^3 + \cdots + a_nX^n.$$

The various coefficients are then held by the parameters $a_0, a_1, a_2, \ldots, a_n$. Similarly, the domain of real polynomials in m indeterminates would be written $\mathbb{R}[X_1, X_2, \ldots, X_m]$. We could even imagine a generic system of k polynomial equations in m unknowns:

$$\begin{aligned} P_1(x_1, x_2, \ldots, x_m) &= 0, \\ P_2(x_1, x_2, \ldots, x_m) &= 0, \\ P_3(x_1, x_2, \ldots, x_m) &= 0, \\ &\vdots \\ P_k(x_1, x_2, \ldots, x_m) &= 0. \end{aligned}$$

Here I have named my polynomials $P_1, P_2, \ldots, P_k$, and I am also using the standard notation $P(x_1, x_2, \ldots, x_m)$ to indicate the value of a polynomial P when given the inputs $x_1, x_2, \ldots, x_m$.

This gives us an easy and flexible way to talk about algebra problems in general. We have a set of indeterminates $X_1, X_2, \ldots, X_m$ (with m being the dimension) and a set of polynomials $P_1, P_2, \ldots, P_k$ contained in $\mathbb{R}[X_1, X_2, \ldots, X_m]$. We want to find the simultaneous zeros of these polynomials, and our plan is to form combinations of them that eliminate the indeterminates one at a time.

Let's suppose we can do this. Then at the culmination of this process, we will be left with a set of polynomial equations, each mentioning one fewer unknown than the previous, until we are down to a polynomial equation in a single unknown. This means the entire classical algebra project can be broken into two separate parts:

(1) The theory and practice of elimination
(2) The solution of polynomial equations in one unknown

So our first order of business is to design an elimination procedure that works. We are given a set of polynomials, and we want to form combinations of them that are somehow simpler than the ones we began with. We are hoping to do this using addition and multiplication only, so that we remain in the polynomial realm.

That means we have a new artisanal craft to master. Deciding which indeterminates to eliminate in which order and choosing which polynomials to combine when and with what scaling factors—these are creative decisions that may lead to an elegant and efficient solution or a hopeless computational boondoggle. We may, by virtue of skill or sheer luck, ascend to elimination heaven or else be consigned to polynomial purgatory—endlessly churning out increasingly elaborate and useless combinations and getting nowhere. (There should be a Greek myth of Polynomius, doomed to calculate for all eternity.)

Notice that in a sense we are no longer doing algebra. There are no unknowns or equations, and we aren't trying to deduce the value of anything. We are simply doing arithmetic, only now with polynomials instead of numbers. Our arithmetic game is to form combinations, and the goal is elimination.

The problem, of course, is that there are an infinity of ways to scale and combine a given set of polynomials, so we can't possibly try them all. On the other hand, being abstract thinkers, there is nothing stopping us from *imagining* them all. This is yet another example—quite common in modern algebra—where a finite collection of objects gives rise to an infinite family of offspring.

Thus, given any set of polynomials $P_1, P_2, \ldots, P_k$, we can imagine the set of all possible combinations

$$Q_1P_1 + Q_2P_2 + \cdots + Q_kP_k,$$

where the scaling factors $Q_1, Q_2, \ldots, Q_k$ can be any polynomials whatsoever. This complete set of such combinations is known as the *ideal* generated by the polynomials $P_1, P_2, \ldots, P_k$. (The term comes from the arithmetic of algebraic number fields and is short for "ideal number.")

So what we are really doing when we are solving an algebra problem is rooting around inside this vast polynomial ideal, examining each possible combination, and searching for the simplest and most convenient ones we can find. Just as we replaced physical reality with idealized mathematical reality, quantities with entities, and algebraic equations with polynomials, we can now go another step further: *an algebra problem is an ideal in a polynomial domain.* We are given a certain set of generators for this ideal, but once the ideal is generated we no longer need them. In fact, our entire strategy is to replace them with a better set of generators more convenient for our purposes.

Given polynomials $P_1, P_2, \ldots, P_k$, let's write $(P_1, P_2, \ldots, P_k)$ for the ideal they generate. (Hopefully, using parentheses in this way won't be too confusing; in any case it's perfectly standard.) The nice thing about ideals is that they form a closed system: any multiple or sum of multiples of any elements of an ideal is already there. Ideals are "closed under combination." So every problem in classical algebra leads to an ideal, and if two puzzles lead to the same ideal, it means the puzzles were essentially the same, only using different sets of generators.

I must say I have somewhat mixed feelings about piling on yet another layer of abstraction. The last thing I want is for you to get frustrated or bleary-eyed staring at indexed indeterminates and trying to imagine infinite sets of polynomial combinations. The good news is that we don't really need any of this architecture in actual practice. In fact, you are more than welcome to stick to equations and unknowns if you prefer that language. We will still be multiplying by this and combining with that, so it's not really much of a change computationally.

But of course I also want to do my best to convey the modern mathematical aesthetic, and I find it quite elegant that we have managed to encapsulate all possible algebra problems in one simple construct. The notion of a polynomial ideal—although infinite and admittedly somewhat unwieldy—replaces our naive idea of unknowns and equations and allows us to classify algebra problems in a more nuanced and sophisticated way. Rather than sorting puzzles by the number of equations or by their various degrees, we instead classify problems by the ideals they generate. Given a polynomial ideal, the problem is then to describe its set of simultaneous zeros.

Let's start by looking at ideals in a somewhat more elementary setting. As I mentioned earlier, there is a nice analogy between polynomials and integers in that they are both integral domains—places where we can freely add, subtract, and multiply but not necessarily divide. The notion of an ideal survives perfectly well under this generalization. If we have any collection of elements of an integral domain, we can speak of the ideal they generate in the same way as before: the set of all possible sums of multiples of them. In particular, we can examine the various ideals of $\mathbb{Z}$.

Suppose we have a collection of integers, say 12, –18, and 30. The ideal (12, –18, 30) then consists of every possible integer that can be manufactured from these three by scaling them (by any integer factor whatever) and adding the results. For example, since two copies of 12 and one copy of 30 make 54, this number must also belong to our ideal.

Since $12 - 30 = -18$, we find that the generator -18 is actually redundant; we can already make it out of 12 and 30. This means that $(12, -18, 30)$ is the exact same ideal as $(12, 30)$. We can go further and observe that $30 - 2 \cdot 12 = 6$, so the number 6 belongs to this ideal as well. But both 12 and 30 are multiples of 6, so we find that our ideal $(12, -18, 30)$ is simply (6), the set of multiples of 6.

In fact, it turns out that all ideals in $\mathbb{Z}$ are of this form. Any collection of integers closed under combination must be the set of multiples of one particular number.

Why is every ideal in $\mathbb{Z}$ generated by a single number?
What are the ideals (0) *and* (1)*?*

Unfortunately, this is not the case for polynomial domains in general. For instance, the ideal (X, Y) in $\mathbb{R}[X, Y]$ consists of all real polynomials with zero constant term, and these are not the multiples of any one particular polynomial. (Luckily, the analogy does go through for polynomial ideals in $\mathbb{R}[X]$, as we shall see.)

As the previous example shows, by playing around with various combinations, we can often improve the set of generators for a given ideal, replacing the original generators with newer versions that are in some sense *smaller.* We were able to do this with integers without too much trouble. Now we have to figure out how to do the same thing with polynomials.

In practice, this turns out to be a fairly technical and computationally demanding enterprise, best relegated to its own separate chapter—one that I highly recommend you skim on first reading. Expect to see multiple pages of almost laughably complicated polynomial arithmetic. I will need to be especially careful with these calculations because I'm actually committing them to paper, but that doesn't mean you need to follow every last computational detail. You can appreciate a tapestry perfectly well without having to keep track of each individual stitch.

ELIMINATION

Now it's time to get into the gory details. The trick is to not get lost in them—it's that forest and trees thing. I want to explain a completely general and generic methodology for polynomial elimination, but in order to really understand the process, I think we'll need to actually undergo the ordeal at least once. So, although we have the conceptual and notational means to support a completely abstract treatment (such as a modern algebraist would undoubtedly prefer), we will instead take the ancient Egyptian approach and carry out our elimination procedure by hand on a specific, concrete example. At the same time, we need to blur our eyes to these details and try to understand our methods in the abstract.

Suppose we are interested in the system of equations

$$\begin{aligned} a^2 + b^2 &= 2ac, \\ ac + bc &= 2, \\ c^2 &= ab. \end{aligned}$$

We will then be forming various combinations of the abstract polynomials

$$X^2 + Y^2 - 2XZ, \quad XZ + YZ - 2, \quad \text{and} \quad Z^2 - XY.$$

At the moment, each of these polynomials involves all three indeterminates. The game is to find clever ways to scale them—that is, to multiply them by whatever numbers or polynomials we see fit—in order to align their terms for elimination. We thus find ourselves faced with the ideal

$$(X^2 + Y^2 - 2XZ, XZ + YZ - 2, Z^2 - XY)$$

in the polynomial domain $\mathbb{R}[X, Y, Z]$. This is the complete set of all possible combinations of these generators, and we're looking for the simplest elements of this ideal we can find—an alternative set of generators that does not alter our problem in any substantive way but changes its outward form to allow for systematic elimination. What a bizarre strategy game! Of course, it's really a distribution game when you get right down to it; we're scaling polynomials by other polynomials, so it's all about the way the various terms are produced and the coefficients they obtain as a result of multiplying and collecting.

The art is in designing combinations that help us eliminate, so the first thing we need is some way to measure complexity so we can tell if we're making progress. Suppose we are trying to eliminate the indeterminate X. Naturally, we would prefer that our polynomial contain an X^2-term rather than an X^3, but what about X^2Y^3 versus X^3Y—which is more complicated? We need a sensible rating system for terms so we can decide if our moves are making things better or worse.

The usual plan is to choose an elimination order (e.g., first eliminate X, then Y, then Z) and to rank the various terms by *multidegree* in the following way. The term X^2Y^3Z has degree 6, but if we focus on one symbol at a time, we could say it has multidegree 2, 3, 1—meaning it has degree 2 in X, degree 3 in Y, and is linear (degree 1) in Z. (If an indeterminate is absent from a term, we would say it has degree 0.) Now we can rank the various possible terms *lexicographically*, as we do with words in a dictionary. The highest ranked terms would be the ones with the most X; if there is a tie, then we look at the amount of Y, and so on.

In this way, our elimination order is telling us what we care most about. In the present case, it happens to be X. We are ranking the terms of our polynomials by how much X they contain, providing us with a rough measure of how much further we have to go to eliminate X entirely. Thus, the term X^3Y would rank higher than X^2Y^3, even though the latter has higher total degree. This means our elimination procedure will to a large extent depend on the elimination order we have chosen. There is a certain amount of art and intuition involved in choosing the most efficient elimination order. Fortunately, this choice does not affect the eventual success of our procedure, only the amount of time and computation involved.

Assuming we have adopted a particular elimination order, we can then write the terms of each of our polynomial generators in decreasing order of importance so that we can more easily see the lay of the land. In the present case, using the elimination order X, Y, Z, our generators would be written

$$\boldsymbol{X^2} - 2XZ + Y^2,$$
$$\boldsymbol{XZ} + YZ - 2,$$
$$\boldsymbol{XY} - Z^2.$$

Notice especially the highest-ranking term of each polynomial, which I have set in boldface type. This is known as the *leading term*. Here we are lucky in that these terms are all *monic*, meaning they have a coefficient of 1, so there's no pesky scaling factor to deal with. Even if there were, we could simply divide it out, so we can always assume that our polynomials have their terms arranged in rank order, with a monic leading term.

Now the game begins: scale, align, combine. We want to do this in such a way that the rank of the leading terms is reduced, until finally there is no X left at all. This is not to be mindless meandering, forming random combinations in hope of accidentally reducing complexity; this is strategic elimination by design. In the linear case, our method was to align the coefficients by scaling appropriately. The system

$$3a - 5b = 1,$$
$$4a + 2b = 7,$$

would be handled by multiplying the first equation by 4 and the second by 3 so that we get $12a$ in both equations, ready for elimination.

So the plan is to do the same thing now, only with polynomial multipliers. One way to keep things simple (although somewhat less efficient) is to work with only two polynomials at a time and to scale by single terms, as opposed to more complicated polynomials composed of multiple terms. In practice, an experienced algebraist working on a particular problem will usually choose an approach to elimination that is tailored to the specific features and symmetries of that problem. The question is whether a systematic, general-purpose method exists. Is there a way to tell which combinations to make and in what order?

It turns out that there is such a procedure, and fortunately it is reasonably simple and straightforward. In particular, such an algorithm can be easily programmed and implemented on computers. This means we don't need to concern ourselves too much with streamlining our methods for maximum speed and efficiency. Whether we require twenty steps or twenty thousand is irrelevant, since both will be accomplished in a matter of milliseconds. The main issue is the existence of a procedure that is guaranteed to work; how efficient or wasteful it may be is a secondary matter.

Now I have to level with you. Truth be told, polynomial elimination is a total nightmare. Generally speaking, if you slap down a random set of polynomials—with whatever degrees and coefficients they may happen to have—you will quickly find yourself in arithmetic hell. Although our procedure is quite simple in theory, it does involve a fair amount of actual arithmetic in the form of addition and multiplication of coefficients. This tends to produce large numbers rather rapidly, typically in the form of fractions with extremely ugly numerators and denominators. A computer may not care about this sort of trivia, but we sure do. In fact, the number of individual arithmetic calculations can get so large that one rapidly loses any feeling of confidence in the results.

Thus, in order to illustrate the general method, I have chosen a set of polynomials designed to keep the computational complexity under control. First of all, I decided to make them quadratic to keep the degree from getting too high. (As it is, our rule of thumb predicts an eventual equation of degree $2 \times 2 \times 2 = 8$.) This is also why the coefficients of my polynomials are mostly zeroes (so that the terms are sparse) or possibly ones or twos (so that their products don't get too big). Even with these safeguards in place, this will undoubtedly be the lengthiest and most involved computation most of you will ever see. Just be glad you live in a time when these sorts of things can be done by machines, so we don't have to fill up notebooks with calculations the way our predecessors did.

Let's begin by writing down our list of generators, with their terms arranged in rank order:

$$\boldsymbol{X^2} - 2XZ + Y^2 \qquad [1]$$
$$\boldsymbol{XZ} + YZ - 2 \qquad [2]$$
$$\boldsymbol{XY} - Z^2 \qquad [3]$$

(I have also numbered them for ease of reference.) Suppose we select our first two polynomials, [1] and [2]. Their leading terms do not agree, so they are not ready to be subtracted. To get them to match, we need to multiply them each by the appropriate factor. Our leading terms are X^2 and XZ. We need a common multiple of these two, and we want it to have the smallest degree possible. In this case it would be X^2Z, so the plan would be to multiply the first polynomial by Z and the second by X. Then we can subtract them and eliminate their common leading term. Let's write $Z[1] - X[2]$ to indicate this combination (in other words, Z times polynomial [1] minus X times polynomial [2]). We have

$$\begin{aligned} Z[1] - X[2] &= Z(X^2 - 2XZ + Y^2) - X(XZ + YZ - 2) \\ &= -XYZ - 2XZ^2 + 2X + Y^2Z. \end{aligned}$$

This new polynomial has a leading term that is still larger (i.e., higher in rank) than some of our original equations. In particular, it is divisible by the leading term XY of equation [3]. So we can get rid of it by adding on the appropriate multiple of this polynomial, namely $Z[3]$:

$$\begin{aligned} Z[1] - X[2] + Z[3] &= -XYZ - 2XZ^2 + 2X + Y^2Z + Z(XY - Z^2) \\ &= -2XZ^2 + 2X + Y^2Z - Z^3. \end{aligned}$$

Our leading term continues to drop, and we can reduce it even further by adding on another $2Z[2] = 2XZ^2 + 2YZ^2 - 4Z$ to get

$$Z[1] - X[2] + Z[3] + 2Z[2] = 2X + Y^2Z + 2YZ^2 - Z^3 - 4Z.$$

Now we're starting to make some progress. We've managed to obtain a combination whose leading term is smaller than any of the polynomials we started with. In fact, it is even *linear* in X. The only slight annoyance is the leading coefficient of 2. So we scale the whole thing by $\frac{1}{2}$ to make it monic, and then add it to our list as our fourth polynomial:

$$\boldsymbol{X} + \tfrac{1}{2}Y^2Z + YZ^2 - \tfrac{1}{2}Z^3 - 2Z \qquad [4]$$

This is what I'll call the "standard maneuver": given any two polynomials, we look to their leading terms to determine the appropriate scaling factors (to bring them both up to their lowest common multiple), and then we subtract. If the new leading term is divisible by any of the leading terms we already have, then we continue to reduce it by subtracting off multiples until we obtain a new polynomial whose leading term is a new low. Then we add it to our list and continue.

Let's try this maneuver using polynomials [1] and [3] instead. The leading terms are X^2 and XY, so the combination we want is

$$\begin{aligned} Y[1] - X[3] &= Y(X^2 - 2XZ + Y^2) - X(XY - Z^2) \\ &= -2XYZ + XZ^2 + Y^3. \end{aligned}$$

The leading term is now divisible by that of our third polynomial, so we can continue to eliminate terms by adding on another $2Z[3]$ to obtain

$$\begin{aligned} Y[1] - X[3] + 2Z[3] &= -2XYZ + XZ^2 + Y^3 + 2Z(XY - Z^2) \\ &= XZ^2 + Y^3 - 2Z^3. \end{aligned}$$

Now we can subtract off $Z[2]$ to get

$$\begin{aligned} Y[1] - X[3] + 2Z[3] - Z[2] &= XZ^2 + Y^3 - 2Z^3 - Z(XZ + YZ - 2) \\ &= Y^3 - YZ^2 - 2Z^3 + 2Z. \end{aligned}$$

We have now eliminated X altogether, and this polynomial is our new leader in the race for lowest leading term. So we certainly want to add it to our list:

$$\mathbf{Y^3} - YZ^2 - 2Z^3 + 2Z \qquad [5]$$

Notice how the elimination of X has caused the degree to increase. The more we scale our polynomials for alignment of X terms, the larger the degrees of the Y and Z terms become. Of course, we are not even close to being done, since we will require at least one more polynomial in Y and Z in order to eliminate Y.

To save space (not to mention sanity), I will spare you the sordid arithmetic details from now on and simply write down the results of our various combinations. (If you would like some arithmetic practice, please feel free to check my calculations!) For our next move, we will now perform the standard maneuver on equations [2] and [3]. The leading terms are XZ and XY, so we make the combination

$$Y[2] - Z[3] = Y^2Z - 2Y + Z^3.$$

This is already a winner, so let's add it to our list:

$$\mathbf{Y^2Z} - 2Y + Z^3 \qquad [6]$$

So far our list of polynomial combinations has done nothing but grow. We started with three generators for our ideal, and now we have manufactured three more. The benefit of lowering the leading terms is that we can now use these new polynomials to reduce the complexity of our original generators. In particular, our fourth polynomial has leading term X, so we can use it to eliminate the X-terms of all the other entries. We can start by combining generators [1] and [4] to get

$$[1] - X[4] = -\tfrac{1}{2}XY^2Z - XYZ^2 + \tfrac{1}{2}XZ^3 + Y^2.$$

Our leading term is divisible by XY, so we can add on an additional $\frac{1}{2}YZ[3]$ to obtain

$$-XYZ^2 + \tfrac{1}{2}XZ^3 + Y^2 - \tfrac{1}{2}YZ^3.$$

We still have an XY in the leading term, so we can toss on another Z^2 copies of [3] to get

$$\tfrac{1}{2}XZ^3 + Y^2 - \tfrac{1}{2}YZ^3 - Z^4.$$

Finally, we can subtract off $\frac{1}{2}Z^2[2]$ and crown a new champion:

$$\mathbf{Y^2} - YZ^3 - Z^4 + Z^2 \qquad [7]$$

Notice that since this polynomial arose from the combination

$$[7] = [1] - X[4] + \tfrac{1}{2}YZ[3] + Z^2[3] - \tfrac{1}{2}Z^2[2],$$

we can then turn this around and write

$$[1] = [7] + X[4] - \tfrac{1}{2}YZ[3] - Z^2[3] + \tfrac{1}{2}Z^2[2].$$

This means polynomial [1] is now redundant and unnecessary, since it can be made from a combination of the others.

In the same way, we can combine [2] and [4] to get

$$[2] - Z[4] = -\tfrac{1}{2}Y^2Z^2 - YZ^3 + \tfrac{1}{2}Z^4 + 2Z^2 + YZ - 2.$$

To this we can add on another $\frac{1}{2}Z[6]$ to obtain a combination with an even smaller leading term, $-YZ^3 + Z^4 + 2Z^2 - 2$. We can then negate this (to make it monic) and append it to our list:

$$\mathbf{YZ^3} - Z^4 - 2Z^2 + 2 \qquad [8]$$

Here we notice that since $[8] = [2] - Z[4] + \frac{1}{2}Z[6]$, we can write $[2] = [8] + Z[4] - \frac{1}{2}Z[6]$ so that [2] is now unnecessary. I will leave it to you to show that

$$[3] = Y[4] - \tfrac{1}{2}Z[5] - Z[6],$$

which means [3] is also redundant. Our ideal, which began as ([1], [2], [3]) now has generators [4], [5], [6], [7], and [8]. The leading terms are X, Y^3, Y^2Z, Y^2, and YZ^3, respectively.

We now have a systematic, step-by-step procedure for reducing the complexity of a given set of generators. We simply perform the standard maneuver repeatedly, shedding redundant generators as we go. This process must eventually come to an end when the degrees can be lowered no further. In many ways, this procedure is reminiscent of long division and other such algorithms. We are shuttling symbols around mechanically in order to create certain alignment patterns, just as a weaver does when using a loom.

At this point we are finished with X (it now appears only in [4]), and we can focus on eliminating Y. Combining [5] and [7], we obtain the combination

$$[5] - Y[7] = Y^2Z^3 + YZ^4 - 2YZ^2 - 2Z^3 + 2Z.$$

We can now subtract $Z^2[6]$ to get

$$YZ^4 - Z^5 - 2Z^3 + 2Z.$$

The leading term is now divisible by YZ^3, the leading term of [8]. But when we subtract off $Z[8]$, something remarkable happens: every term cancels and we get zero! This means our clever combination $[5] - Y[7] - Z^2[6]$ is actually *equal* to $Z[8]$, allowing us to write [5] as a combination of [6], [7], and [8], thus making it redundant.

Proceeding with [6] and [7], we get

$$[6] - Z[7] = YZ^4 - 2Y + Z^5.$$

Subtracting off Z[8], we get $-2Y + 2Z^5 + 2Z^3 - 2Z$. Now we can divide through by –2 to obtain the monic polynomial

$$\mathbf{Y} - Z^5 - Z^3 + Z \qquad [9]$$

This is our new leader, and it also allows us to discard [6]. Thankfully, this polynomial is also linear in Y, so we don't have much further to go.

Our ideal is now generated by [4], [7], [8], and [9]. Since [4] is the only one that so much as mentions X, it will no longer play a role in the elimination process. All that remains is to combine [7], [8], and [9] in order to eliminate Y.

Let's start by combining [7] and [9]. Performing the standard maneuver, we find that

$$[7] - Y[9] - Z^2[8] + Z[9] = 0,$$

which means that [7] is now obsolete and can be discarded. The remaining generators are [8] and [9], so let's combine them and see what happens: $[8] - Z^3[9] = Z^8 + Z^6 - 2Z^4 - 2Z^2 + 2$. At long last, Y has been eliminated, and we have a new grand champion:

$$\boldsymbol{Z^8} + Z^6 - 2Z^4 - 2Z^2 + 2 \qquad [10]$$

We're finally down to only one indeterminate, and we even ended up with a polynomial of degree 8, as predicted. We also see that [8] has become redundant.

Our ideal is now generated by the polynomials

$$X + \tfrac{1}{2}Y^2Z + YZ^2 - \tfrac{1}{2}Z^3 - 2Z \qquad [4]$$
$$Y - Z^5 - Z^3 + Z \qquad [9]$$
$$Z^8 + Z^6 - 2Z^4 - 2Z^2 + 2 \qquad [10]$$

Although these may appear messy and complicated compared to our original generators, we are in fact much better off. Rather than three quadratic polynomials featuring multiple occurrences of X, Y, and Z, we have instead one equation that is linear in X, one involving only Y and Z (and linear in Y), and an equation of degree 8 in Z alone. If we like, we can even simplify further by using [9] to eliminate the Y-terms in [4]. The result is an alternative version in X and Z alone:

$$X + Z^7 + 2Z^5 - Z^3 - 3Z \qquad [4']$$

The generators [4′], [9], and [10] constitute what is known as a *reduced Gröbner basis* for our ideal. This is a set of generators for which no further progress can be made—it is as eliminated as it can get. What makes such a generating set especially interesting is that it turns out to be *unique*. Once you select an elimination order (so that the terms can be ranked) then the reduced Gröbner basis is determined. No matter how we select our generators or in what order we perform the standard maneuver, we will always end up at the same place. The minute we wrote down our generators, all of the reduced bases for all possible elimination orders were decided. The only question is the path we choose to take.

How would this example have gone if we had chosen
the elimination order Z, Y, X instead—
pretty much the same, a bit better, or hellishly worse?

In terms of our original system of unknowns and equations, what these polynomials are saying is that

$$a = -c^7 - 2c^5 + c^3 + 3c,$$
$$b = c^5 + c^3 - c,$$

and c satisfies the eighth-degree equation

$$c^8 + c^6 - 2c^4 - 2c^2 + 2 = 0.$$

The unknowns a and b are now expressed directly in terms of c and are therefore uninteresting. It's as if we've freed two strands of our knot, and all the remaining entanglement has been pushed onto the final strand.

Sorry to say, we won't always be so lucky. We were very fortunate that two of our reduced generators happened to have linear leading terms. This is not at all typical; the leading terms will usually have much higher degree. Luckily, this does not affect our overall strategy, and we can still reduce all of our problems to the one-dimensional case. For example, suppose we ended up with the reduced Gröbner basis

$$(X^5 + 2X^3Y^2 - YZ,\ Y^3 - 3YZ^2 + 2Z,\ Z^4 - 2Z^3 + 7),$$

corresponding to the system of equations

$$a^5 + 2a^3b^2 - bc = 0,$$
$$b^3 - 3bc^2 + 2c = 0,$$
$$c^4 - 2c^3 + 7 = 0.$$

Now nothing is linear, but we still have a sequence of equations that eliminates each unknown in turn. The third equation involves c alone. Assuming that we can somehow solve this equation, we can then substitute any such solution c into the second equation, leading to a cubic equation in b alone. Any solution to this can then be fed to the first equation (along with the current value of c) to obtain an explicit quintic equation in a alone. Once again, it all comes down to solving polynomial equations in one unknown, only now it is not just a single equation but rather a series of them, each determined by the solutions to the previous equation.

The upshot is that *all of our algebra puzzles can be reduced to solving a sequence of polynomial equations in a single unknown*. Assuming the original problem is well posed (meaning it has only finitely many solutions), our elimination procedure is then guaranteed to be successful. If the system is underdetermined, then of course we will be unable to eliminate completely. If, on the other hand, our equations happen to be contradictory, then a nonzero constant polynomial will always be produced, leading to the equation $1 = 0$.

As a somewhat bizarre example, consider the system

$$p^2 + q^2 = 0,$$
$$pq = 0.$$

Of course, it's easy to see that the only possible solution is for both p and q to equal 0. If we now proceed to form the ideal $(X^2 + Y^2, XY)$ and then carry out our standard procedure (with elimination order X, Y), we arrive at the reduced Gröbner basis $(X^2 + Y^2, XY, Y^3)$. Thus, we get the same equations as before, along with the additional constraint $q^3 = 0$. This means that q must equal 0, and then our first equation tells us that $p = 0$ as well.

What is interesting (or perhaps annoying) about this example is that our second generator (namely, XY) never came into play. This means that the equation $pq = 0$ is redundant as information, even though the polynomial XY is not redundant in the sense of elimination theory—there is no way to express XY as a polynomial combination of $X^2 + Y^2$ and Y^3. So there is a bit of blurriness in the correspondence between algebra puzzles and polynomial ideals. The ideal $(X^2 + Y^2, XY, Y^3)$ is essentially the same puzzle as $(X^2 + Y^2, Y^3)$, and both are really just the ideal (X, Y) in disguise, so to speak. Another way to say it is that the equation $x^2 = 0$ is logically equivalent to $x = 0$, but the polynomials X^2 and X are not equal and do not generate the same ideals.

This means there is more to investigate about the relationship between ideals and their simultaneous zeros. In fact, this would be a good time for us to jettison our original conception of a mystery number puzzle as necessarily having only a finite number of discrete solutions. The larger and more sophisticated question concerns the classification of polynomial ideals in general, and we can continue to pursue this inquiry even if the set of simultaneous zeros is infinite, or possibly empty. In other words, now that we have gained a little more perspective, we can drop our prejudice against systems that are underdetermined or contradictory and include their ideals in our general classification project as well.

Suppose, for instance, that we had a system of equations that was internally inconsistent and contradictory. Then there would be no solutions, and the corresponding polynomial ideal would have no simultaneous zeros. Which ideals have this property? What if we have two ideals that share the same infinite set of solutions—must the ideals be the same?

The study of polynomial ideals and their simultaneous zeros is known as *algebraic geometry*. The idea is to view a system of algebraic equations as describing a certain set of points in space, the dimension of this space being the dimension of the system. We discover a surprising and beautiful correspondence between polynomial ideals and the shapes determined by their simultaneous zeros. We will take up this intriguing subject in our final chapter.

In the meantime, let's suppose we are dealing with a system with only a finite number of discrete solutions. We now have a practical way to reduce any such problem to the one-dimensional case (especially if we happen to have a computer equipped with elimination software). The entire classical algebra project then comes down to solving the general polynomial equation in one unknown:

$$a_nX^n + a_{n-1}X^{n-1} + \cdots + a_1X + a_0 = 0.$$

Naturally, we will classify such equations by their degree. In the case of linear and quadratic equations, we have explicit general formulas for their solution. The general linear equation (degree 1) is $Ax + B = 0$, and its solution is given by $x = -\frac{B}{A}$. In the quadratic case (degree 2), we have

$$Ax^2 + Bx + C = 0 \Rightarrow x = \frac{-B + \sqrt{B^2 - 4AC}}{2A}.$$

Can we do something similar for polynomial equations of higher degree? For cubic and quartic equations (degrees 3 and 4), it turns out that we can.

CUBIC AND QUARTIC

It is a bit unclear when human beings first began seeking exact algebraic solutions to cubic equations. The Babylonian and Egyptian scribes certainly knew how to calculate cube roots to any desired degree of accuracy, and approximate numerical solutions to certain cubics appear in several ancient Chinese and Indian texts. The classical Greeks approached such problems geometrically, obtaining solutions to cubic equations as intersections of various curves and surfaces. The Persian mathematician and poet Omar Khayyam published a study of cubic equations in 1079 that provided the first complete classification of cubics as well as general (geometric) methods for their solution.

The question of whether there exists a *solution by radicals*—that is, an explicit formula for the solutions to a cubic equation in terms of its coefficients, involving only the operations of addition, subtraction, multiplication, division, and the extraction of roots—does not seem to have been asked until a few centuries later, though it's hard to believe that this was not the ultimate goal of the ancients as well, however they may have conceptualized it.

The truth is, it's almost impossible for a modern mathematician to fully understand the ancient, medieval, or even Renaissance mathematical mindset. Mathematics in the sixteenth century was essentially regarded as an occult science, right next door to alchemy and astrology. Reason and mysticism coexisted, and the distinction between mathematical and physical reality was blurry at best. (I suppose that for many people, even today, the line between idealism and religious belief is rather thin.) The reliance on a geometrical interpretation of quantity is another strange and awkward feature of the classical approach.

In particular, the geometric treatment of purely algebraic questions feels unnecessarily complicated and contrived to modern eyes, and the abhorrence of negative quantities can't help but seem quaint and naive. Although it's quite clear that the algebraists of the distant past were searching for what we would call a cubic formula, it is unlikely that we would agree as to its ultimate meaning or significance. I, for one, do not believe it to be a deep mystical secret of the universe, nor do I feel that it necessarily pertains to three-dimensional geometry. In any event, I certainly have no intention of classifying cubic equations into types depending on which coefficients happen to be positive or negative.

The modern point of view is that we are asking a linguistic question: we're asking whether the language of radicals is powerful enough to explicitly express any number given implicitly by a cubic equation. There are simply no angels or demons (or even geometric solids) involved. This is a purely algebraic question that can be meaningfully posed in any field environment.

There are actually two subtly different questions here. The first is whether each individual cubic has the property that its solutions can be expressed in this way. The stronger version is whether there is a one-size-fits-all formula that solves *all* cubic equations simultaneously—as we found in the quadratic case. This is the distinction between it just happening to be true, as opposed to having a systematic method that is guaranteed to work.

By the year 1500, algebraists were actively seeking such a method, having failed for centuries to discover a solution. Some were becoming convinced that the quest was hopeless. Might it be that the telluric and chthonic mystical forces do not *want* us to obtain such arcane wisdom?

The true story is in some ways even weirder. At some point in the early 1520s, Scipione del Ferro, a lecturer in mathematics at the University of Bologna, discovered a systematic general method for solving cubic equations. In those days it was customary to keep such discoveries secret—the exact opposite of today's "publish or perish" academic mandate. One's reputation was made by solving problems that others could not. In fact, it was quite common for algebraists of the Italian Renaissance to issue challenges to each other (typically by letter) in the form of a set of problems to be solved—the winner receiving a prize from the loser in the form of ducats and dinners. In this way, a mathematician possessed of secret methods could earn a comfortable living.

Although del Ferro never published his results, he did impart the secret to his student Antonio Maria Fior sometime before his death in 1525. Fior seems not to have been a very talented or original mathematician himself, merely the keeper of del Ferro's secret. Most of Fior's challenge problems are confined to the same narrow class of cubics (e.g., $x^3 + x = 6$

or $x^3 + x = 20$), which suggests that his knowledge and skill did not extend much further than mimicking his teacher. (We've all had grad students like that, I'm afraid.)

In any event, Fior's luck ran out in the winter of 1535 when he made the mistake of challenging the self-taught Venetian mathematician Niccolò Fontana, known as Tartaglia ("the stammerer"). Tartaglia accepted the challenge—consisting almost entirely of cubic equations to be solved—and, late in the evening of February 12, discovered his own independent solution. This not only matched del Ferro's achievement but superseded it, extending the method to any cubic, regardless of its particular form. Tartaglia had finally discovered the solution to the general cubic equation, a problem that (in one form or another) had remained unsolved for over four thousand years.

Of course, Tartaglia had no desire to publish his discovery either. In those days, knowledge was money, and there was no concept of an international scholarly community; it was every intellectual for himself. So it was only by repeated pleas and promises (and being sworn to secrecy) that the Milanese mathematician Girolamo Cardano was able to convince Tartaglia to divulge his methods in 1539. Here is the first stanza of the poem (translated loosely from the Italian) that Tartaglia recited to Cardano:

When the cube and things together are equal to a number,
Find two others of which it is the difference.
Then you will keep this custom
That their product should be equal
To the cube of one-third of the things.
The remainder of their cube roots, well subtracted,
Will be the value of your principal thing . . .

He then goes on to describe the necessary modifications to this scheme when the various terms are on different sides of the equation. (If this confusing and almost indecipherable description doesn't convince you of the value of modern notation, nothing will.) Finally, he sums up his achievement with a flourish:

These things I found, and not with sluggish steps,
In the year one thousand five hundred, four and thirty
With firm and strong foundations
In the city girded by the sea.

(Note that the old calendar was still in effect in Venice at that time; the new year began in March, so it was still 1534 to Tartaglia.)

A few years later, Cardano and his student Ludovico Ferrari discovered del Ferro's unpublished manuscript and decided to publish the method themselves, despite the promises made to Tartaglia. (Tartaglia never forgave Cardano for this and wasted a decade in fruitless lawsuits.)

Cardano's book *Ars Magna* ("The Great Art") was published in 1545. It is generally regarded as one of the most important and influential scholarly works of the Renaissance era. Not only does it contain Tartaglia's solution to the general cubic equation but also Ferrari's discovery of the solution to the general quartic as well. Despite his somewhat unsavory reputation as a rake and a gambler, Cardano does indeed give del Ferro, Tartaglia, and Ferrari full and complete credit for their work. (Incidentally, *Ars Magna* also contains the earliest known mention of complex numbers.) In 1547, Ferrari issued a challenge to Tartaglia and won the contest; Tartaglia's career as a duelist was over.

So that's the story, and I hope you enjoyed it. (Algebra is not exactly brimming over with romantic episodes, so we have to take what we can get!) Now it's time to discuss the mathematics itself. As always, I will try to explain Tartaglia's and Ferrari's ideas and methods, while taking full advantage of the massive simplifications afforded by the modern viewpoint. Essentially, we will be putting sixteenth-century art in a modern frame.

In particular, we can free ourselves entirely from the need to distinguish between positive and negative terms. The general cubic equation (over any field) takes the form

$$Ax^3 + Bx^2 + Cx + D = 0.$$

Here A, B, C, and D are parameters, and there is no reason to inquire any further as to their sign. We do not need to distinguish between "cube and things equal number" (e.g., $x^3 + 5x = 7$) and "cube equals number and things" (e.g., $x^3 = 5x + 7$) because we do not mind negative coefficients. The equation $x^3 - 5x - 7 = 0$ is perfectly valid and meaningful, and we needn't freak out because there is no stick of length −5.

This is already a major improvement. Instead of several different types of cubic, we now have only one. (By contrast, Omar Khayyam's classification included nineteen separate cases!) We can further simplify matters by assuming that our equation is monic. Since we can always begin by dividing everything by A, we may as well suppose from the start that $A = 1$. Thus, our cubic equation becomes

$$x^3 + Bx^2 + Cx + D = 0,$$

and we have lost no generality. Now we have only three parameters to deal with instead of four.

In fact, we can go even further and eliminate the x^2-term as well. The idea is to mimic what we did in the quadratic case. But instead of completing the square to eliminate the linear term, we will attempt to "complete the cube" in order to swallow up the quadratic term. Essentially, we're shifting our unknown by an additive constant to reduce the complexity of the entanglement.

Let's focus on the cubic and quadratic terms, leaving the lower degree terms to take care of themselves. We begin by setting $x = y + \square$, where $\square$ is a parameter to be determined shortly. Our equation then takes the form

$$(y + \square)^3 + B(y + \square)^2 + \cdots = 0,$$

where the ellipsis represents linear and constant terms. Recalling the general expansion $(a + b)^3 = a^3 + 3a^2b + 3ab^2 + b^3$, this becomes

$$y^3 + (3\square + B)y^2 + \cdots = 0.$$

Now the plan is to choose $\square = -\frac{1}{3}B$ so that the second term vanishes. In this way, we can reduce all of our problems to the case of the so-called *depressed cubic*

$$y^3 + Py = Q,$$

where P and Q are parameters that depend upon the original coefficients B, C, and D. (I have also moved the constant term Q to the right side to match Tartaglia's format.)

Show that $P = C - \frac{1}{3}B^2$ *and* $Q = \frac{1}{3}BC - \frac{2}{27}B^3 - D.$

Perhaps a concrete example will help clarify this reduction process. Suppose we have the cubic equation

$$2x^3 - 12x^2 + 4x - 4 = 0.$$

Dividing through by 2, we get the monic version,

$$x^3 - 6x^2 + 2x - 2 = 0.$$

Now we write $x = y + 2$ (since, according to our prescription, we should shift by $-\frac{1}{3}(-6) = 2$) and then substitute into the above equation:

$$(y + 2)^3 - 6(y + 2)^2 + 2(y + 2) - 2 = 0.$$

Expanding each term, we get

$$(y^3 + 6y^2 + 12y + 8) - 6(y^2 + 4y + 4) + 2(y + 2) = 2.$$

Now we can collect (notice the cancellation in the y^2-terms) to obtain the depressed cubic

$$y^3 - 10y = 14.$$

Assuming we can solve this equation for y, we then easily obtain the solution $x = y + 2$ to our original equation. The upshot is that with no loss of generality, we may assume that our cubic is depressed. (Maybe solving it will cheer it up!)

Let's restate our problem in this new simplified format: we seek a general formula for the solution to any depressed cubic $x^3 + Px = Q$. As we have seen, this would be sufficient to allow us to solve any cubic equation. This elementary reduction strategy was well known to the sixteenth-century algebraists as well; the solution to the depressed cubic was their primary goal.

The advantage of this approach is that we now have only *two* parameters to deal with, P and Q. Unfortunately, however, this is where all of our customary methods run aground. Any further shifting will only reintroduce the quadratic term, taking us backward. So a radical new idea is really needed. We have two (thankfully not three) occurrences of our unknown but no obvious way to eliminate further. This is why the problem is so hard and remained unsolved for so many centuries.

The new idea is not only radical and surprising but also ironic and even a bit subversive. Contrary to every classical algebraic instinct, the plan is to *increase* the number of unknowns in order to provide what is now known as *parametric freedom*. In other words, we will up the dimension in exchange for flexibility. This is a bold and brilliant maneuver—a breathtaking work of Italian Renaissance art. Both del Ferro and Tartaglia must have felt divinely inspired.

Given a depressed cubic equation $x^3 + Px = Q$, we begin by writing x as the sum of two other unknowns, say u and v. (Tartaglia uses either the sum or difference depending on which side the terms are on, in order to avoid a negative sign.) Writing $x = u + v$, our equation becomes

$$(u + v)^3 + P(u + v) = Q.$$

Now we have a cubic equation in two unknowns. On the face of it, this looks rather counterproductive, not to say asinine. We have both increased the number of unknowns *and* made our equation more complicated. Nevertheless, if we expand and rearrange terms, we can (very cleverly) rewrite this equation in the form

$$(u^3 + v^3) + (P + 3uv)(u + v) = Q.$$

Verify that these two equations are identical.

Now here is the beautiful observation. We have asked that our two unknowns u and v add up to x, a solution to our original equation. But we never made any other demands on them, so they are free to vary, as long as their sum remains constant. In particular, we never said what their *product* must be. That means we get to play around with the values of u and v in order to make their product especially convenient. Here, the prettiest choice is to demand that $3uv = -P$ so that the second term disappears. (This is the minus sign that Tartaglia didn't like but that we don't mind at all.)

We are thus led to the system of equations

$$u^3 + v^3 = Q,$$
$$3uv = -P.$$

If u and v are solutions to this system, then $x = u + v$ will be a solution to our original cubic, $x^3 + Px = Q$. We have traded a single cubic equation in one unknown for a system of two equations in two unknowns, one cubic and the other quadratic. Again, this does not appear to be a very good trade. We seem to have made matters strictly worse in every respect. What, if anything, did we gain from this exchange?

The payoff comes in the special form of these equations. For one thing, they are symmetrical in u and v. More importantly, they are both very sparse and simple. In fact, if we divide the second equation by 3 and then cube both sides, we get $u^3v^3 = -(\frac{P}{3})^3$, which means that now both equations involve u^3 and v^3 *only*. This clearly calls for some abbreviation. Let's write $\alpha = u^3$ and $\beta = v^3$. (The numbers u and v are then given by $u = \sqrt[3]{\alpha}$ and $v = \sqrt[3]{\beta}$.) Our equations now read

$$\alpha + \beta = Q,$$
$$\alpha\beta = -\left(\frac{P}{3}\right)^3.$$

Magically, our system has become quadratic! Somehow our trade has managed to increase the dimension while also *lowering the degree*—the exact opposite of elimination. As an added bonus, the symmetry of the system means that α and β cannot be distinguished, so we don't have to worry about which is which.

The solution to our depressed cubic $x^3 + Px = Q$ is now at hand. All we need to do is to find α and β—two numbers whose sum is Q and whose product is the (negative) cube of one-third of *P*. Then our solution is simply $x = \sqrt[3]{\alpha} + \sqrt[3]{\beta}$. (We've altered the signs a little, but Tartaglia's poem should now make a bit more sense.)

Let's test our method on a specific example. Suppose we are faced with the depressed cubic equation $x^3 + 3x = 2$. In this case, we have $P = 3$ and $Q = 2$. So we need to solve the system

$$\alpha + \beta = 2,$$
$$\alpha\beta = -1.$$

This quadratic system has the solutions $1 + \sqrt{2}$ and $1 - \sqrt{2}$ (regardless of which you call α and β).

Show that these are in fact the correct values.

Thus, we find our solution to be

$$x = \sqrt[3]{1+\sqrt{2}} + \sqrt[3]{1-\sqrt{2}}.$$

Naturally, Tartaglia and his contemporaries would have written this as the difference between two positive numbers

$$x = \sqrt[3]{\sqrt{2}+1} - \sqrt[3]{\sqrt{2}-1},$$

rather than as a sum of a positive and a negative.

If we like (and we probably don't), we can substitute this expression into the equation $x^3 + 3x = 2$ and check that it does indeed come out exactly right. As an easier (though possibly less convincing) alternative, we can simply take out a calculator and see what the approximations look like:

$$\sqrt[3]{1+\sqrt{2}} \approx 1.3415,$$
$$\sqrt[3]{1-\sqrt{2}} \approx -0.7454.$$

Our solution x is then approximately equal to 0.5961. This means that $x^3 + 3x \approx 2.0001$, which is rather excellent agreement, I would have to say.

What actually happens when we calculate $x^3 + 3x$ exactly?

At this point, there is nothing stopping us from obtaining a general formula. The solution to the quadratic system

$$\begin{aligned} \alpha + \beta &= Q, \\ \alpha\beta &= -\left(\tfrac{P}{3}\right)^3 \end{aligned}$$

is easily found to be

$$\alpha, \beta = \tfrac{Q}{2} \pm \sqrt{\left(\tfrac{Q}{2}\right)^2 + \left(\tfrac{P}{3}\right)^3}.$$

Here I am using the *plus or minus* sign to denote that one of α and β takes the plus and the other the minus, but it doesn't matter which. We thus obtain a formula for the solution to the general depressed cubic equation $x^3 + Px = Q$, taking the striking form

$$x = \sqrt[3]{\tfrac{Q}{2} + \sqrt{\left(\tfrac{Q}{2}\right)^2 + \left(\tfrac{P}{3}\right)^3}} + \sqrt[3]{\tfrac{Q}{2} - \sqrt{\left(\tfrac{Q}{2}\right)^2 + \left(\tfrac{P}{3}\right)^3}}.$$

The answer to our linguistic question is *yes*—the language of roots is indeed powerful enough to allow us to untangle not only any given cubic equation but all of them simultaneously. Notice that to solve the general cubic we require not only cube roots but square roots as well. (There is an additional modern technical issue regarding division by 2 and 3, making our formula problematic for fields where either $2 = 0$ or $3 = 0$, but of course this was no concern of the classical algebraists, who worked exclusively over $\mathbb{R}$.)

The equation $x^3 + x = 2$ clearly has $x = 1$ as a solution.
What does our cubic formula say in this case?

Before we get too excited, it turns out that some pretty strange and unsettling things can happen, especially when the parameter P is allowed to be negative—as it is in the "cube equals things and number" case. Let's look at $x^3 = 3x + 1$, for example. (This corresponds to $P = -3$; $Q = 1$ in our format.) There certainly is a real number that satisfies this equation. One simple way to see this is to use the fact that $\mathbb{R}$ is a continuum. If we imagine the number x as a *variable*—that is, a numerical entity whose value can vary continuously—we see that as x runs from 1 to 2, the left side of the equation (namely x^3) starts out smaller than the right side ($3x + 1$) but ends up larger. So somewhere in between, the two sides have to match exactly, and that is the x we seek. (This same technique can be used to show that *every* cubic equation must have a solution in $\mathbb{R}$.) So there is a certain real number out there satisfying the equation $x^3 = 3x + 1$, having the approximate value $x \approx 1.88$. For the exact value, we can turn to our cubic formula:

$$x = \sqrt[3]{\frac{1}{2} + \sqrt{-\frac{3}{4}}} + \sqrt[3]{\frac{1}{2} - \sqrt{-\frac{3}{4}}}.$$

In this case, with $P = -3$ and $Q = 1$, the expression $(\frac{Q}{2})^2 + (\frac{P}{3})^3$ is now *negative* and equal to $-\frac{3}{4}$. This means our formula requires us to take the square root of a negative number! In Cardano's day, this was considered nonsensical. On the other hand, if one overlooks this absurdity and simply plugs this expression for x into the equation, it works perfectly. What a bizarre situation!

This number x is a perfectly normal and healthy real number between 1 and 2 with the interesting feature that its cube is one more than three times itself. That makes it one of those rare numbers that is not only describable but algebraically describable. On top of that, our number x can be explicitly named using the language of roots—but weirdly, in order to do this, we require the use of numbers that are *not* real.

In some ways, this is reminiscent of our experience with negative quantities in arithmetic. Although there may not be such a thing as a negative number of lemons or a negative amount of money, it is still very convenient to *act* as if there were—to pretend, if you like, that negative numbers make sense and to use them accordingly. The fact that the arithmetic all works out (e.g., accounting for profits and losses) suggests that there may be a certain reality to this notion—a reality beyond the one that we naively perceive. (Now I'm starting to sound like a mystic!) Whenever a new, imaginary construct exhibits patterned behavior, especially if it extends a notion already in place, we should always pay attention. It usually means we are about to make a major breakthrough—an expansion of our consciousness, no less.

The Renaissance algebraists were perplexed and intrigued by this discovery. We have the ability to untie any cubic knot, but to do so we occasionally need to pass one end of our string into an alternate dimension (so to speak) and then pull it out again. Another way to say it is that we have discovered our string $\mathbb{R}$ to be inadequate, even on its own terms. A cubic equation can have positive real coefficients and a positive real solution explicitly given in the language of roots, and yet we still need to jump off the real line in order to say what it is. This was really the last straw, as far as $\mathbb{R}$ was concerned. It's one thing to have quadratic equations with no solution (due to an imaginary square root), but it's quite another to have a real solution to a cubic equation and not be able to write it down because $\mathbb{R}$ is too restrictive an environment.

It took several decades for all of this to sink in. The construction and acceptance of the complex numbers did not happen overnight. (For us, it will happen in the next chapter.) If we step back and consider the three historically important fields $\mathbb{Q}$, $\mathbb{R}$, and $\mathbb{C}$, we see that all three are responses to our idealistic and intellectual choices and demands: $\mathbb{Q}$ is created by arithmetic; $\mathbb{R}$ is born of our desire for an ideal geometry. The field of complex numbers, on the other hand, does not originate in physical reality, idealized or otherwise: $\mathbb{C}$ is the offspring of algebra itself. We did not ask for square roots of negative numbers; we did not demand a number plane incapable of being ordered. These are things we neither sought nor wanted. Instead, they were thrust upon us by our own discoveries. We need complex numbers in order to make sense of what we have done.

Initially, the entire situation was terribly frustrating and disheartening. We thought we were dealing with perfectly reasonable sticks and squares and cubes, and suddenly we find ourselves having to face $\sqrt{-\frac{3}{4}}$. What does it mean? We know that every cubic equation $y^3 + Py = Q$ has a real solution, so why is everything so nice when $(\frac{Q}{2})^2 + (\frac{P}{3})^3$ is positive—giving us two perfectly sensible real numbers whose cube roots also make complete sense—but when it is negative (the so-called *casus irreducibilis*), everything goes kerflooey?

Another related issue has to do with the number of solutions. We saw before that a quadratic equation can have 0, 1, or 2 real solutions and no more. Similarly, as we shall see, a cubic equation can have at most three solutions. Thus, there are cubic equations over $\mathbb{R}$ with exactly one real solution, some with two solutions, and others with three distinct real solutions. For a depressed cubic $y^3 + Py = Q$, this turns out to correspond exactly to whether the expression $(\frac{Q}{2})^2 + (\frac{P}{3})^3$ is positive, zero, or negative. In the first case, the cubic formula gives us an exact solution in real number terms; in the second, we get a formula for one of the solutions but not the other; and in the third case, we get utter nonsense. Something is clearly missing from our analysis because in the quadratic case, we derived a formula for *both* of the solutions, the result of an ambiguous choice of sign in the square root. Shouldn't there be a similar ambiguity regarding cube roots? In the case where our cubic equation has three solutions, don't we want a formula that will produce all three?

So we find that our celebrated cubic formula is both incomplete as well as potentially meaningless. To make sense of it all, let's jump ahead a century or so and suppose that we're completely comfortable with complex numbers and that we understand $\mathbb{C}$ to be a perfectly sensible number environment both extending and improving upon $\mathbb{R}$. Again, the point is that algebra makes different demands than geometry. In retrospect, it was silly of us to interpret algebraic equations as geometric statements confined to positive real numbers. This view led to an awkward and confusing situation. With the benefit of hindsight, we can see that the fear and mistrust of negative and imaginary quantities was the real problem.

What would it be like not to have this prejudice? Rather than viewing $\mathbb{R}$ as the "real" number system and $\mathbb{C}$ as some sort of utopian fantasyland, we take the more enlightened view that $\mathbb{C}$ is the actual field of interest and $\mathbb{R}$ is merely a special subfield for which we have some fond historical regard. In particular, it means we have to start over with our thinking about both cubic and quadratic equations.

Luckily, the quadratic situation requires no new ideas. Given a quadratic equation $Ax^2 + Bx + C = 0$, where the coefficients A, B, and C are now arbitrary *complex* numbers, we can still complete the square as before, since the only move we require is distribution, and this holds for any field. As we shall see, every (nonzero) complex number has exactly two square roots, so our quadratic formula

$$x = \frac{-B \pm \sqrt{B^2 - 4AC}}{2A}$$

remains perfectly valid, and we get a description of both solutions simultaneously. In the event that the coefficients A, B, and C happen to be real, then the solutions will also be real, provided that the discriminant $B^2 - 4AC$ is positive. (When it is zero, the two solutions collapse to the same value—namely, $-\frac{B}{2A}$.) When the discriminant is negative, we still have two solutions, only they are both nonreal complex numbers.

The crux of the matter here is that every number has at most two square roots, and these are negatives of each other. In particular, the two square roots of 1 are the numbers 1 and –1. This is not a special property of $\mathbb{R}$ or $\mathbb{C}$ but is true for any field. The reason is that we always have the difference of squares factorization $x^2 - 1 = (x + 1)(x - 1)$. Thus, in order to satisfy $x^2 = 1$ (i.e., to qualify as a square root of 1), a number must then satisfy $(x + 1)(x - 1) = 0$, and this forces either $x + 1 = 0$ or $x - 1 = 0$, leading to $x = \pm 1$.

What happens if we look for the cube roots of 1? As any classical algebraist would tell you, the only real number satisfying $x^3 = 1$ is 1 itself. But what if we no longer demand that our numbers be confined to the real line? Here we can take advantage of the corresponding *difference of cubes* formula:

$$a^3 - b^3 = (a - b)(a^2 + ab + b^2).$$

Verify that this holds for any numbers a and b in any field.
Does it generalize to higher powers?

We thus obtain the factorization

$$x^3 - 1 = (x - 1)(x^2 + x + 1).$$

This is valid over any field, so a cube root of 1 must either satisfy $x - 1 = 0$ (in which case $x = 1$), or else we must have $x^2 + x + 1 = 0$. This cannot happen in $\mathbb{R}$, but so what? We can still solve this using our quadratic methods. In fact, we get the two complex solutions

$$x = \frac{-1 \pm \sqrt{-3}}{2}.$$

So it turns out there are *three* complex cube roots of 1, and only one of them is real. The other two are a bit difficult to distinguish, but they certainly have the property that when cubed they equal 1. It would therefore be algebraically irresponsible to ignore them. Let's refer to them as ω and ω'. (Here I'm using a lowercase omega, the final letter of the Greek alphabet.)

In addition to being cube roots of *unity* (the fancy old-fashioned term for the number 1), these numbers ω and ω' also satisfy a number of amusing relationships. Adding them together, we find $\omega + \omega' = -1$, and multiplying gives us $\omega\omega' = 1$. This means that ω' is also the reciprocal of ω. Another nice feature is that each of these numbers is the square of the other: $\omega' = \omega^2$ and $\omega = \omega'^2$.

Why is this true?

In fact, we may as well call the complex cube roots of unity 1, ω, and ω^2. (I like how we simply multiply by ω as we go through this list, with the final product taking us back to $\omega^3 = 1$.)

What are the cube roots of unity in $\mathbb{F}_7$?

The cubic equation $x^3 = 1$ now has three distinct complex solutions, as opposed to only one real solution. These three complex numbers need to be understood as equally valid and respectable cube roots of unity. I'm sorry that you cannot hold ω lemons in your hand or pay a bill for ω dollars; I'm sorry there are no pentagons of side ω either. That is simply not ω's job. We hired ω (as well as her friend ω^2) to be complex cube roots of 1, not to be fashionable in the realms of arithmetic and geometry.

One consequence is that *every* real number now has three complex cube roots (except zero, which has only itself). For instance, if $\sqrt[3]{2}$ represents the positive real cube root of 2 (approximately 1.26), then $\sqrt[3]{2} \cdot \omega$ and $\sqrt[3]{2} \cdot \omega^2$ are the two others. So just as with square roots, we'll need to be a little careful about how we use the word *the* when speaking of cube roots as well. In fact, it turns out that every nonzero *complex* number z also has exactly three cube roots, making the expression $\sqrt[3]{z}$ rather ambiguous. If one particular meaning is selected, then the other two possibilities are given by multiplying it by ω and ω^2. Cube roots always come in threes, just as square roots always occur in pairs. So that's pretty, and it also explains where we went wrong with our cubic formula.

The plan, as you may recall, was to write the solution to a depressed cubic as a sum $\sqrt[3]{\alpha} + \sqrt[3]{\beta}$, where α and β satisfy a certain quadratic system. But now there are three different possible interpretations for both $\sqrt[3]{\alpha}$ and $\sqrt[3]{\beta}$. So these must be the solutions we have been missing. In the *casus irreducibilis* (i.e., when there are three real solutions), it means that all three solutions will be represented as combinations of complex numbers. Whether we like it or not, even if our sole interest is in real equations with real solutions, we still require $\mathbb{C}$ in order to talk about the real numbers we want to talk about.

So let's go back to our analysis of the general depressed cubic equation $x^3 + Px = Q$. We began by writing $x = u + v$, where u and v satisfy the system of equations

$$\begin{aligned} u^3 + v^3 &= Q, \\ 3uv &= -P. \end{aligned}$$

The next step was to cube the second equation and write $\alpha = u^3$, $\beta = v^3$. This led to the system

$$\begin{aligned} \alpha + \beta &= Q, \\ \alpha\beta &= -\left(\tfrac{P}{3}\right)^3, \end{aligned}$$

whose solutions are

$$\alpha, \beta = \frac{Q}{2} \pm \sqrt{\left(\frac{Q}{2}\right)^2 + \left(\frac{P}{3}\right)^3}.$$

Due to the symmetry of the equations, there is no way to tell which of these numbers is α and which is β. Fortunately, it doesn't matter. (In fact, in some ways it's even simpler, just as it's easier to put on socks than shoes—there's no wrong way!) Let's suppose (for the sake of definiteness) that we make a particular choice of $\sqrt[3]{\alpha}$ and $\sqrt[3]{\beta}$. Then the options for the number u are $\sqrt[3]{\alpha}$, $\sqrt[3]{\alpha} \cdot \omega$, and $\sqrt[3]{\alpha} \cdot \omega^2$. Meanwhile, v can take on the values $\sqrt[3]{\beta}$, $\sqrt[3]{\beta} \cdot \omega$, and $\sqrt[3]{\beta} \cdot \omega^2$. Our desired solution x is then given by the sum $u + v$.

On the face of it, this would seem to represent *nine* possible choices. But in fact only three are viable. This is because of the constraint $3uv = -P$. When we cubed this equation (in order to rephrase it in terms of α and β), we actually lost a bit of information. Since two different numbers can have the same cube, cubing is in fact a nonreversible act, just like squaring. So we can't just pick any cube roots of α and β. We have to choose them so that their product is $-\frac{P}{3}$ *exactly*, as opposed to being off by a factor of ω or ω^2.

This constraint cuts down our options dramatically. For any given choice of u, there is now only one possible v and vice versa. Supposing that our initial choices of $\sqrt[3]{\alpha}$ and $\sqrt[3]{\beta}$ satisfy this demand (i.e., $\sqrt[3]{\alpha}\,\sqrt[3]{\beta} = -\frac{P}{3}$), then the complete set of possible values for u and v would be:

$$\begin{aligned} u &= \sqrt[3]{\alpha}\,; & v &= \sqrt[3]{\beta}, \\ u &= \sqrt[3]{\alpha} \cdot \omega; & v &= \sqrt[3]{\beta} \cdot \omega^2, \\ u &= \sqrt[3]{\alpha} \cdot \omega^2; & v &= \sqrt[3]{\beta} \cdot \omega. \end{aligned}$$

Whichever cube root of unity we decide to attach to $\sqrt[3]{\alpha}$, we need to choose its reciprocal for $\sqrt[3]{\beta}$ so that the product uv remains constant. Now we can put these together to obtain the three solutions to our cubic equation:

$$x = \begin{cases} \sqrt[3]{\alpha} + \sqrt[3]{\beta}, \\ \sqrt[3]{\alpha} \cdot \omega + \sqrt[3]{\beta} \cdot \omega^2, \\ \sqrt[3]{\alpha} \cdot \omega^2 + \sqrt[3]{\beta} \cdot \omega. \end{cases}$$

This is the full and complete version, valid for any depressed cubic equation with complex coefficients. As we shall see, every (nonzero) complex number does indeed have three distinct cube roots, and thus every cubic equation has three complex solutions (although in extreme cases some of these can be equal, so it then depends on how you wish to count them). This is clearly a much simpler and more symmetrical approach, and the only real difficulty is overcoming the mental hurdle of accepting these new numerical entities in the first place.

In our first example, $x^3 + 3x = 2$, we used Tartaglia's method to obtain the real solution

$$x = \sqrt[3]{1+\sqrt{2}} + \sqrt[3]{1-\sqrt{2}}.$$

Now we know there are two other (complex) solutions, given by inserting the other cube roots of unity:

$$\omega \cdot \sqrt[3]{1+\sqrt{2}} + \omega^2 \cdot \sqrt[3]{1-\sqrt{2}} \quad \text{and} \quad \omega^2 \cdot \sqrt[3]{1+\sqrt{2}} + \omega \cdot \sqrt[3]{1-\sqrt{2}}.$$

These were simply invisible to Tartaglia and Cardano—which is ironic, because it shows that for all the mystical and quasi-religious ideas regarding alternate planes of existence, here the occultists and philosophers had an actual example right under their noses, and it took them centuries to finally lift the veil.

In our second, more confusing example, $x^3 = 3x + 1$, we obtained the values $\alpha, \beta = -\frac{1}{2} \pm \sqrt{-\frac{3}{4}}$. These two numbers are complex, as are their cube roots. Our three solutions for x are all real, but they can only be expressed (at least in the language of roots) using these complex intermediaries. How funny that we need to pop out of the system to find the means to describe things that were already there!

By the way, it just so happens that we've run into these two numbers α and β before—they are precisely ω and ω^2. In other words, in this special example, α and β turn out to be cube roots of unity themselves. This means that $\sqrt[3]{\alpha}$ and $\sqrt[3]{\beta}$ are in fact *ninth* roots of 1, and certain sums of these happen to be real, and those are our solutions. So if we can somehow get truly comfortable with the arithmetic of complex numbers, all the clouds of mystery and confusion will dissipate, and the beautiful symmetry and simplicity will be revealed.

As satisfying as our new and improved cubic formula may be, there is a certain lack of *tangibility* to these creatures, the complex numbers. It's all very well and good to accept $\sqrt{-3}$ as an abstract entity whose square is negative three, but it's quite another to have a concrete and intuitive mental picture. It's nice to be able to write such numbers down and to calculate with them, but I want to be able to see them and to feel them interacting, the way I can with the real number line. So that will be our next big project.

In the meantime—since I haven't written that chapter yet—I want to tell you about Ferrari's solution to the general quartic equation and the clever way he made use of parametric freedom. We will again ignore the sixteenth-century bigotries and proceed with the assumption that all of our coefficients and unknowns belong to the complex number field from the beginning.

The general quartic equation takes the form

$$Ax^4 + Bx^3 + Cx^2 + Dx + E = 0.$$

As usual, we may assume that $A = 1$ for simplicity. We can also perform the customary shifting maneuver, only this time we make the substitution $x = y - \frac{1}{4}B$. Thus, we may assume our equation is monic with no cubic term.

Show that this substitution does the trick.

Moving the lower degree terms to the other side, we will assume that our quartic comes in the depressed form

$$x^4 = Px^2 + Qx + R,$$

where P, Q, and R are parameters that depend on the original coefficients A, B, C, D, and E. Our problem comes down to x^4 = quadratic in x. Now what?

Fortunately, there is one special situation we can handle easily, and that is when the quadratic expression on the right happens to be a perfect square. For example, the quartic equation

$$x^4 = x^2 + 6x + 9$$

can be rewritten as

$$x^4 = (x + 3)^2.$$

In this (admittedly unlikely) event, we can then simply take the square root of both sides (using the word *the* in the professional sense, of course) to get

$$x^2 = x + 3 \quad \text{or} \quad x^2 = -(x + 3).$$

Our quartic thus splits into two separate quadratics. This can also be seen from the factorization

$$x^4 - (x + 3)^2 = (x^2 + x + 3)(x^2 - x - 3)$$

given by the difference of squares formula. The point being that we are all set if the quadratic on the right-hand side just magically happens to be a perfect square.

Ferrari's idea was to employ parametric freedom in order to *make* it be a square. The plan goes like this. We start with our depressed quartic $x^4 = Px^2 + Qx + R$. Now we give ourselves some latitude by introducing a parameter t. The trick is to replace x^4, which can be viewed as $(x^2)^2$, by the slightly modified version $(x^2 + t)^2$. Expanding this, we get

$$\begin{aligned}(x^2 + t)^2 &= x^4 + 2tx^2 + t^2 \\ &= (Px^2 + Qx + R) + 2tx^2 + t^2.\end{aligned}$$

Collecting terms, we arrive at the modified quartic equation

$$(x^2 + t)^2 = (2t + P)x^2 + Qx + (R + t^2).$$

This is what our mystery number x does—regardless of the value of our parameter t. So we get a more complicated equation, but we also gain the freedom to set t to any desired value. Rather than choosing t to eliminate a term, however, Ferrari's idea—his masterpiece, really—was to choose t so that the quadratic expression on the right is a perfect square! If we can do this, then we can square root both sides and we will be done.

Fortunately, it's easy to tell when a quadratic polynomial is the square of a linear polynomial: $AX^2 + BX + C$ is a perfect square precisely when its discriminant vanishes—that is, when $B^2 = 4AC$.

Why is this true?

Here I have done something quite common but perhaps a little confusing: I have labeled my coefficients using prior labels. These new A, B, and C are *not* the same numbers as the original A, B, and C. I don't really care about the general quartic anymore, so I am recycling my symbols to save mental energy. I'm referring now to a general quadratic polynomial. The symbols A, B, and C have temporarily put aside their meanings as coefficients of our quartic equation and are serving my purpose (for the time being) in helping me talk about when quadratic polynomials are perfect squares. Expressed in words, the condition is that the square of the linear coefficient is equal to four times the product of the other two coefficients.

In the present instance, our coefficients are $2t + P$, Q, and $R + t^2$. That means the condition we need to satisfy is: $Q^2 = 4(2t + P)(R + t^2)$. If we can find a value of t that does this, then we can square root both sides of our equation as planned. Ferrari's solution to the quartic comes down to solving this equation for t. Expanding and collecting, our equation becomes

$$8t^3 + 4Pt^2 + 8Rt + (4PR - Q^2) = 0.$$

This is known as the *resolvent cubic* arising from the depressed quartic equation. To find the right value for the parameter t, we need to solve a cubic equation—which we fortunately now know how to do. In other words, the effect of Ferrari's method is to reduce the problem of the general quartic to the (lower degree) case of the general cubic. (This was an additional incentive for Cardano to obtain Tartaglia's secret.)

Of course, I have no intention of writing down a complete quartic formula. As you can see, even in the simplified case of a depressed quartic with only three parameters, we would need to express t explicitly in terms of these coefficients using the cubic formula (having suitably shifted and scaled the resolvent equation to put it in depressed form), then split our quartic into two ghastly quadratics with coefficients that are themselves nightmarish combinations of roots—forget it.

The point is not what the cubic and quartic formulas look like in detail—you can look them up if you want—but rather that they exist and that we can know that. The question concerns the expressive power of a certain language, not precisely what is being said when. The language of radicals is indeed strong enough to allow us to untangle cubic and quartic equations over the complex numbers, and we can even (in principle) write down a general formula, if we so desire.

As you might expect, a generic quartic equation will tend to have *four* distinct complex solutions, some of them coinciding in certain special cases. In the simplest instance of $x^4 = 1$, we expect there to be four complex fourth roots of unity, and indeed there are: 1, –1, and the two different values of $\sqrt{-1}$. If we choose one of these two and name it i (as has become customary), then the complex fourth roots of unity can be written as 1, i, i^2, and i^3. In general, the solutions to a given quartic equation will be certain sums of roots (with constraints on these choices) involving expressions made of pieces involving roots of things made of roots. It's pretty gruesome, but it is sayable—and more importantly, that fact is *knowable*.

Show that $\frac{1+i}{\sqrt{2}}$ is an eighth root of unity.
Can you find the other seven?

It is now clear what direction the classical algebra project must take. First, we need to develop the complex number concept and put it on some kind of solid footing so that we are not merely inventing hopeful truths but are making definitive statements about logically coherent entities. The goal is then to expand our methods to include equations of degree 5, 6, and beyond. Ideally, we could even dream of a solution to the general polynomial equation of degree n—a single, master formula for all algebraic equations. Obviously, finding a quintic formula is the natural next step. This problem would occupy Cardano's followers for the next 250 years.

COMPLEX NUMBERS

Now we get to do something really fun. We get to bring a whole new world of numbers into existence and then play around in it and see how it works. We already know what we want: a field extension of $\mathbb{R}$ that also contains numbers like $\sqrt{-1}$ that we currently lack. In addition to this purely algebraic demand, we're also hoping to construct some sort of intuitive model of this new environment akin to the number line for $\mathbb{R}$. We want to be able to calculate and do arithmetic, but we also seek to gain intuition and understanding—to see and feel the ways that our new numbers behave and interact. This means we're looking for both symbolic as well as visual (and even tactile) representations.

Let's suppose for the moment that there is such a realm. In particular, let's imagine that there is some entity i that plays the role of $\sqrt{-1}$. The only thing we know about this number is that $i^2 = -1$. Since we want $\mathbb{C}$ to be a field (meaning that it satisfies our nine demands), it must contain the number $-i$ (the additive inverse of i), and this will also be a square root of -1. That means these two values of $\sqrt{-1}$, despite being unequal, are impossible for us to tell apart—we'll simply have to call one of them i and the other $-i$. (We will return to this fundamental ambiguity later.)

So our first plan will be to define the complex number field $\mathbb{C}$ to be $\mathbb{R}(i)$, the field extension of $\mathbb{R}$ generated by throwing in the new number i and forming every possible combination of sums, products, negations, and reciprocals. If $\mathbb{C}$ is going to be a field environment, it must be a closed system under all these operations. Thus, $\mathbb{C}$ will need to include such numbers as $2 + i$, $5 - \frac{3}{i}$, and even

$$\frac{1 + 7i}{\frac{i}{2} - \left(\frac{3}{i-4}\right)^2}.$$

Fortunately, this collection of numbers is not as vast and unruly as it may appear. In fact, every element of the field $\mathbb{R}(i)$ can be written in the simple form $a + bi$, where a and b are real numbers.

To see this, let's first notice how this particular type of combination behaves with respect to our field operations. Suppose we have two such numbers, say $a + bi$ and $x + yi$. Adding and multiplying them, we get

$$(a + bi) + (x + yi) = (a + x) + (b + y)i,$$
$$(a + bi) \cdot (x + yi) = (ax - by) + (ay + bx)i.$$

(Here I have used distribution, along with $i^2 = -1$.) We see that any two numbers of this special form combine to make another number of the same form, so these numbers already form a closed system under addition and multiplication. The special numbers 0 and 1 are also included in this family, since any real number r can always be written $r + 0i$. The additive inverse of $a + bi$ is clearly $(-a) + (-b)i$, so the only thing left to show is that every (nonzero) number $a + bi$ must have a reciprocal of the same form.

There are several ways to demonstrate this, but my favorite is to use the difference of squares formula to write

$$\begin{aligned}(a + bi)(a - bi) &= a^2 - (bi)^2,\\ &= a^2 + b^2.\end{aligned}$$

The peculiar property of i (i.e., that its square is -1) means we get a factorization formula for the *sum* of two squares as well. The consequence is that every complex number $a + bi$ has a partner $a - bi$ so that when multiplied together they make the positive real number $a^2 + b^2$. (The expression $a^2 + b^2$ can also be zero, but only if both a and b are zero to begin with.) For example, the number $3 + 2i$ can be multiplied by $3 - 2i$ to make $3^2 + 2^2 = 13$.

Now it's easy to construct the reciprocal of any complex number $a + bi$ (other than $0 + 0i$, of course). We're looking for a complex number that when multiplied by $a + bi$ makes 1. The number $a - bi$ almost works, since multiplying $a + bi$ by $a - bi$ produces the real number $a^2 + b^2$. So all we need to do is divide by this number to get 1. Thus, the reciprocal of $a + bi$ is given by

$$\frac{a - bi}{a^2 + b^2} = \left(\frac{a}{a^2 + b^2}\right) + \left(\frac{-b}{a^2 + b^2}\right) i.$$

In the case of our example $3 + 2i$, we find its reciprocal to be

$$\frac{1}{3+2i} = \frac{3-2i}{13} = \frac{3}{13} - \frac{2}{13}i.$$

Another way to think of it is that we are simply rescaling the fraction $\frac{1}{3+2i}$, multiplying both the top and bottom by $3 - 2i$ in order to make the denominator real.

The upshot is that the sum, difference, product, and quotient of any two numbers of the form $a + bi$ is again of that same form, so these numbers constitute the desired extension field $\mathbb{C} = \mathbb{R}(i)$. We now have our first model of the complex numbers: they are the expressions of the form $a + bi$, where a and b are real numbers and $i^2 = -1$. It's bit abstract, but at least we know exactly what entities we've got and how to work with them.

What is $(7 + i) \div (3 + 4i)$?

There is something a little unsettling about simply assuming that such a thing as i exists and that we get to work with it as if it enjoys all the usual field properties. How do we know there isn't some sort of internal contradiction here? It would be nice to have a model of $\mathbb{C}$ that wasn't based so much on wishful thinking.

One way to construct the complex number system—without any such assumptions—is to view a complex number $a + bi$ as being simply a *pair* of real numbers (a, b). That is, we let the number pair hold the information, and this then becomes our concrete representation. Thus, the pair (3,2) refers to the desired entity $3 + 2i$, and the number i itself would be encoded as (0,1).

So the proposal is to define $\mathbb{C}$ as the set of all pairs of real numbers. We can then define addition and multiplication of such pairs by mimicking the patterns that we want our complex numbers to obey:

$$(a, b) + (x, y) = (a + x, b + y),$$
$$(a, b) \cdot (x, y) = (ax - by, ay + bx).$$

Essentially, we are training number pairs to behave exactly like the complex numbers we are hoping to create. So we let the pairs *be* the numbers! Notice that the pair (0,0) behaves exactly as zero should behave (i.e., when added, it does nothing), and that (1,0) is multiplicatively inert as well. These pairs are the complex numbers 0 and 1. It is now a straightforward (but somewhat tedious) matter to check that these operations on pairs do indeed satisfy our nine demands for a field. In particular, the reciprocal of the pair (a, b) is given by the same pattern as before:

$$\frac{1}{(a,b)} = \left(\frac{a}{a^2+b^2}, \frac{-b}{a^2+b^2}\right).$$

Now we have a new number system—a perfectly well-defined and legitimate field—whose elements are number pairs. The elements of the form $(r, 0)$ form a subfield that is indistinguishable from the real numbers, and the special pair $i = (0,1)$ possesses the very property that we require: $i^2 = (0,1) \cdot (0,1) = (-1,0)$, which is precisely the -1 in this pair world, since $(1,0) + (-1,0) = (0,0)$. At this point, we are free to return to the $a + bi$ format if we like because we now have a concrete model that shows that these numbers exist and constitute a field.

One nice feature of number pairs is that they are especially easy to visualize. The usual idea is to imagine a pair of number lines, one horizontal and the other vertical.

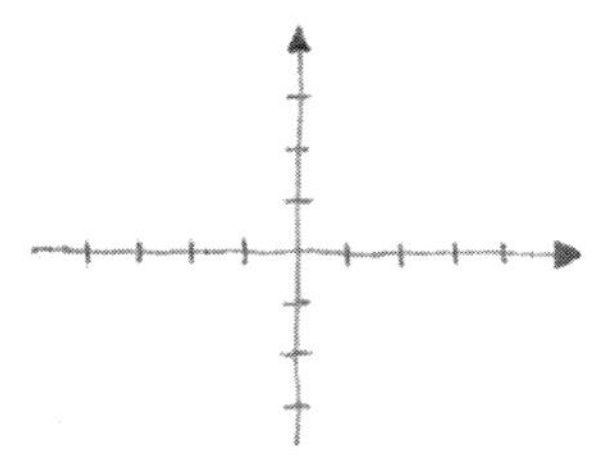

What we are doing is making a *map* of the plane, using this pair of lines as a reference system. The customary choice is to make the first direction horizontal, pointing to the right, and the second direction vertical, pointing up. (These are purely cultural and typographical decisions and have nothing to do with the intrinsic geometry of the plane itself.)

For simplicity, it is also customary to choose our axes so that their crossing point is the origin of both (i.e., the point labeled 0), and the units are also the same.

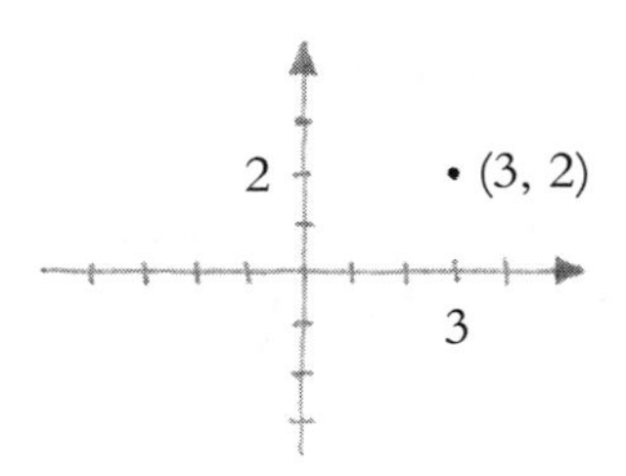

In this way, a number pair like (3,2) can be seen as a point in the plane—namely, the point with horizontal position 3 and vertical position 2. I like to imagine two dials (or maybe sliders) that control the location of the point. As I adjust the horizontal control, the point moves side to side; the vertical slider raises the point up and down. When the horizontal is set to 3 and the vertical to 2, the point then sits at position (3,2).

We now have a simple way to visualize complex numbers: we view the complex number $a + bi$ as sitting at the map location (a, b). Now every complex number has not only a name but also a *place*. This means we are once again making a map of the plane, only this time rather than labeling each point with a pair of real numbers, we are instead using a single complex number.

Our *complex number plane* now looks like this:

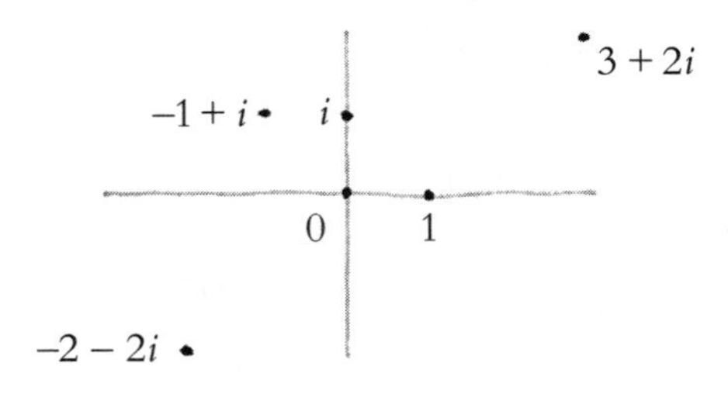

The horizontal axis is simply the real number line itself, whereas the vertical axis is labeled with the "pure imaginary" numbers of the form bi. Now we finally have a *picture* of $\mathbb{C}$, and that is worth some number of somethings, so they say. The complex number $z = a + bi$ lives at the address (a, b) on our map. These two numbers a and b are known as the *real part* and the *imaginary part* of the complex number z. (Ironically, the imaginary part of z is in fact a real number—namely, the amount of i involved.)

The good news is that we now have a simple geometric model of our field $\mathbb{C}$. In fact, it's a perfectly natural two-dimensional analog of the real number line. However, the ultimate value of this model will depend on how well our field operations interact with the geometry. In the case of $\mathbb{R}$, addition and multiplication correspond to natural geometric activities like shifting and scaling. Is this still true for the number plane? If not—if multiplication, for instance, has the effect of rearranging the points of the plane in some unimaginably complicated way—then our model will have limited utility.

Luckily for us, this is not the case. Part of the great beauty of the complex number field is the way it so nicely engages with the geometry of the plane. Not only do the field operations correspond to simple and natural geometric transformations, but in fact all of classical Euclidean plane geometry can be rephrased—often more simply—in terms of complex arithmetic as well.

What number is located at the midpoint
between the complex numbers z and w?

Let's start by figuring out what addition looks like. Suppose $z = a + bi$ is a complex number. What happens to the points of the number plane when we add z to everybody? It's pretty clear what happens if we add the number 1—the complex number $3 + 2i$ becomes $4 + 2i$, increasing its real part (or horizontal coordinate, if you prefer) by one unit. In other words, adding 1 has the same effect on the plane as it does on the line: a shift of one unit to the right. Similarly, adding the number i increases the vertical position by 1, shifting the plane upward by one unit. Of course, subtracting 1 (i.e., adding -1) shifts the plane one unit to the left, and the addition of $-i$ moves everybody down by one.

This means that adding a complex number like $4 - 3i$ would have the effect of sliding the plane four units to the right and three units down. If we like, we can think of this as one total shift but in a slanted direction:

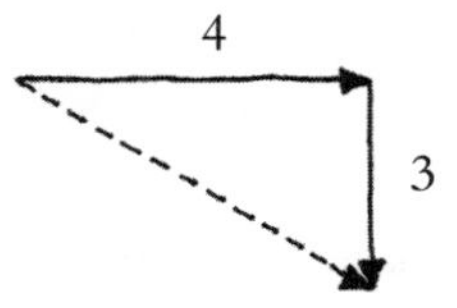

The nice thing about shifting (also known as *translation*) is that when you perform any sequence of shifts (in any order), the result is the same as one single shift. In other words, translations possess a natural additive structure. Clearly, the opposite of a given shift is the shift by the same amount in the opposite direction. The zero shift, of course, is that most beautiful and elegant of activities, doing nothing at all.

One convenient way to think of a translation is to imagine it as an arrow of a certain length pointing in a certain direction. This arrow then acts as an instruction: shift the plane this much in this direction. Such an arrow is known as a *vector* (Latin for "carrier"). A vector is simply an information holder, carrying both distance and direction information. Since a vector represents a shift of the entire plane, the vector itself has no specific location. We're not telling a particular point how to move; we're moving *everyone*.

So if we like, we are free to interpret a complex number $a + bi$ as a vector—namely, the shift that moves everyone over a and up b (or back and down if these numbers happen to be negative). In other words, the addition of the complex number z shifts the plane by the vector that points from 0 to z.

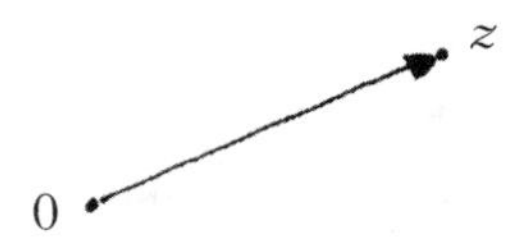

To understand the effect of adding a complex number, we need only watch what happens to one particular point. Whatever way it gets translated is the same for everyone.

This makes it very easy to add and subtract vectors visually. Given two vectors v and v' of whatever lengths and directions, we simply place them end to end to form a new vector $v + v'$ that represents the combined shift.

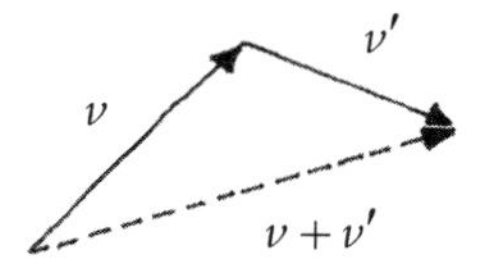

If we choose to think of complex numbers as vectors—that is, if we imagine the number z as the vector taking 0 to z—we then obtain a simple geometric representation of complex addition.

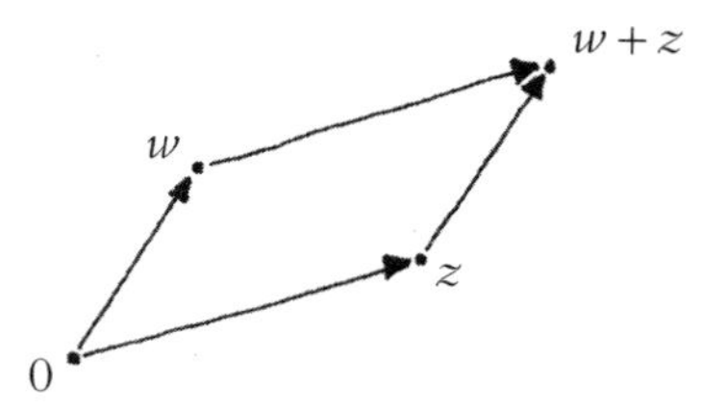

Given two complex numbers w and z represented in this way, their sum $w + z$ then appears as the fourth vertex of the parallelogram formed by 0, w, and z. The symmetry of addition is reflected in the symmetry of the parallelogram—it makes no difference whether we first shift by w and then z, or vice versa.

So the additive structure of the number plane is about as simple and pretty as we could wish. We are already used to thinking of addition and subtraction in $\mathbb{R}$ as shifts of the number line, and this generalizes perfectly—the only modification being the jump from one dimension to two.

In fact, there is nothing stopping us from extending these ideas to *any* dimension. We can easily design a map of three-dimensional space using coordinate triples (a, b, c) to represent points, and we can certainly view these as vectors and shifts if we so desire. In higher dimensions, although we lose the ability to *visualize* this sort of structure (at least in the conventional sense), there's no reason why we can't view quintuples $(a_1, a_2, a_3, a_4, a_5)$ of real numbers as constituting a map or model of five-dimensional space.

While we're at it, we may as well expand our notion of dimension a bit. Suppose we have a continuum of some sort. Let's call it a "space of possibilities." For example, a chemist performing an experiment might keep track of a few different quantities—time, volume, temperature, and pressure, say. Each of these can be measured on a numerical scale, and the state of the experiment at any moment can then be recorded as a quadruple of numbers (t, V, T, p). The complete space of possible states is then a *four-dimensional* space because it requires exactly four pieces of information to determine a single point. (The experiment itself can be seen as tracing out a continuous path through this space.) So the dimension of a space of possibilities is a measure of how much room there is for variation—how many independent coordinates are needed to make a map of that space.

We thus obtain a notion of *n-dimensional Euclidean space* for every natural number n. (We can even include a single point as our model of zero-dimensional space, if we like.) Each of these spaces then comes equipped with the structure of a vector space. The shifts combine in the same way as always, and we do not need to be able to see them to understand the pattern: if $(a_1, a_2, \ldots, a_n)$ and $(b_1, b_2, \ldots, b_n)$ are two vectors in n-dimensional space, then their sum is the vector $(a_1 + b_1, a_2 + b_2, \ldots, a_n + b_n)$.

Returning to the complex number plane, the question now is what happens with multiplication and division. Let's begin by looking at doubling. What is the effect of multiplying a complex number by 2? If $z = a + bi$ is a complex number, then $2z = 2a + 2bi$, so both the real and imaginary parts get doubled. This means that if we view our number z as a vector, then $2z$ is the vector that points in the same direction and is twice as long.

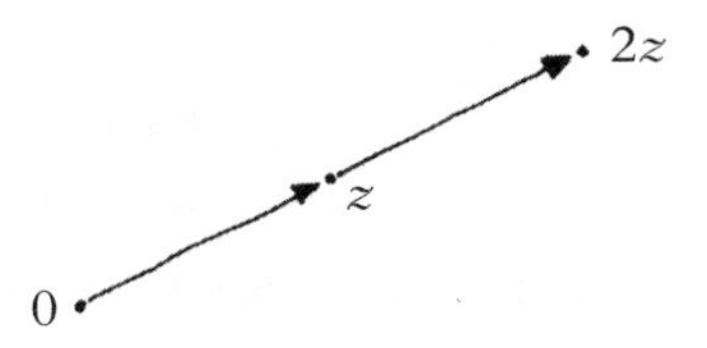

We can also see this by viewing $2z$ as $z + z$. In other words, multiplication by 2 has the same effect in $\mathbb{C}$ that it does in $\mathbb{R}$—it moves everyone further away from 0, magnifying the distance by a factor of 2. So it is a stretch in all directions at once.

Of course, the same goes for any positive real number. Multiplication by π or by $\frac{1}{7}$ simply stretches or contracts the plane, multiplying all distances by a factor of π or dividing them by 7. So in the case of positive real numbers, we see that multiplication is the same as scaling, just as before. (We can also extend this idea to higher-dimensional vector spaces.)

What about multiplication by negative numbers? In $\mathbb{R}$ we are accustomed to viewing multiplication by -1 (otherwise known as negation) as performing a *flip*—a reversal of direction, changing the orientation of the line. Let's see what happens in $\mathbb{C}$. The complex number $z = a + bi$ becomes $-z = -a - bi$. Thus, the negation of a complex number has the effect of reversing both coordinates, flipping the point to the other side of zero, just as in the real case. If we want to think of z as a vector, then we are simply reversing its direction, keeping its length the same.

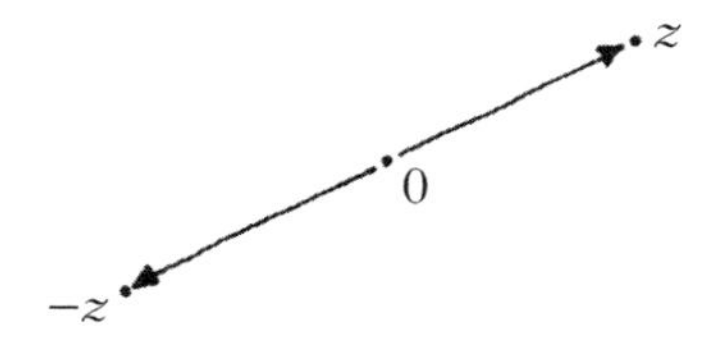

One way to view the negation activity is that we are *inverting* the complex plane, sending each point to its opposite. This is very much in keeping with our idea of flipping in $\mathbb{R}$—and indeed, the real line in $\mathbb{C}$ certainly does get reversed. But there is another way to think about this same activity: we are performing a *half-turn rotation*. Negation in $\mathbb{C}$ has the effect of turning the entire plane 180 degrees, so that each vector now points in the opposite direction.

We now understand the effect of multiplying by any real number r. If r is positive, we are simply stretching or contracting, depending on whether r is greater or less than 1. (Of course, multiplication by 1 does nothing, because that is its sole mission in life.) Multiplication by 0 is completely destructive as usual, collapsing the entire complex plane to a single point. If the number r is negative, then multiplication not only scales the plane by the numerical value of r but also performs an inversion, or half-turn rotation, if you prefer. This generalizes easily to higher dimensions as well, allowing us to both scale and negate in any vector space environment.

Is negation in three dimensions also a rotation?
What about in four dimensions?

The real question, however, is not about doubling or negation; it's about what happens when we multiply by a complex number in general. We already understand doubling and negating in $\mathbb{R}$, and the effects of these multiplications in $\mathbb{C}$ are utterly analogous. The new twist (literally, as it turns out) comes when we multiply by numbers like $3 + 2i$. What on earth does this do to the number plane?

Let's start by examining the effect of multiplication by i itself. A generic complex number $z = a + bi$ will then get sent to the point $iz = i(a + bi)$, or $-b + ai$. In other words, the point at map location (a, b) moves to location $(-b, a)$. If we imagine the point (a, b) as the corner of an $a \times b$ rectangle (I'm assuming for the moment that both a and b are positive), then $(-b, a)$ is the corner of the same rectangle, only rotated 90 degrees counterclockwise:

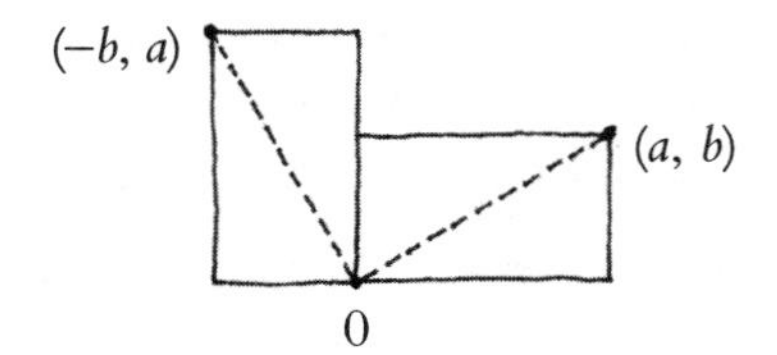

Show that this is true no matter where the point is located.

This means the effect of multiplication by i is to rotate the plane by a quarter turn!

Let's let this sink in for a minute. We began with a vague and romantic dream of extending the real numbers to include square roots of negative quantities so that our number environment would be richer and better able to support the kind of algebraic activities we enjoy. The trouble with a number like $\sqrt{-1}$ is that it makes no sense from a classical geometric perspective. There is no length that when squared can produce a negative amount of area. This is why i and its fellow complex numbers seem impossible or nonsensical, or at any rate hopelessly abstract.

Now, however, we have a new interpretation of i (or at least multiplication by i) that is about as concrete and tangible as anyone could want—namely, a 90-degree rotation of the plane. To add i is to shift up one unit and to multiply by i is to rotate a quarter turn counterclockwise. If that ain't a visual and tactile representation of a purely abstract mathematical notion, then I don't know what is!

We now have a simple geometric way to interpret the algebraic statement $i^2 = -1$. This equation says that if we multiply by i twice, it's the same as negation. In other words, two quarter turns make a half turn! Of course, division by i (which is the same as multiplication by its reciprocal, $-i$) has the effect of rotating by a quarter turn *clockwise*, in order to undo multiplication by i itself.

So far, we have found that multiplication by positive real numbers corresponds to scaling, and multiplication by -1, i, and $-i$ perform rotations. That means a number like $-3i$ will have a combined effect: multiplication by $-3i$ is the same as multiplying by 3 and by $-i$, so it scales by a factor of 3 and also rotates clockwise 90 degrees. It thus performs a kind of "stretch-turn," reminiscent of throwing pizza dough—it rotates and expands at the same time. Let's call this sort of activity a *twist*.

A twist consists of a stretch and a turn, and different numbers will perform different amounts of each. Multiplication by 3 is all stretch and no turn; multiplication by i is pure turn and no stretch. Multiplication by 1 does none of both. (I suppose we could say that multiplication by 0 performs an extreme contraction—by a factor of zero—but then the question of how much it turns becomes meaningless.)

So what is the effect of multiplying by a complex number such as $3 + 2i$? It turns out this also performs a twist. For any (nonzero) complex number z, multiplication by z will always rotate and stretch the plane by a certain turn and a certain stretch factor. In fact, the precise twist is quite easy to determine—we just look at the relationship between the vectors 1 and z.

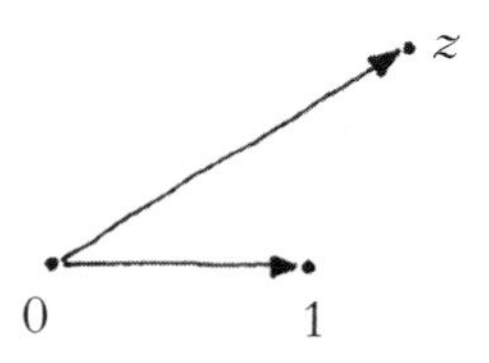

Multiplication by z clearly sends 1 to z, so the turn is just the angle between these vectors and the stretch is just the length of z itself.

This is quite beautiful, actually. Here's what we're essentially saying. The addition of a complex number z performs a shift, and we can tell which shift it is by watching what happens to the additive identity element 0. Adding z sends 0 to z, so that is the precise shift we want. Multiplication by z, on the other hand, performs a twist—that is, a stretch-turn—and exactly which twist is determined by watching what happens to 1, the multiplicative identity element. The number 1 is sent to z, and this shows us the precise twist that multiplication by z performs.

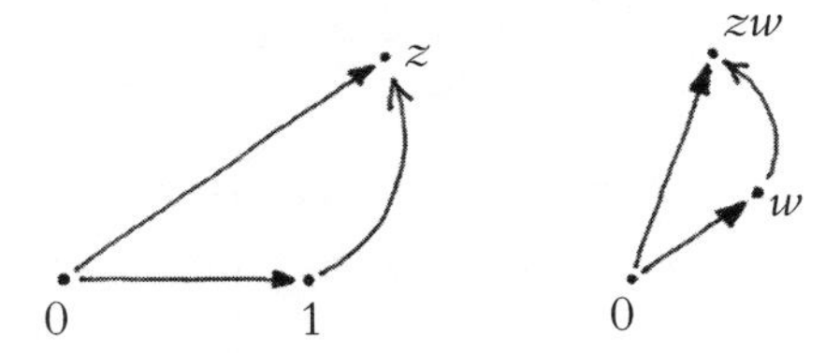

Of course, I still have to explain *why* multiplication is the same as twisting; I don't get to just say that it is, no matter how much I might want it to be. Just because numbers like 2, –1, and i happen to have this effect doesn't mean that all complex numbers must behave in this twisty way. Inductive reasoning (i.e., jumping to conclusions based on patterns) can be very helpful in making good guesses and getting ideas, but true mathematical knowledge comes from deductive argument, not wishful thinking.

So let's see if we can understand why multiplication by a generic complex number z must have the effect we are hoping it does—namely, that it rotates and stretches in the same way that the number 1 does when it becomes z. Let's call that particular twist T. (Now I am using a symbol to denote a geometric transformation rather than a number.) For any complex number w, let's write $T(w)$ for the effect of the twist T on w. This means that $T(w)$ is the complex number obtained by twisting w in the exact same way that 1 twists to make z. What we are trying to prove is that $T(w)$ is the same as zw, the product of z with w. This is what it means to say that multiplication by z is the same as twisting by T.

Of course, we already know this is true for $w = 1$, since that is how we defined T in the first place: $T(1) = z \cdot 1$ because T is the very twist that takes 1 to z. But that does not necessarily mean that $T(w) = zw$ for every complex number w; we need to understand why the nature of multiplication in $\mathbb{C}$ *forces* this to be true.

As a warm-up, let's see what happens for $w = i$. We want to know if zi is the same as $T(i)$. Luckily, we already know that $zi = iz$, which is the same as z rotated a quarter turn counterclockwise.

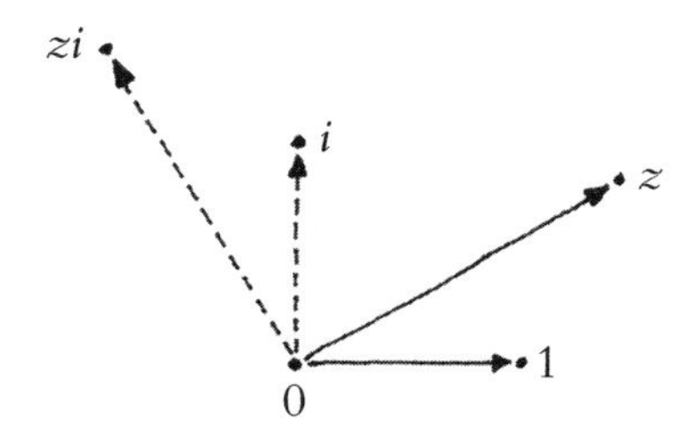

This means that the relationship between 1 and z is the same as that between i and zi, since multiplication by i simply rotates the picture 90 degrees. So in this case, we see that we're right—the number zi is the same as $T(i)$.

More generally, we can imagine any two complex numbers, say w and w', along with their sum $w + w'$.

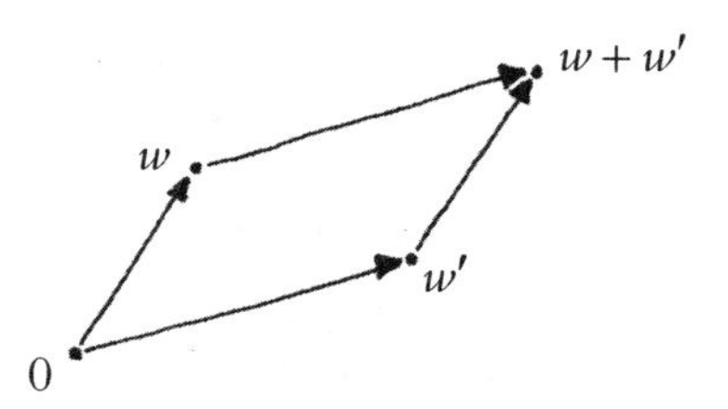

As we have already seen, these points (or vectors, if you like) form a parallelogram rooted at 0. Now, suppose we apply the twist T. This will perform a stretch-turn of the entire plane, scaling and rotating everyone together. In particular, the parallelogram will stretch and turn as well. But a rotated and scaled parallelogram is still a parallelogram. This means that the twisted $w + w'$ must be the sum of the twisted w and the twisted w'. In other words, $T(w + w') = T(w) + T(w')$.

What we're saying is that twisting gets along very nicely with vector addition. You can add and then twist or twist first and then add—either way, you get the same thing. A modern algebraist would say that twisting and adding *commute*. Another way to say this is that twisting is *distributive*, meaning that it flows though sums just as multiplication does. This is in fact the crux of the matter: the reason that multiplication and twisting are so intimately connected is that they are both distributive activities with respect to complex addition.

Now we can see why these two activities must in fact be the same. Given any complex number $w = p + qi$, we see that

$$\begin{aligned} T(w) &= T(p + qi) \\ &= T(p) + T(qi). \end{aligned}$$

The distributive property of twists means we get to split $T(w)$ into two parts, $T(p)$ and $T(qi)$. Of course, p and qi are just scaled versions of 1 and i, so $T(p) = p \cdot T(1)$ and $T(qi) = q \cdot T(i)$. (In other words, twisting also commutes with scaling.) This means that $T(p) = pz$ and $T(qi) = qzi$. Putting these pieces together, we get

$$\begin{aligned} T(w) &= T(p) + T(qi) \\ &= pz + qzi \\ &= z(p + qi) = zw. \end{aligned}$$

Since twisting and multiplying are distributive and have the same effect on both 1 and i, they are forced to agree everywhere.

Now at last we have an intuitive visual model of the complex number field. Algebraically, $\mathbb{C}$ is the extension field $\mathbb{R}(i)$, whose elements can be written in the form $a + bi$. Geometrically, these numbers may be viewed either as coordinate pairs (a, b) or as the corresponding vectors. Addition then corresponds to shifting (the number 0 being our helpful guide) and multiplication becomes twisting (watching 1 for our cue). Every complex number has a horizontal and vertical component (its real and imaginary parts), and if we like, we can break any shift into the sum of a horizontal and vertical part accordingly. Similarly, we can factor any twist into the product of a pure rotation and a pure scaling factor. Arithmetic in $\mathbb{C}$ can then be viewed geometrically, as a sequence of natural geometric transformations of the plane.

There are a few questions remaining, however. We now know that multiplication by z performs a twist, and we also understand geometrically which twist we are talking about—the one that turns 1 into z.

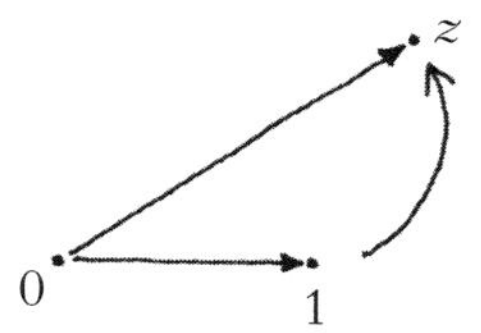

The turn portion of this twist is just the angle between these two vectors, and the stretch part is simply the length of z. But suppose we have a complex number $z = a + bi$ given by two real numbers a and b. What are its length and direction?

The length of a complex number z (i.e., its distance from zero) is denoted by the symbol $|z|$. (This is also known as the *modulus* or *absolute value* of z, but I prefer to simply call it the length.) Fortunately, we know enough about the geometry of the plane to figure this out. The complex number $z = a + bi$ sits at the corner of an $a \times b$ rectangle, so we're really just talking about the diagonal measurement.

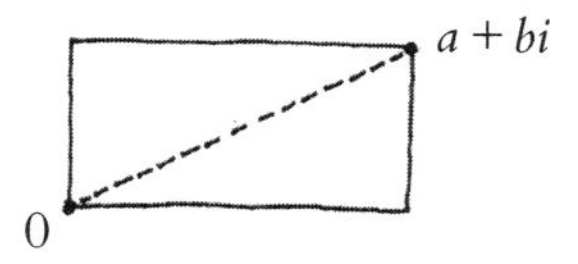

This is exactly what the Pythagorean relation is all about. We find that the length $|z|$ of this diagonal, when squared, is equal to $a^2 + b^2$. This means that for any complex number $z = a + bi$, its length is given by

$$|z| = \sqrt{a^2 + b^2}.$$

Thus, the length of a complex number can be algebraically expressed in terms of its real and imaginary parts.

Unfortunately, the same cannot be said of its direction. For example, our friend $3 + 2i$ forms a 3×2 rectangle, and the diagonal makes a certain angle A with the horizontal.

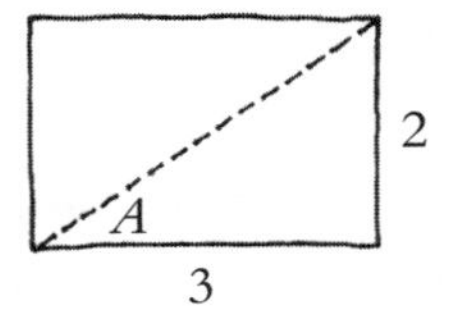

We can assign a numerical value to this angle, either in degrees or as a portion of a full turn, or in whatever units you prefer. This means that A is now a real number that depends somehow on the numbers 2 and 3.

In general, the angle formed by the diagonal of a rectangle depends only on the *shape* of the rectangle, not its size. So it's really only the proportions that matter. We have a rectangle of proportion 3:2, and we want to explicitly measure the angle A. To obtain the diagonal length, all we need to do is form the Pythagorean combination $\sqrt{2^2 + 3^2} = \sqrt{13}$. But what is the corresponding algebraic formula for the angle?

Well, it turns out that there is no such formula, and there's never going to be one. The angle measurement is generically transcendental, meaning that almost all such numbers (including our particular angle A) are simply not algebraically describable. There certainly is a relationship between the sides of a rectangle and the angle formed by the diagonal, but it is not an *algebraic* relationship, so there is (in general) no way to express the angle in the language of field operations and roots. While there are special rectangles (e.g., the square) whose diagonal angles can be algebraically measured (for a square it is one-eighth of a turn, or 45 degrees), such rectangles are exceedingly rare.

Can you find another rectangle whose sides
and diagonal angles can all be measured explicitly?

This is one of the great tragedies of plane geometry—or perhaps one of its most fascinating marvels, depending on how you want to look at it. We can always specify a particular rectangle by giving its sides, and we can then easily (and algebraically) determine its area and diagonal, but not the angles made by its diagonal. They simply are what they are. We can give them names—even fancy names like $\arctan(\frac{2}{3})$—and we can approximate them, but in general we cannot explicitly refer to them in the way we would wish.

So that's a bit of a blow. The cozy relationship between algebra and geometry is not quite as cozy as we thought. The central issue is the interaction between translation and rotation. The $a + bi$ format is great if we intend to do a bunch of adding and subtracting. However, it is not so convenient in terms of twisting. Conversely, suppose we start with the number 1, and we rotate it some amount, say one-seventh of a full turn. Where exactly are we in the over and up, $a + bi$ sense?

The relationship between angles and sides, as well as rotations and translations, forms part of the study of *trigonometry* (Greek for "triangle measurement"). Although we usually cannot explicitly name both the sides and angles of a triangle, we can still discover interrelationships among such measurements, and many of these are even algebraically describable. But the fact remains that our description of complex multiplication via twisting requires a bit of an asterisk next to it.

One way to sidestep this annoyance is to modify the way we specify directions in the plane. Rather than thinking in terms of angles or turns, we can instead use the points of a *circle* to indicate direction, as with a compass. Given any (nonzero) complex number z, we can write $z = ru$, where $r = |z|$ is the length of z (a positive real number) and u is the complex number of length 1 pointing in the same direction as z.

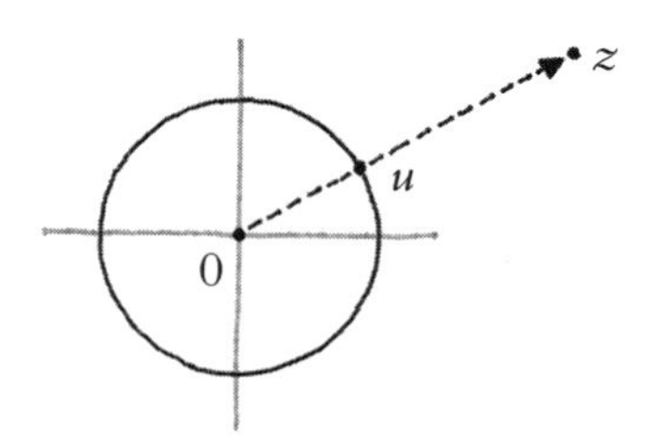

In other words, u is just z scaled down by its own length, thus making it a *unit vector*—that is, a vector of length 1. The collection of unit vectors in the plane forms the *unit circle* (i.e., the circle whose radius is our unit).

The advantage of the factorization $z = ru$ is that it breaks the twisting into separate parts. Multiplication by z is then a pure stretch by a factor of r, combined with a pure rotation given by multiplication by u. Thus, the number $u = z/|z|$ contains all of the angle information, and if we like, we can simply choose to call u itself the "direction" of z, leaving angles to the trigonometers. It's a bit of a cop-out, but it does allow us to skirt the transcendence issue.

So what we're saying is that multiplication by $3 + 2i$ acts as a twist, the stretch factor being the length $\sqrt{13}$ and the turn being represented by the unit vector $\frac{3}{\sqrt{13}} + \frac{2}{\sqrt{13}}i$ (whose angle is whatever it is). If we have two complex numbers, say $z = ru$ and $z' = r'u'$, then their product can be written $zz' = rr' \cdot uu'$, and we see that the stretch factors get multiplied as well as the directions. This means that length and multiplication are on very friendly terms: $|zz'| = |z| \cdot |z'|$. It also means that the unit circle forms a closed system under multiplication—the product of any two unit vectors is again a unit vector. If we want to think in terms of rotation, then the amount of turn produced by uu' will be the combined rotation coming from u and u'. In other words, when we multiply two complex numbers their lengths get multiplied and their turns get combined (or their angles *added*, if you want to use that language).

The upshot is that we now have a fully realized algebraic and geometric structure for $\mathbb{C}$. We can do arithmetic symbolically as well as visually, and our complex entities are now every bit as "real" as $\mathbb{Q}$ or $\mathbb{R}$ or any other number environment with which we are familiar. Not only do we obtain a new realm of numbers in which we can explore and discover unexpected patterns, but we also gain a fresh perspective on $\mathbb{R}$ itself, seeing what we once thought was the entire universe as merely a thin slice of a much wider world.

Which brings me to an important philosophical point that I alluded to earlier. When we began on this mission to construct $\mathbb{C}$, we first imagined it as the extension field $\mathbb{R}(\sqrt{-1})$. We then chose to give one of the square roots of -1 the name i, so that the other was $-i$. We also decided on a pictorial convention whereby we locate i at $(0,1)$ with the second coordinate oriented upward on the page. All of these are completely arbitrary decisions. But suppose for the moment that you and I made *opposite* choices, so that your i is my $-i$. Could we tell if this had happened? Is there any true statement about i that isn't also true for $-i$?

We can tell the numbers 1 and -1 apart because $\mathbb{R}$ is an *ordered* field: one number is positive and the other is negative. (Or, if you prefer a more algebraic distinction, one of them is a square in $\mathbb{R}$ and the other isn't.) However, $\mathbb{C}$ cannot be ordered in the usual way (i.e., so that positive times positive is positive) because *every* number is a square, –1 in particular. The number i is neither positive nor negative because these words have no meaning in $\mathbb{C}$. A plane is not a line. The "unit circle" in $\mathbb{R}$ consists only of the points 1 and –1, so there are only two directions —hence the ordering. But in $\mathbb{C}$ we have an entire circular continuum of directions, and there is no way to consistently order the points of a circle. We lose a very desirable feature of $\mathbb{R}$ when we extend to $\mathbb{C}$, and that means different problems will demand one setting or the other depending on whether ordering is required. For this reason, one often sees the words *real* and *complex* tacked on to the names of mathematical structures to indicate which environment the numbers are coming from (e.g., real polynomials versus complex polynomials).

The point is that even though i and $-i$ are distinct complex numbers, there's no way to tell them apart. This means there is a fundamental symmetry to the complex plane: if we switch the meanings of i and $-i$, all the arithmetic stays the same. Algebraically, we are sending each complex number $z = a + bi$ to its partner, $\bar{z} = a - bi$. This is known as the *complex conjugate* of z (from the Latin *coniunx*, meaning "spouse"). The complex conjugate of a number is usually denoted by a *macron*, or overline. Thus, we have $\overline{3+2i} = 3 - 2i$ and $\overline{3-2i} = 3 + 2i$.

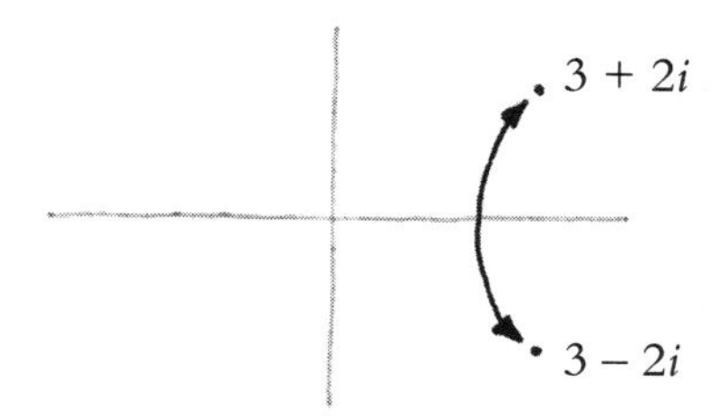

Visually, this swapping of identities results in a *reflection* of the plane across the real line. This is because conjugation affects the imaginary part only. Since reflection preserves parallelograms, conjugation must commute with addition: for all complex numbers z and w, we have

$$\overline{w+z} = \overline{w} + \overline{z}.$$

This is quite clear algebraically as well. Rotation and scaling are also preserved under reflection, only the direction of the rotation gets reversed. This means that conjugation also commutes with multiplication:

$$\overline{wz} = \overline{w} \cdot \overline{z}.$$

Verify this directly, writing $w = p + qi$ *and* $z = a + bi$.

Of course, the real reason that conjugation preserves the field operations is that if it didn't, then i and $-i$ would behave differently and we could tell them apart. (It's sort of like how you and your "mirror self" both think the other is the reflection.) Complex conjugation is a *field symmetry*, rearranging the points of $\mathbb{C}$ while preserving the algebraic structure.

So now we have a new operation—complex conjugation—that gets along famously with our field operations and also has a very natural geometric interpretation. The subfield $\mathbb{R}$ now stands out as the set of *fixed points* of the conjugation symmetry. A complex number z is real precisely when $\bar{z} = z$ (i.e., when it is married only to itself).

Show that the real and imaginary parts of z *are*
$\frac{1}{2}(z + \bar{z})$ *and* $\frac{1}{2i}(z - \bar{z})$.

Now it's all very well and good that we have this field $\mathbb{C}$ and that it does all these nice things, but the whole point was to improve upon $\mathbb{R}$—which, for one thing, lacks square roots for all of its members—by moving to a wider realm that is more symmetrical and better behaved. In particular, it would be rather embarrassing if it turned out that some elements of $\mathbb{C}$ also lack square roots. So before we uncork the champagne, we'd better make sure that $\mathbb{C}$ does in fact have the properties we want, or else $\mathbb{R}$ is going to be really mad and call us hypocrites. We need to make sure that $\mathbb{C}$ is actually *better*, not just bigger.

Let's start by thinking about squaring. What happens to a complex number z when we square it? We're multiplying z by z, which means the lengths will multiply, so the length of z^2 will be the square of the length of z: $|z^2| = |z|^2$. The turns will also combine, so whatever angle

z makes with 1 will get *doubled.* Squaring therefore squares the length and doubles the angle. That means the square root operation must square root the length and cut the angle in half!

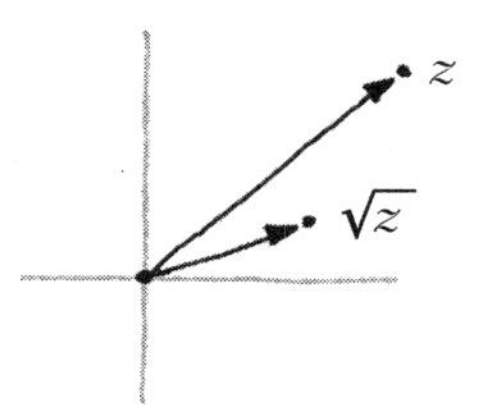

Now we can see that every complex number does indeed have a square root. But wait! What about the other one? Aren't we supposed to have *two* square roots?

What happened is that we got a little sloppy with our logic. (Notice how I'm including you as one of the culprits!) We saw that squaring doubles angles and we deduced that square root must then cut them in half, but it turns out that there are two equally legitimate ways to cut a turn in half. This is because any turn can be viewed as either clockwise *or* counterclockwise. So when we cut the turn in half, we get two diametrically opposed possibilities, corresponding to the two different square roots, which are of course negatives of each other.

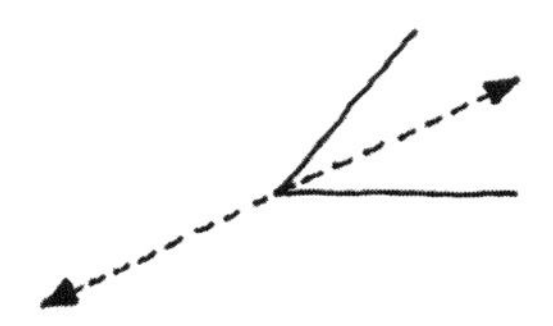

Another way to think of this is that we have a full circle of possible directions. That means performing a full turn and doing nothing have the same effect—in other words, an angle of 0 and an angle of 360 degrees are really the same. But although these two turns have the same effect, when we cut them in half they do not. The numbers 1 and –1 have the same square but point in opposite directions. Their angles are 0 and a half turn, and when doubled these become equal and point in the same direction. In general, two numbers will have the same square if they are off by a half turn—that is, they are negatives of each other.

When doubled, the half turn discrepancy between them becomes a full turn, and we can then no longer tell the difference.

Now we can see why the two-dimensional plane is superior to the one-dimensional line: we have enough directions so that every angle has two distinct half angles. We have two values of $\sqrt{1}$ in $\mathbb{R}$ because we can cut the angle 0 in half both ways, but we can't cut a half turn in half to form $\sqrt{-1}$ because the real line does not allow for a quarter turn. So our continuum $\mathbb{R}$ is not quite continuous enough, in the sense that it contains only a pair of discrete directions rather than a full circle.

What about cube roots? Since cubing a number triples its angle, we'll need to cut angles into thirds in order to cube root. How many ways are there to cut a given turn into thirds? The full-turn ambiguity leads to three different ways. If we have three turns equally spaced by one-third of a turn (i.e., 120 degrees), they will all triple to the same thing. In particular, our old friends 1, ω, and ω^2 (the cube roots of unity) form the vertices of an equilateral triangle.

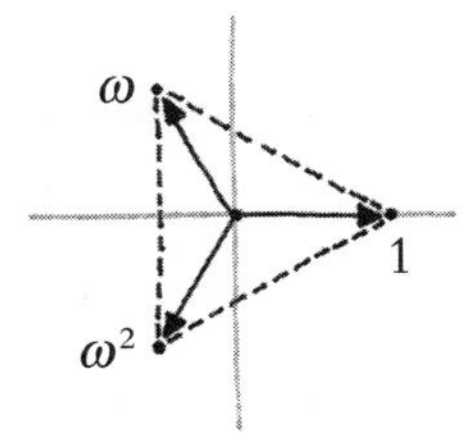

We see that every (nonzero) complex number z has exactly three complex cube roots, and these are symmetrically arranged around a circle of radius $\sqrt[3]{|z|}$.

What are the three complex cube roots of −1?
How about the cube roots of i?

Now it's clear what happens in general. Raising a complex number to the nth power causes the angle to get multiplied by n, so taking the nth root must cut the angle into n equal parts. These will be equally spaced around a circle, forming the vertices of a regular polygon with n sides. In this way, we see that $\mathbb{C}$ contains exactly n n^{th} roots of every nonzero number.

Thus, it is really the two dimensionality of the complex plane that guarantees every number its full complement of roots. Despite the fact that $\mathbb{C}$ arose purely from our desire to include certain square roots, we nonetheless also attain all of the nth roots as well. What is even more surprising is that not only do simple equations like $z^5 = 2 + 3i$ always have a solution, but in fact *every* polynomial equation with coefficients in $\mathbb{C}$ must have a solution in $\mathbb{C}$. To be precise, for any complex coefficients $a_0, a_1, \ldots, a_{n-1}$, the monic polynomial equation of degree n,

$$z^n + a_{n-1}z^{n-1} + \cdots + a_1 z + a_0 = 0,$$

always has a complex solution z.

This means that $\mathbb{C}$ is a closed system not only in the usual sense of being a field but also in the larger linguistic sense: there is no number that can be described as the solution to an algebraic constraint that we do not already possess. The complex field $\mathbb{C}$ is algebraically closed, and this makes it the ideal environment for classical algebraists. Every number we could ever (algebraically) dream of is already present. We never need to make another extension! The path of desire that led us from $\mathbb{N}$ to $\mathbb{Z}$ to $\mathbb{Q}$ to $\mathbb{R}$ to $\mathbb{C}$ has finally come to an end. We now have every number we could ever wish for—at least until our wishes become considerably more abstract and modern.

So how do we prove that every polynomial equation over $\mathbb{C}$ must have a solution? The field of complex numbers (in the form of the number plane) began to gain acceptance in the early 1600s. However, this beautiful and important feature of algebraic closure, although widely assumed to be true, was not fully proved—and therefore not fully understood—until two centuries later.

I want to show you a very beautiful argument that I have always loved. It is a simplified version of a proof first published in 1799 by the German mathematician Carl Friedrich Gauss. To explain the idea, I thought it might be best to work with a specific (though perfectly arbitrary) polynomial equation as opposed to one with symbolic coefficients. The advantage is that we won't have to stare at a pageful of indexed parameters; the disadvantage is that the argument might not be completely general. We need to be able to work with a particular equation but also blur our eyes to the details—making sure that we never require the coefficients or the degree to have any specific values or properties.

Let's choose a reasonably simple but generic equation, say

$$z^6 + 3z^4 - z^3 + 5z + 1 = 0.$$

We want to show that this equation has a solution, but we do not want to make any substantive use of its particular degree or coefficients. I have chosen integer coefficients (including zero) for simplicity, but they can be arbitrary complex numbers.

The plan is to exploit the fact that $\mathbb{C}$ is a continuum to force the existence of a solution. Since addition and multiplication in $\mathbb{C}$ are given by continuous geometric transformations such as shifting and twisting, we know that any polynomial process—for instance, the one that sends every complex number z to the destination $z^6 + 3z^4 - z^3 + 5z + 1$—must also be continuous. This means that if we regard z as a *complex variable* roaming around in the plane, then as z traces out a path in $\mathbb{C}$, so does its "image" $z^6 + 3z^4 - z^3 + 5z + 1$. If we like, we can view this process as mapping an input plane to an output plane:

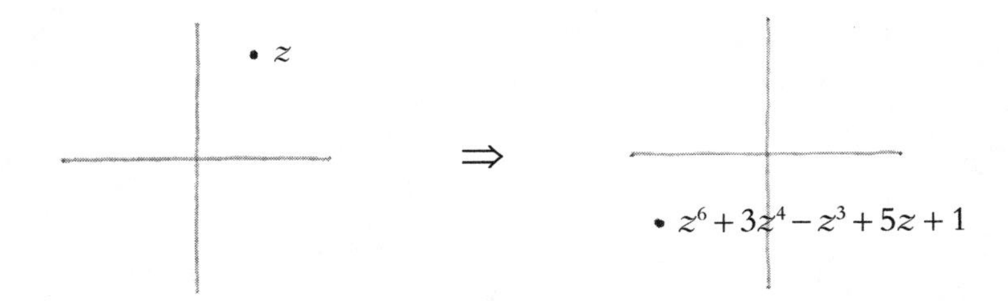

You may recall how in the previous chapter we used a continuity argument to show that a certain cubic equation had a real solution. In fact, we can use this same technique to show that every cubic equation over $\mathbb{R}$ must have a solution in $\mathbb{R}$. The idea is that any (monic) cubic polynomial must send large positive inputs to large positive outputs because the cubic term will dominate when the input gets big. For the same reason, large negative inputs will be sent to large negative outputs. Thus, every cubic polynomial with real coefficients will attain both large positive and large negative values. Since a polynomial process is continuous, at some point (i.e., for some value of the input) the output must be zero exactly, and therefore the cubic equation must have a solution.

Now we need a two-dimensional version of this argument, and this requires a bit of ingenuity. The trick is to look at circles. Rather than viewing our input point as moving along the real line, we will instead imagine it describing a circular loop of a certain radius r, centered around 0. This will constitute our input circle.

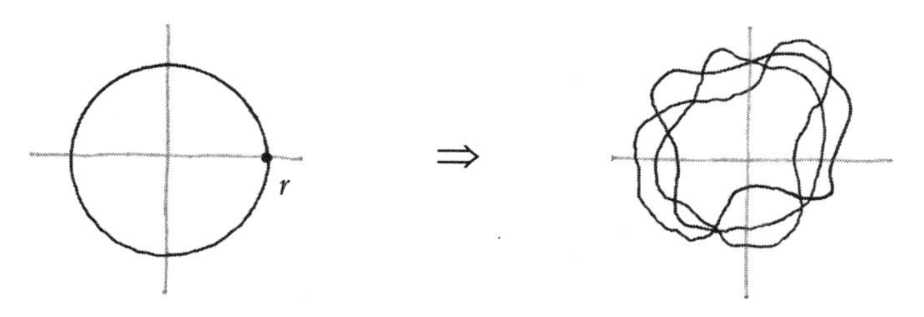

As our input point z traces out one complete revolution, the output point will traverse a certain complicated closed curve. The plan is to watch what happens to this tangle as the input circle grows and shrinks.

When the radius r is extremely large, the highest degree term—in our case, z^6—will dominate the other terms in size. The second term, $3z^4$, will certainly get quite large, but it will still be dwarfed by z^6 once r (i.e., the length of z) gets large enough. One way to see this is to rewrite our polynomial expression in the form

$$z^6 + 3z^4 - z^3 + 5z + 1 = z^6\left(1 + \frac{3}{z^2} - \frac{1}{z^3} + \frac{5}{z^5} + \frac{1}{z^6}\right).$$

When $|z|$ becomes enormous, the terms in parentheses become tiny, except for the initial 1 coming from the leading term. Thus, for very large r, our process resembles $z \to z^6$, with a certain small relative error coming from the nonleading terms.

Happily, we already understand the effect of raising a complex number z to the sixth (or n^{th}) power. As z travels once around a circle of radius r, z^6 will trace out a circle of radius r^6—only it will wrap around six times instead of once! This is because raising to the nth power multiplies turns by n. What this means is that our somewhat more elaborate process $z \to z^6 + 3z^4 - z^3 + 5z + 1$ must also behave in a similar manner, only with proportionally small perturbations as we go around.

As z goes once around the (large) input circle, the output traces out a (very large) wiggly path that goes around six times before it closes up again. We can't say much about this output path in general, other than that it wraps around n times (n being the degree of the polynomial) and that the points on this path are extremely large in absolute value. In particular, the output path must contain the point 0 in its *interior.*

But what happens at the other extreme, when the input circle is very small? When the radius r is very small, our complex variable z is very close to 0, so the output curve will stay in the vicinity of wherever the number 0 is sent. This is simply the constant term of the polynomial. In our case, it's the number 1, but the only thing that really matters is that it be nonzero. (If the constant term were absent, then $z = 0$ would already be a solution to the equation.) Thus, as z traverses a tiny enough circle around 0, the output will trace out some small, snarled up path near the point 1. In particular, the point 0 will lie *outside* this closed path.

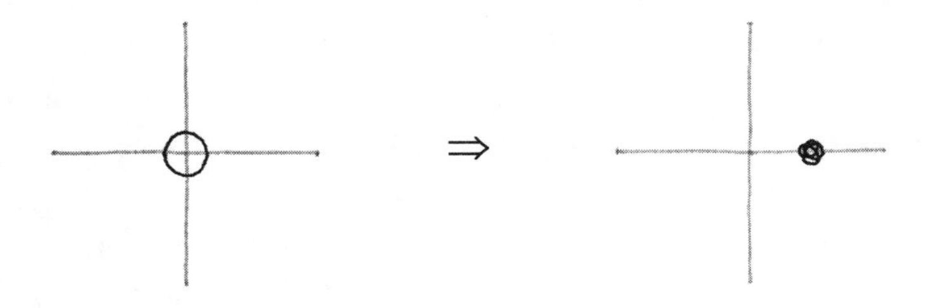

Maybe you can already see the punchline coming. As we continuously vary the radius of our input circle from giant to tiny, our output path smoothly morphs from a curve that wraps around the point 0 to one that has 0 on the outside. This means that at some particular moment in this morphing process, the number 0 must actually lie *on* our output curve, meaning there is an input z that produces an output of 0. This z is then a solution to our polynomial equation.

What I love about this proof is its simplicity—we don't need to get into the mucky details of exactly how our polynomial process transforms the plane; we need only understand roughly what happens at the extremes of size. I also love how continuity comes into play. This is not the kind of argument one can apply to a field like $\mathbb{Q}$ or a domain like $\mathbb{Z}$. These sorts of environments are much too grainy and full of holes.

Continuity arguments have the great advantage that they allow us to prove the *existence* of certain special values (e.g., solutions to polynomial equations) but also the frustrating disadvantage that they are *nonconstructive* —we can prove that the thing we seek is out there, but we have no idea where.

This result, the algebraic closure of the complex numbers, is known as the *fundamental theorem of algebra.* It is rightly regarded as one of the greatest achievements of mathematics (and of the human imagination in general, if you ask me). To a classical algebraist, it means that $\mathbb{C}$ is clearly the place to do business, the ultimate and final piece of string, amenable to any algebraic entanglement and lacking nothing that we are capable of describing in the language of knots.

The introduction of complex numbers by Cardano and their development by Rafael Bombelli and others in the late 1500s ushered in a new era of *complexification,* where all of classical mathematics—from algebra and geometry to the study of differential equations and mathematical physics—was gradually reimagined, rephrased, and reconstituted in complex number terms. In particular, there is nothing stopping us from imagining and investigating *complex n-*dimensional space consisting of all possible n-tuples $(z_1, z_2, \ldots, z_n)$ with complex entries. In this way, we can see the classical complex plane as being *one dimensional* (in the complex sense). This is why modern geometers often refer to $\mathbb{C}$ as the *complex line.*

The nice thing about an algebraically closed field is that all the numbers you might ever want are already there. Even if your sole field of interest is $\mathbb{R}$, it can't hurt to regard it as a subfield of $\mathbb{C}$ so that you have the flexibility to jump out if you need to, even if you intend to eventually jump back in. The point being that $\mathbb{R}$ itself is not algebraically closed (e.g., it lacks a solution to $x^2 + 7 = 0$), but it has an extension field—namely, $\mathbb{C}$—that is. One of the great results of modern algebra is the discovery (and proof) that *every* field has such an extension. Even the tiny field $\mathbb{F}_2$ has an *algebraic closure* $\overline{\mathbb{F}}_2$, the smallest algebraically closed field containing $\mathbb{F}_2$. (The bar in this context means algebraic closure, not complex conjugation.) In the case of the real numbers, we have $\overline{\mathbb{R}} = \mathbb{C}$, and the algebraic closure is a relatively small extension, generated by the single element $\sqrt{-1}$. For a finite field like $\mathbb{F}_2$, on the other hand, the extension $\overline{\mathbb{F}}_2$ requires infinitely many generators.

Whenever a modern algebraist is working in a given field environment, the typical first step is to extend the field to its algebraic closure. This provides a large ambient space of numerical entities in which the original field now sits. In a way, it's like placing a sculpture in the center of a gallery so that we can walk around it and see it from all possible perspectives. Once you have placed $\mathbb{R}$ inside $\mathbb{C}$, you can never again look at the real line in quite the same way.

FACTORIZATION

I want to return to a subject we were discussing a little while back: the strong analogy between integers and polynomials. Suppose $\mathbb{F}$ is a field (you can think of it as $\mathbb{R}$ or $\mathbb{C}$ if you like). We saw before that both $\mathbb{Z}$ and $\mathbb{F}[X]$ are integral domains—realms in which we can freely add, subtract, and multiply, but not divide. In particular, they satisfy eight of our nine postulates for a field, the existence of reciprocals being replaced by the weaker demand that the product of any two nonzero elements be nonzero.

What makes an integral domain like $\mathbb{Z}$ such a fascinating environment is that because we lack the reciprocals necessary for division, what we have instead is an issue of *divisibility*. In the field $\mathbb{Q}$, we have the freedom to divide anybody by anybody and still remain in $\mathbb{Q}$. But $\mathbb{Z}$ does not afford us this luxury. Some integers divide others; some do not. In place of complete freedom and flexibility (division by zero being the only taboo), we have instead a divisibility *relationship*, and the question then becomes: Who divides whom, and how can we tell?

Let's first think about what divisibility means in the abstract. For any two elements a and b in an integral domain, we'll say that *a divides b* (written $a \mid b$) if b is a multiple of a—that is, if there is an element q with $b = aq$. For example, since $6 = 2 \times 3$, we see that $2 \mid 6$ and $3 \mid 6$ in $\mathbb{Z}$. In the polynomial domain $\mathbb{Q}[X]$, we have $X^2 - 9 = (X + 3)(X - 3)$, which means that $X + 3$ and $X - 3$ both divide $X^2 - 9$. We thus acquire a new verb—one number divides another—and that means more language and more descriptive power and more beautiful patterns to be discovered and explained.

Show that $a \mid b$ precisely when the ideal (b) is contained in (a).

Another way to think of a product like $6 = 2 \times 3$ is that we are building the number 6 from the parts 2 and 3. In this way, we can see each of these parts as being a *factor* (Latin for "maker") of the number 6, which is then the *product* of this manufacturing process. So if we like, we can view the divisibility relationship as telling us which numbers can be used to build which. As we shall see, it is a lot easier to build numbers up than it is to tear them down.

The product of two natural numbers is 437. *Who are they?*

The study of $\mathbb{Z}$ and its arithmetic is called *number theory*. Despite their apparent simplicity, integers are surprisingly subtle creatures, and no small part of this is due to divisibility. To clarify matters, let's restrict ourselves to an even simpler environment and start by investigating the arithmetic of the natural numbers $\mathbb{N}$. Here, we can freely add and multiply, we have identity elements 0 and 1, and all the usual associative, commutative, and distributive patterns are in effect. What we don't have are negatives and reciprocals. So we can build larger whole numbers by adding and multiplying smaller ones, and we won't have to worry about any obnoxious negative signs.

One amusing metaphor is to think of natural numbers as substances akin to chemical compounds. If we combine two numbers to make a larger one—such as 3 and 5 adding to make 8—we can then view 8 as a compound consisting of the molecules 3 and 5 attached together via "additive bonding." Similarly, we can think of the number 60 as being the result of the multiplicative bonding of 5 and 12.

The process of breaking a substance down into its component parts is known as *analysis* (Greek for "loosening up"). This is often achieved by using a catalyst of some kind to break apart the bonds, allowing us to get at the various individual constituents. So let's imagine a kind of "additive acid" that we can pour on numbers to break them down into sums of smaller pieces. For example, if we pour some of this acid on the number 13, it might dissolve into 5 and 8. If we keep pouring acid over these numbers, they will continue to break down, and the end result will be the complete dissolution of the number 13 into a sum of ones:

$$13 = 1 + 1 + 1 + 1 + 1 + 1 + 1 + 1 + 1 + 1 + 1 + 1 + 1.$$

This means the additive structure of $\mathbb{N}$ is quite simple. Every number is just a sum of ones, and addition is nothing more than counting. If we want to think of numbers as chemical substances, then the corresponding periodic table is very small: there is only one irreducible element, and it's the number 1. Every substance in $\mathbb{N}$ is simply a long molecule of ones linked together by additive bonds, so the number 1 is the only building block needed. (I suppose we could also include the number 0, corresponding to nothingness, but as we've noticed before, its only value would be to build itself.)

Things get much more interesting when we consider multiplication instead. When we pour multiplicative acid on a number, it breaks down into a product of smaller pieces. For example, if we pour this acid on the number 90, it might split into the factors 6 and 15, or it may decompose into 9 and 10, or possibly 45 and 2. Whichever way the breakdown occurs, we can continue pouring acid on the pieces until they break down as well. When completely dissolved, the number 90 appears as the product $2 \cdot 3 \cdot 3 \cdot 5$. Each of these parts is irreducible—multiplicative acid cannot break down 2, 3, or 5 any further.

Now it's the number 1 that corresponds to nothingness, whereas 0 is a sort of black hole that sucks everything else into it. Every other number is either irreducible (an element, so to speak) or a product of such. No matter what number we start with, the breakdown process must eventually come to an end because at each stage the pieces get *smaller*, and this cannot continue indefinitely.

Thus, in both the additive and multiplicative arenas, we end up with a collection of primordial irreducible elements out of which everything can be made. In the additive case, we have only one element (namely, 1), whereas in the multiplicative version, we find quite a bit of residue in the bottom of the crucible—an entire sequence of irreducible elements: 2, 3, 5, 7, 11, 13, and so on. These are of course the infamous prime numbers about which we know so little. What we do know is that they are the fundamental building blocks of natural numbers under multiplication: every natural number (putting 0 and 1 aside) is either prime itself or a product of primes.

So who are these strange irreducible numbers? They are certainly anything but random. After all, these are the very rare and special numbers that are immune to multiplicative acid. Of course they have a pattern—namely, the pattern of being prime. It's the least random sequence there could be. And yet, there they sit, looking sporadic and irregular and certainly appearing quite random to *us*. What does it mean when something completely deterministic appears random? It means we lack understanding and need to think harder. We may even need to gain altitude so that we can see better.

One of the few things that we do know about the sequence of prime numbers is that it goes on forever—there are infinitely many primes out of which all natural numbers can be built. The earliest known proof of this fact is due to Euclid. In many ways it resembles Cantor's diagonal argument, though it is more than two thousand years older.

The idea is to show that no *finite* list of primes could ever be complete. Suppose we have a finite collection of prime numbers, say $p_1, p_2, \ldots, p_n$. Euclid's plan is to use these very primes to construct a new prime number not on the list, just as Cantor uses a given list of bead strings to build a new string that it does not contain. First, imagine that we multiply all of our primes together to make one large number, $p_1 p_2 \cdots p_n$. This number is divisible by every one of our primes. That means if we alter it slightly—by adding 1, say—then our new number, $p_1 p_2 \cdots p_n + 1$, will now leave a remainder when divided by any of the primes on our list, making it divisible by none of them. (The addition of 1 in order to foil divisibility is analogous to the switching of colors in Cantor's argument.) Now we have a natural number that is not divisible by any of the primes on our list. But this number must be divisible by *some* prime, and since this prime cannot be on our list, our list must therefore be incomplete.

So in contrast to the additive situation, in which there's only one irreducible element, our multiplicative periodic table is infinite. We know that every number is a product of primes (I'm including primes themselves here), but we don't really know the primes, so it's a mixed blessing. On the other hand, divisibility is much simpler when our numbers are expressed in this way. For instance, it's not immediately clear whether 1001 is divisible by 91. But if we write these numbers in factored form, we find that $1001 = 7 \cdot 11 \cdot 13$ and $91 = 7 \cdot 13$, and it becomes patently obvious.

This often makes it convenient—especially when dealing with problems of a multiplicative nature—to think of our positive integers as products of primes, as opposed to sums of ones or a string of decimal digits. Given a number expressed in a certain form—the decimal number 1200, say—we are then faced with the problem of determining its prime factorization. As the numbers get larger, this becomes an increasingly difficult and time-consuming process. But once it is done, it never needs to be done again (assuming we can store the information in some way). We have $1200 = 2 \cdot 2 \cdot 2 \cdot 2 \cdot 3 \cdot 5 \cdot 5$, and this fact is not subject to change.

One way to help simplify matters is to make use of exponents by writing $1200 = 2^4 \cdot 3 \cdot 5^2$. In this way, we are expressing our numbers as products of *prime powers*—numbers of the form p^n, where p is prime and n counts the number of occurrences of p in the prime factorization. If we like, we can also arrange these factors in order of increasing p. Thus, $140 = 2^2 \cdot 5 \cdot 7$ and $150 = 2 \cdot 3 \cdot 5^2$.

We can push this idea even further by generalizing our notion of exponent to include zero and negative values as well. Our original motivation for writing 5^3 instead of $5 \cdot 5 \cdot 5$ was to save space; the exponent notation began as a simple abbreviation. But exponents are not just abbreviations; they are abbreviations *with a pattern*. Each time we multiply by 5, the exponent increases by 1: $5 \times 5^3 = 5^4$. Likewise, every time we divide by 5, the exponent must *decrease*: $5^4 \div 5 = 5^3$. This gives us a natural way to extend the meaning of our exponents. Since $5^2 = 25$, we would want 5^1 to mean $25 \div 5 = 5$. This makes a certain sense, if we want to say that "multiplying one 5 together" is the same as doing nothing. The pattern would then suggest that $5^0 = 5 \div 5 = 1$, which is a little weirder. Are we saying that multiplying *no* fives together somehow makes 1?

Yes and no. It's not that it makes sense that multiplying no fives together should be 1; it's more like we are ceasing to rely on "number of copies" as our meaning of exponent and are instead listening to what $\mathbb{N}$ and $\mathbb{Z}$ have to say. In this way, we let the pattern tell us what we want our symbols to mean. We find that 5^0 should be taken to mean 1, despite being meaningless under our original interpretation. Essentially, we are allowing the pattern to *create* the meaning. The symbol 5^{-1} now means $1 \div 5 = \frac{1}{5}$, and 5^{-2} is $\frac{1}{25}$, not because that's what we intended these symbols to mean but because that's what *math* wants them to mean.

Of course, none of this has anything to do with the number 5, or even natural numbers. If w is an element of any field, we can define w^n in the usual way for positive integers n, then extend into the negatives as before. Thus, $w^3 = w \cdot w \cdot w$ and $w^{-3} = \frac{1}{w} \cdot \frac{1}{w} \cdot \frac{1}{w}$. (We should also remember to be careful when $w = 0$, since numbers like 0^{-3} do not exist.) One beautiful consequence of this choice of meaning is that the simple pattern $w^m \cdot w^n = w^{m+n}$ now holds for *any* integers m and n, positive, negative, or zero. For example, we have $5^{-3} \cdot 5^2 = 5^{-1}$.

Show that $w^m \cdot w^n = w^{m+n}$ *and* $(w^m)^n = w^{mn}$,
for any nonzero w *and all integers* m *and* n.

This notational device allows us to extend the prime factorization idea to fractions as well. Suppose we have a fraction $\frac{a}{b}$, where a and b are positive integers. Since a and b themselves are products of primes raised to various exponents, their quotient will also be a product of prime powers—only with possible negative exponents. For example, since $60 = 2^2 \cdot 3 \cdot 5$ and $99 = 3^2 \cdot 11$, we could say that their ratio $\frac{60}{99}$ has the prime factorization

$$\frac{60}{99} = 2^2 \cdot 3^{-1} \cdot 5 \cdot 11^{-1}.$$

If we like, we can even use the fact that $5^1 = 5$ and $7^0 = 1$ to write this product in the form $2^2 \cdot 3^{-1} \cdot 5^1 \cdot 7^0 \cdot 11^{-1}$.

The upshot is that every (positive) fraction can be expressed as a product of prime powers. For any fraction r and any prime number p, we can refer to the exponent of p that occurs in the prime factorization of r. I suppose we could even call this "the number of times p occurs in r." For example, in the fraction $\frac{8}{5}$, the prime 2 occurs three times, the prime 5 occurs negative one times, and the prime 3 occurs zero times. When our number is an integer, these exponents simply tell us how many times we can divide by a given prime.

One rather pretty idea is to assign to any number the complete list of all its prime exponents in order. Thus, to the number $140 = 2^2 \cdot 3^0 \cdot 5^1 \cdot 7^1$, we would attach the infinity-tuple (2,0,1,1,0,0,0, ...) that stores this information prime by prime. The first entry tells us how many factors of 2 are involved, then how many 3s (in this case none), and so on. The value of doing this is not only to clarify divisibility, but also to simplify multiplication itself.

Suppose we have two positive integers a and b, and we construct the corresponding exponent sequences $(a_2, a_3, a_5, a_7, \ldots)$ and $(b_2, b_3, b_5, b_7, \ldots)$. In other words, our numbers have prime factorizations

$$a = 2^{a_2} \cdot 3^{a_3} \cdot 5^{a_5} \cdot 7^{a_7} \cdot \cdots$$
$$b = 2^{b_2} \cdot 3^{b_3} \cdot 5^{b_5} \cdot 7^{b_7} \cdot \cdots$$

(Here I have chosen to index the exponents with the corresponding prime for simplicity.) Naturally, all of these exponents will be zero from some point on, so these are really finite products. Now if we multiply these numbers together, we get

$$ab = 2^{a_2 + b_2} \cdot 3^{a_3 + b_3} \cdot 5^{a_5 + b_5} \cdot 7^{a_7 + b_7} \cdot \cdots.$$

This means that the exponent sequence for ab is just the *sum* of the two original sequences: $(a_2 + b_2, a_3 + b_3, a_5 + b_5, a_7 + b_7, \ldots)$.

Let's try this out on a concrete example. Using ordinary decimal arithmetic, we can certainly multiply 240 by 98 to get 23520, but it's an ugly business full of carrying and place shifting. But suppose we encoded our numbers as prime exponent sequences instead. Then 240 would be written (4,3,1,0,0, ...), and 98 would be represented as (1,0,0,2,0,0, ...). The sequence for their product would then be obtained by adding these together: (5,3,1,2,0,0, ...). Depending on what we are doing, this may be vastly preferable to the usual Hindu-Arabic decimal format.

In particular, if we express our numbers in this way, it becomes almost effortless to tell if one positive integer is divisible by another. The number b is divisible by a precisely when each entry in the exponent sequence for b is at least as big as the corresponding entry for a. (Another way to say this is that the fraction $\frac{b}{a}$ would then have only non-negative entries, corresponding to a whole number.)

Thus, the entire divisibility issue—along with the complete multiplicative structure of the positive rational numbers—can be transferred over to the addition and comparison of integer sequences. The number 1, being the multiplicative identity element, would have the exponent sequence (0,0,0,0,0, ...), corresponding to the additive identity in the world of sequences. What we have here is a machine for converting multiplication and division into addition and subtraction—the price being that our sequences are infinitely long (albeit mostly all zeroes) and involve the mysterious prime numbers.

Any mechanism for converting multiplication into addition is called a *logarithm* (Greek *logos* + *arithmos*, "way of calculating"). In the practical sense (e.g., a slide rule), this can be a clever means of calculating products rapidly, especially approximations. For theoretical purposes, a logarithm can be even more useful in drawing out analogies and exploiting structural similarities. As an example, we are now in a position to understand the irrationality of $\sqrt{2}$ in a much deeper way. Suppose r is a positive fraction with exponent sequence $(r_2, r_3, r_5, \ldots)$. These entries may be positive or negative, and all but a handful of them will be zero. What happens when we square this number? Since multiplication has the effect of adding the exponent sequences, we see that the sequence for r^2 must be $(2r_2, 2r_3, 2r_5, \ldots)$. In other words, squaring corresponds to *doubling*.

We now have an almost babyishly simple way to tell if a given rational number is a square: we simply look at its exponent sequence and make sure that every entry is even. If not, then it cannot be the square of a fraction, and its square root cannot be rational. Since the sequence for 2 is (1,0,0,0, ...) and the first entry is odd, we get an instant proof that $\sqrt{2}$ is irrational.

Show that $\sqrt{6}$, $\sqrt{20}$, *and* $\sqrt{\frac{3}{4}}$ *are irrational.*
Why is $\sqrt[3]{2}$ *irrational?*

So just as a chemist finds it valuable to think of table sugar as $C_{12}H_{22}O_{11}$ in order to understand the way it is built out of the elements carbon, hydrogen, and oxygen, we mathematicians enjoy viewing our numbers—positive integers and fractions—as complex structures manufactured from simpler pieces. In fact, we are in a much nicer position than the chemists because multiplication is a lot simpler than chemical bonding. The properties of a chemical substance depend not only on the elements involved but also on the way they are bonded together. Luckily for us, numbers present no such difficulties. Multiplication is both commutative and associative, so a given product of primes is the same no matter how you choose to arrange the terms. Thus, every exponent sequence corresponds to one and only one number. But does every number have only one exponent sequence? What if there is more than one way to build a certain numerical substance?

We know that every positive integer is a product of prime powers; the question is whether this breakdown must be *unique*. Might it be possible that $11 \times 31 = 17 \times 23$ for some strange reason? Our experience with whole number arithmetic suggests that this never happens, but that's not a terribly satisfying or convincing explanation.

The truth is that every positive integer does indeed factor uniquely into a product of prime powers and thus corresponds to a unique exponent sequence. This means that our logarithm is a perfect correspondence: multiplication and division of positive fractions is exactly the same as addition and subtraction of infinite sequences of integers whose entries are almost all zero.

In effect, unique factorization into primes allows us to split the divisibility issue into separate parts, one for each prime. We can put on our 7 glasses, so to speak, and look only at the exponents of that particular prime. The complicated "global" issues of multiplication and divisibility become simple counting and comparison at each "local" prime, the trade-off being that we have infinitely many such localities to deal with. This technique is so useful (not to mention beautiful) that it behooves us to really understand unique factorization into primes and why $\mathbb{N}$ got so lucky as to have this convenient property.

The polynomial domain $\mathbb{F}[X]$ has its own notion of divisibility, but we can still pour multiplicative acid on polynomials until they are broken down completely into irreducible pieces. Again we can guarantee that the process must eventually end because factorization lowers the *degree*. Both $\mathbb{Z}$ and $\mathbb{F}[X]$ share the valuable feature of having a natural size measurement (i.e., absolute value and degree, respectively) that decreases as the pieces break down so that our factorizations into irreducibles are necessarily finite. Every polynomial can then be shattered into a product of irreducible polynomials, and we can ask if that factorization is unique and in what sense.

Another question is how to extend this notion to larger realms like $\mathbb{Z}$ and $\mathbb{Q}$ that include negative numbers. To what extent can we say that the domain $\mathbb{Z}$ has unique factorization? One thing to notice is that although the numbers 8 and –8 are not the same, they do have the exact same divisibility properties—they divide the same numbers and are divisible by the same numbers. From a divisibility standpoint, they are essentially identical. (In fact, they divide each other.) We could even say that –5 is irreducible because it cannot be factored into numbers that are smaller in size (i.e., absolute value).

The point is that the number –1, like its positive cousin 1, doesn't affect divisibility. The reason is that these are precisely the elements of $\mathbb{Z}$ that have a reciprocal in $\mathbb{Z}$. That means we can both multiply and divide by them, so they can slosh around fluidly among the factors without changing who divides whom. The number 6, for example, can be factored into irreducibles as $(-2) \times (-3)$ as well as 2×3. So if we want any kind of uniqueness statement in $\mathbb{Z}$, we'll need to deal with this redundancy somehow.

In the polynomial realm, the redundancy comes in the form of the nonzero constant polynomials (i.e., those that are just a single number). These are the polynomials that possess reciprocals, so they are in a sense invisible as far as divisibility is concerned. For instance, the polynomials $\frac{1}{2}X^2 + \frac{1}{2}X$ and $3X^2 + 3X$ have the exact same divisibility properties in $\mathbb{Q}[X]$.

In general, an element of an integral domain that has a reciprocal is called a *unit*. These are the numbers that divide everybody, including 1. For divisibility purposes, this makes them boring and redundant—*every* element is divisible by a unit. If we are fortunate enough to be in an environment where things factor into irreducible pieces, then we can always scatter a few units around and it won't affect anything. In particular, a unit multiple of an irreducible is still irreducible.

What are the units of a field?

When it comes to factorization and divisibility, it's really best to put 0 in its own category—it divides no one but itself—and to put the units in their own class as well. In the case of $\mathbb{Z}$, we are fortunate that the only units are 1 and –1. To avoid ambiguity, we can always decide to take the *positive* primes as our irreducible factors. Likewise, in the polynomial case, we can select from among the constant multiples of a given irreducible polynomial the unique *monic* one to be the standard bearer.

Thus, for any (nonzero) integer n in $\mathbb{Z}$, we can write

$$n = \pm\, 2^{n_2} \cdot 3^{n_3} \cdot 5^{n_5} \cdot \cdots$$

for a unique choice of sign and exponents. Similarly, it turns out that every nonzero polynomial in $\mathbb{F}[X]$ can be factored uniquely as a leading coefficient times a product of monic irreducibles. If we like, we can then collect these factors into prime powers in the customary way. In fact, using negative exponents, we can even extend such factorizations to the field of *polynomial fractions* $p(X)/q(X)$, just as we did with the rational numbers.

Show that the collection of polynomial fractions forms a field.
Does this generalize to any integral domain?

Again, the trouble is that the primes—in this case, monic irreducible polynomials—are somewhat mysterious. Depending on the field that we are dealing with, some polynomials will factor and others will not. The polynomial $X^2 + 1$ is irreducible in $\mathbb{R}[X]$ but can be factored as $(X + i)(X - i)$ in $\mathbb{C}[X]$. So it is not always so clear which polynomials are prime and which can still be broken down. One thing we can say for sure is that a polynomial of degree 1 (e.g., $X + 3$) is *always* irreducible since the degree cannot be lowered any further.

So the analogy between $\mathbb{Z}$ and $\mathbb{F}[X]$ is quite strong. Both domains possess a notion of irreducibility as well as a convenient size measurement. In both cases we get a breakdown of each entity into a product of irreducible pieces, and (as we will show) this breakdown is essentially unique.

A domain with the property that every element can be factored into irreducible parts uniquely (disregarding unit multiples and the order of the factors) is called a *unique factorization domain*. Unfortunately, this turns out to be an extremely rare property. Our familiarity with arithmetic in $\mathbb{Z}$ makes it feel almost obvious that factorizations should exist and be uniquely determined, but this is not at all the case for a garden-variety domain environment. For instance, in the imaginary quadratic domain

$\mathbb{Z}[\sqrt{-5}]$ (consisting of all numbers of the form $a + b\sqrt{-5}$ with integers a and b), we have the factorization

$$(1 + \sqrt{-5})\,(1 - \sqrt{-5}) = 2 \cdot 3,$$

and all four factors are irreducible in $\mathbb{Z}[\sqrt{-5}]$.

Even more disturbing, there are many domains that do not even possess a coherent notion of irreducibility—you can keep pouring acid until the cows come home and the pieces will keep breaking down indefinitely. (One amusing example is $\mathbb{Z}[\sqrt{2}, \sqrt[4]{2}, \sqrt[8]{2}, \ldots]$, obtained from $\mathbb{Z}$ by throwing in infinitely many generators, each the square root of the previous.)

So it's the fact that both $\mathbb{Z}$ and $\mathbb{F}[X]$ have a discrete size measurement that keeps divisibility from getting too out of hand. When we factor, the size gets reduced, and since these sizes are non-negative integers, they must eventually hit rock bottom. This is how we can guarantee that our prime factorizations are finite. But even in those domains where irreducibility does make sense, there is nothing saying that our factorizations have to be unique. And in most cases, they're not.

The question is, what special properties do the domains of integers and polynomials possess that endows them with the rare and luxurious feature of unique factorization? If we can isolate the necessary criteria, then we can widen our viewpoint and gain altitude by studying all of the environments that meet these conditions. In other words, our proof of unique factorization will also inform us of the natural level of generality.

Our desire for understanding has once again led us to higher ground. We are now studying integral domains *in general* to gain a clearer grasp of unique factorization. As usual, we have a class of structures (in this case, integral domains) and a pattern (unique factorization), and we are trying to classify those structures that obey the pattern. The question is, at what level of generality does this pattern reside? Is it only $\mathbb{Z}$ and $\mathbb{F}[X]$ that have this beautiful property, or are there other such domains?

Before we begin this investigation, we should generalize our notation a bit to give us more flexibility. Up to now we have been writing $\mathbb{F}[X]$ for the polynomial domain in one indeterminate over the field $\mathbb{F}$. Essentially, we are starting with $\mathbb{F}$, throwing in the indeterminate X, and then forming all possible sums and products. Similarly, we have been writing $\mathbb{F}[X, Y]$ for the polynomial domain in two indeterminates. More generally, for any environment A and any entities b_1, b_2, …, we can write $A[b_1, b_2, \ldots]$ to denote the extension of A obtained by tossing in the b's (which may be any combination of numbers and indeterminates) and forming every possible sum and product.

Thus, the domain $\mathbb{Z}[\sqrt{2}]$ consists of every number that can be made from any combination of integers and $\sqrt{2}$ using addition and multiplication only. (Due to the fact that $\sqrt{2} \cdot \sqrt{2}$ is an integer, this turns out to be simply the numbers of the form $a + b\sqrt{2}$ for integers a and b.) Similarly, $\mathbb{Z}[X]$ would denote the domain of polynomials with integer coefficients.

While we're at it, we may as well generalize our field notation as well. We have been writing $\mathbb{Q}(\sqrt{2})$ to denote the field extension of $\mathbb{Q}$ generated by $\sqrt{2}$. This means we are tossing in $\sqrt{2}$ and then forming all possible sums, products, and *quotients* as well. So the idea is to use brackets to indicate closure under addition and multiplication, and parentheses for when division is also included. In particular, $\mathbb{F}(X)$ now denotes the field of polynomial fractions over $\mathbb{F}$. The polynomial domain $\mathbb{F}[X]$ then sits inside $\mathbb{F}(X)$ in the same way that $\mathbb{Z}$ sits inside $\mathbb{Q}$. In fact, it is not hard to see that every integral domain lives inside its own field of fractions.

Now let's go back to integer arithmetic for a minute. How can we tell if one number is divisible by another? For example, is 423 a multiple of 19 or not? The usual approach is to perform a "division with remainder" process: we start with 423 and then subtract off as many copies of 19 as we can until what's left is as small as possible. If this remainder happens to be zero, then the answer is yes: 423 is divisible by 19; if not, then not.

As it happens, $423 = 22 \cdot 19 + 5$, so the answer is no. The point being that the multiples of 19 are evenly spaced among the integers, and therefore every number must lie somewhere between two consecutive multiples. By subtracting off copies of 19, we are shifting our number so that it lies in the range 0 to 18, allowing us to see more clearly where we are. In this way, we can regard the number 423 as being equivalent to 5, at least from 19's point of view. (We will come back to this idea a little later.)

The most important feature of this division process is that it ensures the remainder will be *strictly smaller* than the number we are dividing by. This holds true for polynomials as well. We can even mimic the usual long division procedure, using the degree of the leading term as our guide. For example, suppose we want to divide $X^3 - 3X^2 + 2X + 1$ by $X^2 + 1$. The leading terms are X^3 and X^2, so we can start by subtracting off X copies. (This should remind you of our elimination procedure.) The calculation would proceed as follows:

$$\begin{array}{r|l} & X \quad - 3 \\ \hline X^2 + 1 & X^3 - 3X^2 + 2X + 1 \\ & X^3 \qquad\quad + X \\ \cline{2-2} & \quad -3X^2 + X + 1 \\ & \quad -3X^2 \qquad - 3 \\ \cline{2-2} & \qquad\qquad X + 4 \end{array}$$

This means that

$$X^3 - 3X^2 + 2X + 1 = (X - 3)(X^2 + 1) + (X + 4),$$

and the remainder $X + 4$ has smaller degree than $X^2 + 1$. We find that $X^2 + 1$ does *not* divide $X^3 - 3X^2 + 2X + 1$ because it leaves a remainder of $X + 4$.

Both $\mathbb{Z}$ and $\mathbb{F}[X]$ thus feature the same simple and convenient division property: for any two elements a and b, we can always find a quotient q and a remainder r so that $a = qb + r$, with r strictly smaller than b. In both cases, if the remainder falls below a certain minimal size, it is forced to be zero. In $\mathbb{Z}$, the smallest size a nonzero element can have is 1, and this is shared by the two units 1 and -1. In the polynomial case, we have the linear polynomials (of degree 1), below which are the nonzero constant polynomials (of degree 0), and these are our units. The only thing smaller is the zero polynomial itself. (Some people like to assign 0 a negative degree for this reason.)

This property can be used in a very pretty way to determine the so-called *greatest common divisor* (or GCD) of two or more numbers. Suppose we take 3234 and 315. What is the largest number that divides both? One way to proceed would be to factor both numbers and see which prime powers they have in common. The downside of this approach is that it becomes increasingly time consuming and laborious as the numbers get larger.

A much simpler and faster procedure was invented by Euclid, using repeated division with remainder. The idea is to first divide the larger number by the smaller one and see what remainder we get. In our case, we find

$$3234 = 10 \cdot 315 + 84.$$

Whatever factors 3234 and 315 may have in common, 84 must share them as well because it is a whole-number combination of them: $84 = 1 \cdot 3234 - 10 \cdot 315$. Another way to say this is that the ideal (3234, 315) in $\mathbb{Z}$ is the same as the ideal (315, 84). The numbers you can make from a combination of 3234 and 315 are the same as the numbers you can make out of 315 and 84.

Show that in general, if $a = qb + r$ then $(a, b) = (b, r)$.

Just as with elimination, we're finding new combinations of our original generators that are in some sense smaller. Now we can continue with 315 and 84:

$$315 = 3 \cdot 84 + 63.$$

This says the ideal (315, 84) is the same as (84, 63). Next, we get

$$84 = 1 \cdot 63 + 21,$$
$$63 = 3 \cdot 21 + 0.$$

We can now see that our original ideal (3234, 315) is equal to (63, 21) = (21, 0) = (21), the multiples of 21. This tells us two things. First, it says that 21 is the GCD of 3234 and 315. If we like, we could even say that what it *means* for a number d to be the GCD of a and b is that the ideal (a, b) is equal to (d).

Show that this agrees with our previous notion of GCD.

The other thing we discover is that since (3234, 315) = (21), the number

21 must be a combination of our two original numbers. Thus, the GCD of two numbers is not only the largest number that divides both, but is also a whole-number combination of the two.

The foregoing procedure is known as the *Euclidean algorithm*. In order for it to work in a more general setting, we need our domain to have a notion of size that allows us to perform division with remainder in the right way. Namely, we need the size of the remainder to be smaller than the size of the divisor. A domain with this property is said to be *Euclidean*. Euclidean domains are thus a natural generalization of both the integer and polynomial realms.

Show that the domain $\mathbb{Z}[i]$ *is Euclidean.*
What are its units?

The practical utility of the Euclidean algorithm is in calculating GCDs rapidly, but its theoretical value is that it guarantees their existence: if a domain is Euclidean, then we know that any ideal (a, b) is just (d), the multiples of a single element. In fact, *every* ideal in such an environment must have this form. Our notion of size—and the way it controls remainders—makes it that every ideal is simply the multiples of its smallest nonzero element. Since there is a smallest, everybody has to be a multiple of it; otherwise, any remainder would be even smaller.

So the ideals in a Euclidean domain like $\mathbb{Z}$ or $\mathbb{F}[X]$ have a particularly simple structure: every ideal is *principal*—that is, generated by a single element. This not only guarantees a GCD for any pair of elements, but also for larger collections: if $a_1, a_2, a_3, \ldots$ is any (finite or infinite) set of elements, then there is a smallest element d of the ideal they generate and $(a_1, a_2, a_3, \ldots) = (d)$. We can then take this element d to be the greatest common divisor. For example, we saw before that the ideal (12, –18, 30) in $\mathbb{Z}$ is equal to (6), so the GCD of 12, –18, and 30 is just 6. (You could also call it –6, if you want to be contrarian.)

What is the GCD of $X^4 - 1$, $X^3 + 1$, *and* $X^2 - 1$ *in* $\mathbb{R}[X]$?

Notice how the meaning of GCD has subtly shifted from being the greatest common divisor of $a_1, a_2, a_3, \ldots$ (in the sense of size) to being a generator of the ideal $(a_1, a_2, a_3, \ldots)$. Of course, in the context of a Euclidean domain, these are equivalent. However, in the case of more general domain environments, we will not have a notion of size at our disposal, yet we can still speak of ideals and generators. The only trouble is that there's no guarantee that a given ideal will be principal, so a GCD may not even exist. For example, the ideal (X, Y) in the polynomial domain $\mathbb{R}[X, Y]$ cannot be generated by a single element.

Why can't (X, Y) be a principal ideal?

A domain with the property that all of its ideals are principal is called a *principal ideal domain*, or PID. This condition ensures that any collection of elements will always have a GCD (in our new, generalized sense of the word). A principal ideal domain is thus the abstract way of saying "a place where greatest common divisor makes sense." Euclidean domains are merely a special case of this wider phenomenon.

This turns out to be just the right level of generality. Our proof of unique factorization will not require the Euclidean algorithm, but we will need a notion of GCD. By generalizing to the level of principal ideal domains, we include not only our original environments, $\mathbb{Z}$ and $\mathbb{F}[X]$, but also many others, such as $\mathbb{Z}[i]$ and $\mathbb{Z}[\sqrt{2}]$. In a nutshell, integers and polynomials possess unique factorization because those domains are Euclidean, and Euclidean domains are PIDs, and PIDs *always* have unique factorization. So our goal is to explain why a principal ideal domain (i.e., a domain possessed of GCDs) must necessarily also be a unique factorization domain.

Let's imagine that we are working in a principal ideal domain such as $\mathbb{Z}$, and for the sake of argument, let's suppose it were possible to have two distinct factorizations of the same number. This is exactly the situation we talked about before, when we were worried that maybe $11 \times 31 = 17 \times 23$. We want to understand why something like this can't happen in a PID. Of course, if this were actually true—that $11 \cdot 31$ were the same as $17 \cdot 23$—it would mean that the product $17 \cdot 23$ is somehow a multiple of 11 without 11 dividing either of the factors! But our everyday experience with prime numbers says that if a product is divisible by a prime, then one of the factors must be also. This is certainly not the case for nonprime numbers like 6, however. In fact, 6 divides $4 \cdot 9$ without dividing either factor.

It is thus a characteristic feature of prime numbers that when a prime divides a product it must divide one of the factors (or possibly both). This is what makes an ambiguous factorization impossible. So what we need to understand is *why* prime numbers (i.e., irreducibles) must behave in this way. To clarify matters, let's start by making a careful distinction between these two different aspects of prime behavior.

Suppose p is a nonzero, nonunit element of a domain. Then p is said to be *irreducible* if for every factorization $p = ab$, one of the factors a or b must be a unit. On the other hand, we will say that p is *prime* if whenever p divides a product ab, then p must divide either a or b. Thus, irreducibility and primality are not quite the same notion. In fact, it turns out that primality is the stronger condition, in that primes are *always* irreducible.

Prove that a prime element must be irreducible.

Our goal is to show that in a principal ideal domain, irreducible elements must also be prime. This means that in the context of a PID, the two terms are synonymous. If we can prove this, it will mean that an irreducible cannot divide a product without dividing one of the factors, and this is enough to prevent ambiguous factorizations from occurring.

So why must an irreducible element have this primality property? Just because your number can't break down anymore doesn't mean it has to do this funny divisibility dance. We need to understand how the existence of GCDs *forces* irreducible elements to act this way.

Let's look at a simple example in the familiar context of $\mathbb{Z}$. The number 7 is irreducible in this domain. Now suppose we have an integer b with the property that $7 \mid 12b$. This means that $12b$ is divisible by 7, and we want to show that this means b itself must be a multiple of 7. The plan is to use the fact that 7 and 12 have no common factors, so their GCD must be 1. This means the number 1 can be written as a combination of 7 and 12. In fact, we have $1 = 3 \cdot 12 - 5 \cdot 7$. Now if we multiply both sides of this equation by b, we get $b = 3 \cdot 12b - 5 \cdot 7b$. This says that the number b is then a combination of $7b$ and $12b$, both of which are divisible by 7—the latter by our initial assumption. This forces b to be divisible by 7 as well. Notice how this argument relies strongly on the existence of GCDs. Fortunately, this same proof can be applied equally well to $\mathbb{F}[X]$ or any other principal ideal domain.

Let's now imagine we are in the setting of an arbitrary PID. Suppose p is irreducible and $p \mid ab$ for some elements a and b. Let's further assume that p does *not* divide a. We want to show that p is then forced to divide b. Again, the idea is to consider the GCD of p and a. Since this must divide the irreducible element p, it must either be a unit or a unit multiple of p itself. (This is, after all, what it means for p to be irreducible.) But this GCD must also divide the number a, and we assumed at the start that a was not divisible by p. So that means we must have $(a, p) = (1)$, and the only common divisors are units. In particular, the number 1 must be a combination of a and p—that is, we can write $1 = ax + py$ for some elements x and y. This is what the existence of GCDs allows us to do that we cannot do in general.

Now here's the punchline. If we multiply both sides of the equation $1 = ax + py$ by the number b, we get

$$b = (ab)x + (pb)y,$$

and both expressions in parentheses are divisible by p. This shows that b is indeed a multiple of p, and we are done. We have just succeeded in proving that in a PID, irreducible elements are always prime.

The defining feature of a prime is that if it divides a product ab, then it must divide either a or b (or possibly both). As a consequence, we can see that if a prime divides the product of any number of elements, then it must divide at least one of the pieces. For example, if p is prime and $p \mid abc$, then either $p \mid a$ or else $p \mid bc$. In the latter case, either $p \mid b$ or $p \mid c$. No matter what, we see that p divides at least one of a, b, or c. This same reasoning clearly generalizes to any number of factors.

Now we can see why a PID must have unique factorization. Suppose, on the contrary, that we had irreducible elements $p_1, p_2, \ldots, p_n$ and $q_1, q_2, \ldots, q_m$ with

$$p_1 p_2 p_3 \cdots p_n = q_1 q_2 q_3 \cdots q_m.$$

Since we can always cancel any common factors, we may assume that none of the p's are equal to (or even unit multiples of) any of the q's. Let's choose one particular p. Being irreducible, this p must also be prime because we are in a principal ideal domain. But p divides the product of the q's, so it must divide some particular q. Since this q is irreducible, it must then be a unit multiple of p, and this contradicts our initial assumption.

We can now see that the reason for unique factorization is that irreducible elements are prime, and the way we explain this for PIDs like $\mathbb{Z}$ and $\mathbb{F}[X]$ is that these are environments in which GCDs exist, so we can then play our little trick of writing 1 as a combination of elements with no common factor.

Show that the rational numbers with odd denominators form a unique factorization domain. What are the units and primes?

The domains $\mathbb{Z}$ and $\mathbb{F}[X]$ are special in that they are principal ideal domains and thereby possess GCDs, whereas $\mathbb{Z}[\sqrt{-5}]$ and $\mathbb{F}[X, Y]$ do not. This is yet another reason why elimination is so valuable; it is only in the one indeterminate case (i.e., $\mathbb{F}[X]$) that a polynomial domain is fortunate enough to be a PID. Luckily, more general polynomial domains (such as $\mathbb{Z}[X, Y]$) still retain unique factorization, but the lack of GCDs makes the arithmetic far more complicated. As an example, suppose we have two integers with no common factors—let's say 26 and 15. Such numbers are said to be *relatively prime* to each other. In this case, the ideal (26, 15) is equal to (1), which is the entire domain $\mathbb{Z}$. This means that every integer can be written as a combination of 26 and 15, including the number 1. This is not the case in $\mathbb{R}[X, Y]$, for instance, because the ideal (X, Y) is not equal to (1), despite X and Y having no factors in common.

Can you find integers a and b so that $1 = 26a + 15b$*?*

Now let's focus on the polynomial setting. For any field $\mathbb{F}$, we know that the polynomial domain $\mathbb{F}[X]$ is Euclidean. That means we have a notion of size (namely, degree) that provides us with a means of measuring remainders: for any polynomials a and b, there are polynomials q and r so that $a = qb + r$, with the degree of r strictly less than the degree of b.

This feature of polynomial division with remainder has a particularly simple and elegant consequence. Suppose α is an element of our field $\mathbb{F}$, and we consider the monic linear polynomial $X - \alpha$. What does it mean to be divisible by this element? Suppose p is a polynomial in $\mathbb{F}[X]$, and we try to divide it by $X - \alpha$. Then we will have a quotient polynomial q and a remainder r, where the degree of r is less than 1, the degree of $X - \alpha$. This means r must actually be a *constant* (including possibly zero). We have

$$p(X) = q(X) \cdot (X - \alpha) + r.$$

Here I am writing $p(X)$ and $q(X)$ to emphasize that these are polynomials involving X, whereas r is simply a number. We can then deduce the value of this remainder r by substituting α for X to cancel the first term on the right. This tells us that in fact $r = p(\alpha)$, the value of the polynomial p at α. Therefore, a polynomial $p(X)$ is divisible by $X - \alpha$ precisely when $p(\alpha) = 0$.

For example, suppose we are interested in the cubic polynomial $p(X) = X^3 - 3X^2 + 4$ in $\mathbb{Q}[X]$. Let's say we've been playing around and we happen to notice that $p(2) = 0$. This means that the polynomial p must be divisible by $X - 2$. We thus obtain the factorization $p(X) = (X - 2) \cdot (X^2 - X - 2)$. Of course, if we had started with this representation of p, it would then be patently obvious that $p(2) = 0$; any polynomial with $X - 2$ as a factor will become zero when we plug in 2 for X. What we're saying is that this is the *only* way this can happen. Finding a linear factor of a given polynomial is thus tantamount to finding a zero.

This means we can reinterpret the classical algebra project yet again! We've already succeeded in reducing all of our algebra problems down to finding the zeros of a polynomial—that is, given a polynomial $p(X)$, we hope to determine the numbers α satisfying $p(\alpha) = 0$. Such a number is also called a *root* of p, generalizing the usual notion to any polynomial process. (The square root of 2 is simply a root of the polynomial $X^2 - 2$).

What we have just seen is that there is a perfect correspondence between roots and linear factors: α is a root of p precisely when $X - \alpha$ is a factor of p. Thus, the problem of finding the roots of a polynomial (or its zeros, if you prefer that language) is the same as finding its linear prime factors. Classical algebra, like much of arithmetic, reduces to the problem of prime factorization.

This correspondence is particularly pretty in the case of complex polynomials. Not only is $\mathbb{C}[X]$ a unique factorization domain, but we also know that $\mathbb{C}$ itself is algebraically closed—every (nonconstant) complex polynomial has at least one root. This means that if p is a complex polynomial with the root α, then we can write $p(X) = (X - \alpha) \cdot q(X)$ for some polynomial q. But this q is also a complex polynomial, so it too must have a root, and we can continue splitting off linear factors until we are left with only a constant.

Thus, every nonzero complex polynomial $p(X)$ can be fully factored into linear prime factors $X - \alpha$ for various complex numbers α. If we like, we can gather these pieces together into prime powers and write

$$p(X) = c(X - \alpha_1)^{n_1}(X - \alpha_2)^{n_2} \cdots (X - \alpha_m)^{n_m},$$

where c is the leading coefficient of p and the α_k are its distinct complex roots. The exponent n_k is called the *multiplicity* of the root α_k. For example, the quartic polynomial $4X^4 - 6X^3 + 2X$ factors as

$$4X^4 - 6X^3 + 2X = 4(X - 1)^2(X + \tfrac{1}{2})X.$$

We see that this polynomial has single roots at 0 and $-\frac{1}{2}$ and a double root (of multiplicity two) at 1.

What is the prime factorization of $X^4 + 2X^2 + 1$?

Factorization of complex polynomials is especially simple because we can describe the irreducible elements so easily: the primes in $\mathbb{C}[X]$ are simply the monic linear polynomials $X - \alpha$, one for each complex number α. So the good news is that we know all the primes, but the trade-off is that there are uncountably infinitely many of them—as opposed to $\mathbb{Z}$, where the primes are listable but mysterious.

Are there infinitely many primes in $\mathbb{F}_2[X]$*?*

This means that the factorization of quadratics, for example, is not as artful or as subtle as it may have appeared. It turns out that *every* polynomial factors completely over $\mathbb{C}$, so the only question is how to describe the factors. If we like, we can rephrase the classical algebra project in this way: How do we factor a given polynomial? I'm not saying that prime factorization is any easier than finding roots or solving equations; all I'm saying is that it offers us a new vantage point—one that may call to mind different ideas and approaches.

I suppose my theme all along has been the value of abstraction and the perspective it provides. Sure, there's been a lot of talk about knots, but what I'm really talking about are *connections*—new ways of seeing that lead to epiphany and deeper understanding. The truth is that we cannot begin to understand classical algebra until we start to leave it behind. The equivalence of roots and prime factors offers us a new perspective not only on classical algebra but also integer arithmetic. For instance, it is not unusual for a modern algebraist to view a number like $60 = 2^2 \cdot 3 \cdot 5$ as having "roots" at 3 and 5 and a "double root" at 2. These are really divisibility statements; having a root simply means being divisible by the corresponding prime.

But what is the ultimate point of all these analogies and metaphors? Aren't mathematical truths simply what they are? Mathematical proofs depend only on logic and formal definitions; they seem to require no colorful imagery or deep metaphorical connections. Why do we bother?

The trouble is that profound and beautiful mathematical arguments do not simply fall from the sky; we poor *humans* have to devise and invent them, and this requires insight and imagination, inspiration and epiphany, creativity and stamina, and all sorts of intangibles like these. We need all the help we can possibly get, and the modern mathematical viewpoint essentially comes down to the belief that the best way to see is to gain altitude. Abstraction and generalization are almost synonymous with making connections, uniting formerly disparate ideas under the same conceptual heading. So the reason for all the metaphors is not that math demands them; it's that *we* require them to make any progress. Essentially, we make analogies in order to connect two trees with a tightrope; if we make enough of these, we can use them as a rope ladder to climb higher.

In any event, we now have two complementary ways to view polynomials. We can think of them additively, as sums of individual terms, or we can view them multiplicatively, as products of irreducible factors. In the complex case, these prime factors are linear, which is especially convenient.

What are the primes in $\mathbb{R}[X]$*?*

I know this has been a lot to take in, but I want to show you just one more example of the close analogy between integers and polynomials. Let's go back to where we divided 423 by 19 and got a remainder of 5. Relative to the multiples of 19, the numbers 423 and 5 are in the same position. They are said to be equivalent *modulo* 19, or (mod 19) in parentheses, for short. Usage varies somewhat, but I like to write $423 = 5 \pmod{19}$ using the equals sign, knowing full well what I mean by it in this context: not equal as integers but having equal *remainders*. In a sense, we are redefining equals so that 19 (and consequently all of its multiples) becomes zero. We are wrapping the system of integers around a circle, like the numbers on a clock. In fact, the usual way we tell time is to do arithmetic (mod 12).

For any integer n, arithmetic (mod n) is just ordinary arithmetic in $\mathbb{Z}$—the exact same addition and multiplication—but with the added demand that $n = 0$. This is not a contradiction but instead leads to an alternative system of numerical entities based on our original integers. This is the system of *integers modulo n*, or $\mathbb{Z}_n$, for short. For example, in $\mathbb{Z}_5$ we have $-1 = 4$ and $\frac{1}{2} = 3$ because $4 + 1 = 0$ and $2 \times 3 = 1$. Arithmetic in $\mathbb{Z}_2$ consists only of the numbers 0 and 1, subject to the condition that $1 + 1 = 0$. In fact, we've been here before—it's the field $\mathbb{F}_2$. It turns out that when p is a prime, the integers (mod p) always form a field, so $\mathbb{Z}_p = \mathbb{F}_p$ is a finite field with p elements.

Why is $\mathbb{Z}_p$ a field when p is prime?
What goes wrong otherwise?

The most beautiful and important feature of modular arithmetic is that it preserves the arithmetic structure of $\mathbb{Z}$. If a statement involving addition and multiplication of integers is true, then it remains true (mod n) for any n. So in a sense, the modular environments $\mathbb{Z}_n$ are like shadows of $\mathbb{Z}$ itself—they preserve the arithmetic relationships but blur the distinctions among the elements themselves. If we are operating (mod 7), then all of the multiples of 7 collapse to a point—the zero of the system. Two integers that differ by a multiple of 7—and thus have the same image in $\mathbb{Z}_7$—are like two objects aligned with the direction of the light: they cast the same shadow.

When we turn on the 7 light, the system of integers $\mathbb{Z}$ casts the shadow $\mathbb{Z}_7$. By turning on multiple lights simultaneously, we can tease out a lot of divisibility information. It turns out that these various colors are not all independent, however. Turning on the 6 light ends up revealing exactly as much information as turning on the 2 and 3 lights together, for example.

Show that if m and n are relatively prime, then every element of $\mathbb{Z}_{mn}$ corresponds to a unique pair of elements in $\mathbb{Z}_m$ and $\mathbb{Z}_n$.

Thus, as a consequence of division with remainder, we get this nice modular equivalence idea. Which means, of course, that we should be able to do the same thing with polynomials. What happens when we "mod out" by a prime polynomial like $X - \alpha$? We saw before that when we divide a polynomial $p(X)$ by $X - \alpha$, the remainder is equal to $p(\alpha)$, the value of the polynomial at α. In modular terms, this means that $p(X) = p(\alpha)$ (mod $X - \alpha$). Essentially, when we mod out by $X - \alpha$, we are treating $X - \alpha$ as zero, and this is the same as setting X equal to α.

For example, since $X^2 - X + 2$ has the value 8 when we plug in 3 for X, we see that $X^2 - X + 2 = 8$ (mod $X - 3$). This means we can think of the plugging-in process as being a kind of shadow: setting X equal to 3 has the same effect on polynomials as turning on the $X - 3$ light. We can even transfer this language back to $\mathbb{Z}$ and view an equivalence like

$12 = 5 \pmod 7$ as saying that the "value" of 12 at the prime 7 is equal to 5.

More generally, we can choose any polynomial q and consider the arithmetic of $\mathbb{F}[X] \pmod q$. As an example, suppose we take the domain of real polynomials $\mathbb{R}[X]$ and we mod out by $X^2 + 1$. This means that $X^2 + 1$ is now to be regarded as zero. Since this polynomial is quadratic (degree 2), any remainders will have degree at most 1. The upshot is that every polynomial now looks like $a + bX \pmod{X^2 + 1}$, and X^2 is now equal to -1. In other words, the system of real polynomials $(\text{mod } X^2 + 1)$ is identical with $\mathbb{C}$!

This provides us with an alternative approach to constructing the complex numbers. Rather than throwing in some imaginary external entity $\sqrt{-1}$, we can instead construct the polynomial domain $\mathbb{R}[X]$ and then mod out by $X^2 + 1$. In fact, this gives us a uniform method for constructing algebraic extension fields in general. If you want to work in the field $\mathbb{Q}(\sqrt[3]{2})$ obtained by adjoining $\sqrt[3]{2}$ to the rational field, one option is to view this environment as being the domain of rational polynomials $\mathbb{Q}[X]$ considered $(\text{mod } X^3 - 2)$. In this way, X itself will now play the role of $\sqrt[3]{2}$. What I like about this approach is that we are building our extensions from within, as opposed to reaching outward for a number we do not yet possess. In this way, every field $\mathbb{F}$ already contains the seeds of all of its algebraic extensions.

SYMMETRY

Here is where things start to get insanely beautiful. The fact that every polynomial factors completely over $\mathbb{C}$ allows us to reimagine our project in an utterly new and striking way. Just as with Tartaglia's solution to the general cubic equation, this will involve making what appears to be a rather pointless and unproductive trade, increasing the number of unknowns dramatically in exchange for a certain technical convenience—in this case, *algebraic symmetry*.

This trade is a lot like Jack selling the family cow for a handful of magic beans. At first it may seem foolhardy (especially to Jack's mother, a classical algebraist), but then symmetry begins to work its subtle magic, and before we know it, an entirely new kingdom emerges in the clouds. Symmetry is our beanstalk, and we will use it in the best way possible: to climb higher.

Let's begin with the familiar quadratic case. For simplicity, we will suppose we have a monic quadratic equation of the form $x^2 - Px + Q = 0$. (I am choosing to call the linear coefficient $-P$ for stylistic reasons that will become clear.) The solutions of this equation are the roots of the polynomial $X^2 - PX + Q$, which I will call α and β. We then have the factorization

$$X^2 - PX + Q = (X - \alpha)(X - \beta).$$

Expanding and collecting the right-hand side, we get

$$X^2 - PX + Q = X^2 - (\alpha + \beta)X + \alpha\beta.$$

Thus, for α and β to satisfy our equation, we must have

$$\begin{aligned} \alpha + \beta &= P, \\ \alpha\beta &= Q. \end{aligned}$$

In other words, every quadratic equation is equivalent to a broken bones puzzle in two unknowns with a prescribed sum and product.

(This is why I chose $-P$ initially, so that this system would not feature an ugly minus sign.) We have sold our cow (the original quadratic equation in one unknown) for a system of two equations in two unknowns, and now we look like fools who have been swindled at the county fair.

But we have actually gained a commodity of great value. This system is not some random collection of clunky and complicated polynomial equations. Not only are the equations themselves as simple as they can be, but they are also *symmetrical* in the unknowns α and β. If we were to switch the meanings of these symbols—exchanging all of the α's and β's in each equation—we could not tell the difference. Our system is *invariant* under the swap $\alpha \leftrightarrow \beta$. This means there is action and energy in this setting. With two unknowns, we have the possibility of interchange, whereas our original equation $x^2 - Px + Q = 0$ just sits there—a lifeless amalgam of terms of various degrees. Our symmetric system, on the other hand, treats α and β identically, and the equations are the epitome of elegance.

So we do gain something from the exchange. We just need to figure out how best to exploit it. Our problem is to solve the symmetric system

$$\alpha + \beta = P,$$
$$\alpha\beta = Q.$$

We could use our classical methods—substitution and elimination—to reduce this to a single quadratic equation in one unknown, but that would merely return us to our starting point. Instead, we will develop a new approach based on symmetry. This means we will always strive to treat α and β equally; we do not want to favor one over the other, as with substitution.

One amusing consequence of algebraic symmetry is that there is no way to tell our solutions apart. If we call the two (typically distinct) solutions α and β, then swapping these labels will also provide a solution. Another way to say it is that we're not trying to determine the particular values of α and β themselves, but rather the value of the unordered pair $\{\alpha, \beta\}$. These two numbers will have the given sum and product that we demand, but there is no way to say which one is α and which is β.

Fortunately, we already know how to solve quadratic equations, and we even have a quadratic formula that tells us the solutions: α and β are the two numbers given by $\frac{1}{2}(P \pm \sqrt{D})$, where D is the discriminant $P^2 - 4Q$. The ambiguity is now contained in the square root symbol—the two solutions correspond to the two choices of sign. Somehow, the swapping of α and β is being converted into the negation of a square root. One way to convey this state of affairs is to write

$$\alpha, \beta = \frac{P \pm \sqrt{D}}{2}.$$

This says that the two numbers on the left are the same as the two numbers on the right, but we can't tell which is which. That is the content of the quadratic formula.

Now I want to show you a different way to obtain this same information from symmetry considerations alone, without the need for completing the square or any other quadratic technique. We are given information about α and β in the form of their sum and product—symmetric combinations $\alpha + \beta$ and $\alpha\beta$. We need to somehow take this symmetrical information and turn it into a description of α and β themselves, or at any rate the pair of them.

The first step is to write α and β as a combination of their sum and difference:

$$\alpha, \beta = \tfrac{1}{2}(\alpha + \beta) \pm \tfrac{1}{2}(\alpha - \beta).$$

(Notice that this is the Babylonian average-and-spread idea once again.) The first term is symmetric and equal to $\frac{1}{2}P$, as it should be. The second term involves $\alpha - \beta$, which is an asymmetric combination of the roots. If we interchange α and β, this term gets negated, leading to the two choices of sign and therefore the two possible values α and β.

Thus, we can express the pair of roots α and β as a sum of two parts, one symmetric and the other *antisymmetric*—meaning that it negates when we exchange symbols. The symmetric part we already know, so everything comes down to the antisymmetric expression $\alpha - \beta$.

The key idea is this: since $\alpha - \beta$ gets negated when we swap α and β, its square $(\alpha - \beta)^2$ must then be symmetric. A number and its negative always have the same square, so it makes no difference if you transpose the symbols or not. The difference $\alpha - \beta$ is thus a nonsymmetric combination whose square is symmetric. In fact, we can write

$$\begin{aligned}(\alpha - \beta)^2 &= \alpha^2 - 2\alpha\beta + \beta^2,\\ &= (\alpha + \beta)^2 - 4\alpha\beta,\end{aligned}$$

and this is nothing other than $P^2 - 4Q$, the discriminant D of our quadratic polynomial. Thus, we discover that $(\alpha - \beta)^2 = D$ and the ambiguous $\alpha - \beta$ (which negates if we swap α and β) is simply $\pm\sqrt{D}$. Notice also that we get a new description of the discriminant as $D = (\alpha - \beta)^2$, which clearly shows why $D = 0$ detects the equality of the two roots.

From a symmetry point of view, a quadratic equation is just a two-dimensional symmetric system with a given sum and product of unknowns. The asymmetric *difference* of the unknowns then becomes the combination of interest, its square being symmetric and expressible in terms of the coefficients. The difference itself can then be described using square roots, with exactly the right amount of ambiguity to match the asymmetry.

In the cubic case, we are given a monic cubic polynomial $X^3 - AX^2 + BX - C$ with roots α, β, and γ. (The symbol γ is gamma, the third letter of the Greek alphabet.) The factorization

$$X^3 - AX^2 + BX - C = (X - \alpha)(X - \beta)(X - \gamma)$$

then leads to the symmetric system

$$\begin{aligned}\alpha + \beta + \gamma &= A,\\ \alpha\beta + \beta\gamma + \gamma\alpha &= B,\\ \alpha\beta\gamma &= C.\end{aligned}$$

The first equation involves the sum of the roots, the second the sum of all pairwise products, and the third the product of all three. We thus have one symmetric equation for each degree, 1, 2, and 3. (Note that in the case of a depressed cubic, we have $A = 0$; when we shift our unknown in order to eliminate the quadratic term, we are in effect shifting the roots so they total to zero.)

The question is how to invert this system. As in the quadratic case, the plan will be to construct clever combinations of the roots α, β, and γ that exhibit various types and levels of symmetry. The hope is that we can express the roots themselves as combinations of asymmetric expressions whose squares and cubes are more symmetrical. We are trying to build a bridge from the asymmetric world of the individual roots to the completely symmetric realm of the coefficients. The information we have—that is, the parameters A, B, and C—consists of symmetric combinations of the roots, and we want to get at the roots themselves. Clearly, our first order of business is to understand the symmetries of three objects.

There are six ways to permute the symbols α, β, and γ. One of these is to do nothing, of course. There are also three different swaps, or *transpositions.* (For example, we can switch α and β, leaving γ alone.) Finally, there are the two "roll" permutations, or 3-*cycles*: we can replace α by β, β by γ, and γ by α, or we can do the reverse. These six permutations form a closed system of invertible activities, and our system of equations is invariant under all of them.

The cubic analog of the difference of two roots is the product of all three such differences, which I will call δ. (This is the Greek letter delta, for "difference.") Thus,

$$\delta = (\alpha - \beta)(\beta - \gamma)(\gamma - \alpha).$$

The beautiful thing about this particular expression is that it is both invariant under rolls and antisymmetric under swaps.

Show that δ has these properties.

This means that δ^2 is invariant under both rolls *and* swaps, and that makes it fully symmetric. In fact, $D = \delta^2$ can be explicitly written in terms of the coefficients A, B, and C.

Show that $D = A^2B^2 - 4B^3 - 4A^3C - 27C^2 + 18ABC.$

The nonsymmetric combination δ—which negates under all three transpositions—can thus be written $\pm\sqrt{D}$, and the ambiguity now resides with the square root.

With three roots, we need to be a bit more creative about how we express them as sums of symmetric and asymmetric parts. In the quadratic case, we used the sum and difference of the roots. The trick now is to employ the complex cube roots of unity 1, ω, and ω^2 to write

$$\alpha = \tfrac{1}{3}(\alpha + \beta + \gamma) + \tfrac{1}{3}(\alpha + \omega\beta + \omega^2\gamma) + \tfrac{1}{3}(\alpha + \omega^2\beta + \omega\gamma),$$

with analogous expressions for β and γ. Notice how the alignment of terms causes the α's to be reinforced and the β and γ terms to cancel out, due to the fact that $1 + \omega + \omega^2 = 0$.

The first term is symmetric (the average of the three roots) and is equal to $\tfrac{1}{3}A$. (This term is absent in the case of a depressed cubic.) Let's give names to the expressions appearing in the other two terms:

$$\eta = \alpha + \omega\beta + \omega^2\gamma,$$
$$\eta' = \alpha + \omega^2\beta + \omega\gamma.$$

These combinations are asymmetric, but they do have one nice feature: if we perform a roll permutation, they both get multiplied by a cube root of unity. For instance, if we turn α into β, β into γ, and γ into α, then η becomes $\beta + \omega\gamma + \omega^2\alpha$, which is the same as $\omega^2\eta$. Likewise, this same roll turns η' into $\omega\eta'$.

As a consequence of this behavior, we see that both η^3 and η'^3 are roll-invariant, as is their product $\eta\eta'$. (When we apply a roll, one of them multiplies by ω and the other by ω^2, so the effects cancel.) As a matter of fact, $\eta\eta'$ turns out to be fully symmetric and expressible in terms of the coefficients.

Show that $\eta\eta' = A^2 - 3B$.

Now we have roll-invariant combinations η^3 and η'^3. These have the convenient feature that when we perform any swap, they are interchanged.

Show that η^3 and η'^3 trade places under transpositions.

This means we can play our clever trick of writing the pair as a sum of symmetric and antisymmetric parts:

$$\eta^3, \eta'^3 = \tfrac{1}{2}(\eta^3 + \eta'^3) \pm \tfrac{1}{2}(\eta^3 - \eta'^3).$$

The first term is now fully symmetrical and can be written in terms of the coefficients A, B, and C.

Show that $\eta^3 + \eta'^3 = 2A^3 - 9AB + 27C$.

The second term is roll-invariant and antisymmetric under swaps, just like our friend δ. In fact, it is a constant multiple of δ.

Show that $\eta^3 - \eta'^3 = 3(\omega^2 - \omega)\delta$.

Now we can understand the cubic formula from a symmetry standpoint. Given the coefficients A, B, and C, we can calculate D and $\eta^3 + \eta'^3$ explicitly, as well as $\eta\eta'$. We then know δ (up to a sign ambiguity) and thus also $\eta^3 - \eta'^3$ (again up to sign). This gives us the pair $\{\eta^3, \eta'^3\}$ and thus η and η'—only now the ambiguity involves possible cube roots of unity. These must be chosen so that $\eta\eta'$ has its known value, and we thus get three possibilities for the pair $\{\eta, \eta'\}$. The roots are then obtained from the prescription $\alpha, \beta, \gamma = \tfrac{1}{3}(A + \eta + \eta')$.

The cubic formula is therefore a direct consequence of the behavior of roll and flip symmetries. The combinations δ and η are both constructed to have the feature that a certain power of them is more symmetrical than they are themselves. Rather than finding clever ways to untangle equations, we're instead finding clever symbolic expressions that straddle different levels of symmetry.

To say that the roots of a polynomial can be expressed in terms of the coefficients using the language of radicals is to say that certain nonsymmetric expressions (namely, the roots themselves) can be written as combinations of various parts whose powers are more symmetrical, and that these new pieces can then be expressed in terms of other combinations whose powers are even more symmetrical, and so on. Our "unknowns and equations" symbol game becomes a "symmetric combinations" symbol game.

To talk about these ideas in general, we'll need to develop some notation for the various permutations of symbols and their effects on polynomial expressions involving several unknowns. As always, the point will be to find a way to write down our ideas as simply and as economically as possible.

The general polynomial equation of degree n can be written

$$x^n - c_1x^{n-1} + c_2x^{n-2} - c_3x^{n-3} + \cdots \pm c_n = 0,$$

where the coefficients are labeled so as to alternate in sign. (The sign of the constant term will depend on whether n is even or odd.) We then have the corresponding polynomial factorization

$$X^n - c_1X^{n-1} + \cdots \pm c_n = (X - r_1)(X - r_2) \cdots (X - r_n).$$

Expanding the right side and equating coefficients, we obtain the symmetric n-dimensional system

$$\begin{aligned} r_1 + r_2 + \cdots + r_n &= c_1, \\ r_1r_2 + r_2r_3 + \cdots + r_{n-1}r_n &= c_2, \\ r_1r_2r_3 + r_1r_2r_4 + \cdots + r_{n-2}r_{n-1}r_n &= c_3, \\ &\vdots \\ r_1r_2 \cdots r_n &= c_n. \end{aligned}$$

These are known as *Viète's equations*, after the sixteenth-century French algebraist François Viète (who is also credited with the invention of modern algebraic notation). Here each coefficient c_k is expressed as the sum of all k-fold products of the roots $r_1, r_2, \ldots, r_n$ with distinct indices. (Notice how I'm calling the coefficients c and the roots r—now that's some creative notation!)

As always, the problem is how to invert these relationships to express the r's in terms of the c's. For each degree n, the plan is to understand the various permutations of n symbols well enough to design special combinations of roots with just the right symmetry properties.

What is so interesting about this approach is that it is entirely symbolic. It doesn't even matter that the r's and c's represent numbers, per se. It's more about symbols and how various combinations of them are affected by their rearrangement.

This allows us to make a very convenient and powerful abstraction. Forget about polynomial equations and roots. Instead, let's recast the whole problem in terms of indeterminate symbols $X_1, X_2, \ldots, X_n$. (These will now assume the role formerly played by the roots.) We will thus be working in the polynomial domain $\mathbb{C}[X_1, X_2, \ldots, X_n]$ rather than the field $\mathbb{C}$ itself. We can then define a set of symmetric polynomials by analogy with Viète's equations:

$$\begin{aligned} s_1 &= X_1 + X_2 + \cdots + X_n, \\ s_2 &= X_1X_2 + X_1X_3 + \cdots + X_{n-1}X_n, \\ &\vdots \\ s_n &= X_1X_2 \cdots X_n. \end{aligned}$$

The polynomials $s_1, s_2, \ldots, s_n$ are known as the *elementary symmetric polynomials* (or ESPs) in n indeterminates. (These will take the place of our coefficients.) We have already met the ESPs in the $n = 2$ case, namely $X + Y$ and XY. When $n = 3$, we have the ESPs $X + Y + Z$, $XY + YZ + ZX$, and XYZ. (I usually prefer to use X, Y, and Z for $n \leq 3$, but this ceases to be an option for large n.) The ESPs in four indeterminates are:

$$\begin{aligned} s_1 &= X_1 + X_2 + X_3 + X_4, \\ s_2 &= X_1X_2 + X_1X_3 + X_1X_4 + X_2X_3 + X_2X_4 + X_3X_4, \\ s_2 &= X_1X_2X_3 + X_1X_2X_4 + X_1X_3X_4 + X_2X_3X_4, \\ s_4 &= X_1X_2X_3X_4. \end{aligned}$$

In general, the degree of the polynomial s_k is k, so we have one ESP for each degree, 1 through n.

Of course, none of this changes the fundamental problem. We still need to design clever combinations and understand their various levels of symmetry, but at least now it's a purely symbolic question, unaffected by whatever values some numbers may happen to have. We're no longer trying to express the roots in terms of the coefficients; instead, we're trying to express a set of indeterminates in terms of their ESPs.

The first important result in this direction is known as the *fundamental theorem of symmetric polynomials*: every symmetric polynomial is a polynomial combination of ESPs. To be precise, if p is a polynomial in $\mathbb{C}[X_1, X_2, \ldots, X_n]$ that is invariant under every permutation of the symbols $X_1, X_2, \ldots, X_n$, then p can be written as a polynomial combination of the elementary symmetric polynomials $s_1, s_2, \ldots, s_n$. Naturally, any combination of ESPs must be symmetrical; the news here is that this is the *only* way that symmetric polynomials can arise. For example, the symmetric polynomial $X^2 + Y^2 + Z^2$ in three indeterminates can be rewritten as $(X + Y + Z)^2 - 2(XY + YZ + ZX)$, so it is equal to the polynomial combination $s_1^2 - 2s_2$.

How can we write $X^3 + Y^3 + Z^3$ as a combination of ESPs?

Thus, the collection of all symmetric polynomials in $\mathbb{C}[X_1, X_2, \ldots, X_n]$ is simply the domain $\mathbb{C}[s_1, s_2, \ldots, s_n]$ generated by the ESPs. We can extend this result to polynomial fractions as well. A polynomial fraction $\frac{p}{q}$, where p and q are polynomials in n indeterminates, is also called a *rational function*—a term I have always despised. The word *rational* rightly refers to ratio and proportion, just as with the rational numbers $\mathbb{Q}$, but the word *function* is, in my view, both inaccurate and misleading. (Unfortunately, the nomenclature is fairly entrenched, so we're probably stuck with it.) The field $\mathbb{C}(X_1, X_2, \ldots, X_n)$ is typically referred to as the *rational function field* in n indeterminates. The set of symmetrical elements of this field is exactly the subfield $\mathbb{C}(s_1, s_2, \ldots, s_n)$ consisting of the rational functions in the ESPs. It actually turns out to be a bit simpler to work in these field environments (as opposed to the polynomial domains) because we then have the freedom to divide.

Our goal is to express the indeterminates $X_1, X_2, \ldots, X_n$ in terms of the ESPs $s_1, s_2, \ldots, s_n$. If we can do this, the resulting expression will then constitute an explicit formula for the solution of the general polynomial equation of degree n. So the plan is to mimic what we did in the quadratic and cubic cases. We'll attempt to construct a sequence of subfields of $\mathbb{C}(X_1, X_2, \ldots, X_n)$ with varying levels of symmetry, each one generated by a cleverly designed polynomial combination, to be determined.

Since the ESPs are playing the role of our coefficients, we should regard the field $\mathbb{C}(s_1, s_2, \ldots, s_n)$ of symmetric rational functions as our starting point, or *base field*. These are the objects we have, whereas the X's are the objects we want. In this way, it is natural to regard the full rational function field as an extension of the symmetric field. The hope is that we can chain our way up by building an *extension tower* consisting of a sequence of field extensions, each one a bit less symmetric than the previous, until we arrive at the fully asymmetric field at the top. We want these to be *radical* extensions, meaning that at each stage they are obtained by adjoining a square root (or a cube or higher root) of an element from below. This is why we seek special combinations whose powers have increased symmetry, thus placing them on a lower level of the tower.

Our quadratic results can now be restated in this setting. We consider the rational function field in two indeterminates, $\mathbb{C}(X, Y)$. The ESPs are $s_1 = X + Y$ and $s_2 = XY$, and the symmetric subfield is $\mathbb{C}(s_1, s_2)$. The extension $\mathbb{C}(X, Y)$ is generated by the single element $X - Y$, whose square lies in the base field and is equal to $s_1^2 - 4s_2$. Thus, the field $\mathbb{C}(X, Y)$ is obtained from $\mathbb{C}(s_1, s_2)$ by the adjunction of a square root of $s_1^2 - 4s_2$, and we obtain the usual quadratic formula in the generic form $X, Y = \frac{1}{2}(s_1 \pm \sqrt{s_1^2 - 4s_2})$.

The cubic formula can be understood in a similar way, only now there are more types and levels of symmetry available. In this case, we obtain a tower of field extensions with $\mathbb{C}(X, Y, Z)$ at the top, the symmetric subfield $\mathbb{C}(s_1, s_2, s_3)$ at the bottom, and the roll-symmetric subfield in the middle. This extension is generated by the roll-symmetric polynomial $\delta = (X - Y)(Y - Z)(Z - X)$, whose square is fully symmetric. The middle extension can thus be obtained from the base field by adjoining $\delta = \sqrt{D}$, where $D = \delta^2$ lies in the base field.

$$\begin{array}{c} \mathbb{C}(X,Y,Z) \\ | \\ \mathbb{C}(s_1,s_2,s_3,\delta) \\ | \\ \mathbb{C}(s_1, s_2, s_3) \end{array}$$

To get from the roll-symmetric subfield $\mathbb{C}(s_1, s_2, s_3, \delta)$ to the top field $\mathbb{C}(X, Y, Z)$, we need only throw in $\eta = X + \omega Y + \omega^2 Z$ (or η', if you prefer; they generate the same field). The number η^3 is roll-invariant and hence already downstairs in the middle field of the tower. The top field is thus obtained by the adjunction of a cube root of this number. All told, we can get from the symmetric subfield $\mathbb{C}(s_1, s_2, s_3)$ to the complete field $\mathbb{C}(X, Y, Z)$ by first making a radical extension of degree 2 (that is, throwing in a square root), followed by a radical extension of degree 3 (adjoining a cube root). This is the meaning of the cubic formula from the symmetry point of view.

When it comes to four or more indeterminates (corresponding to quartic or higher-degree equations), we will need to deal with quite a large number of potential symmetries and symmetry types. With n symbols there are exactly $1 \times 2 \times 3 \times \cdots \times n$ permutations. A consecutive product like this is also known as *n factorial* and is written $n!$ with an exclamation point.

Why are there n! permutations of n objects?

Even in the quartic case, there are already $4! = 24$ symmetries to worry about, and when $n = 5$ we have 120. (I wanted to add an exclamation point for emphasis, but you see the trouble.)

The simplest way to work with permutations is to employ *cycle notation*. Suppose we wish to describe a given permutation of five symbols, for instance:

$$
\begin{aligned}
1 &\to 4 \\
2 &\to 1 \\
3 &\to 5 \\
4 &\to 2 \\
5 &\to 3
\end{aligned}
$$

If we choose a particular element, say 1, then we can list the chain of replacement: 1 becomes 4, which becomes 2, which becomes 1. So the symbols 1, 2, and 4 engage in a 3-cycle that we can denote by (142), telling us who goes to whom, reading left to right and then cycling back. At the same time, we see that 3 becomes 5, which becomes 3, so we have the 2-cycle (35) as well. That means the complete permutation can be encoded as (142)(35). In this way, any permutation of any number of symbols can be succinctly described in one line of type.

I like to think of a permutation like (142)(35) as a machine that takes one symbol as input and produces one symbol as output. I imagine the symbol 4 entering the machine on the left, being converted to 2 by the first cycle, then being unaffected by the second. If 5 walks in, the first cycle does nothing to it, but the second turns it into a 3 and then spits that out. Permutations are activities, and we can perform any number of such activities in succession.

Suppose we have two permutations of five objects, such as $\sigma = (142)(35)$ and $\tau = (1325)$. (I will stick to using lowercase Greek letters for names of permutations. These are sigma and tau.) Notice that in this case the permutation τ *fixes* the symbol 4, meaning it is left unchanged. We could also write $\tau = (1325)(4)$, but it is simpler (and more customary) to simply omit any symbols that remain fixed. Now we can combine these two activities by first performing one and then the other. Let's write $\sigma\tau$ for the permutation achieved by first doing σ and then doing τ. (If we were interested in doing τ first, we would write $\tau\sigma$.) To understand the combined effect, we can send each symbol through the gauntlet and see what happens to it. For instance, if we send the symbol 4 though σ it becomes 2, and then τ turns it into 5. Thus, $\sigma\tau$ turns 4 into 5. Sending each symbol through in turn, we obtain

$$\sigma\tau = (142)(35) \cdot (1325) = (14523).$$

What is the cycle notation for $\tau\sigma$?

We now have a simple and efficient way to calculate the effects of permutations. In a sense, we have a new arithmetic system with its own idea of multiplication: the product of two permutations is their combined effect. Since actions are always associative, this aspect of multiplication is the same. The new wrinkle is that the product is generally *noncommutative*— $\tau\sigma$ is not the same as $\sigma\tau$, so we have to be extra careful about the order in which we write our symbols. Since we are viewing the inputs as coming in from the left, $\sigma\tau$ means first performing σ and *then* doing τ.

One nice feature of cycle notation is that it makes it almost effortless to produce the inverse of a given permutation: just write the cycles in reverse order! The inverse of $\sigma = (142)(35)$ would thus be $(53)(241)$. By analogy with numbers, we can call this σ^{-1}, the multiplicative inverse of σ. I also tend to use the symbol 1 to indicate the trivial or do-nothing permutation. In this way, we have $\sigma \cdot \sigma^{-1} = \sigma^{-1} \cdot \sigma = 1$, and our notation is consistent with our previous usage and expectations.

There is one slightly obnoxious aspect of cycle notation, and that is the built-in redundancy: the cycle (123) is the same as (231) and (312). So we need to read such things cyclically and not put any emphasis on where they happen to start and end. In particular, a transposition (ab) is the same as its inverse (ba). I always try to imagine the symbols of a cycle as being arranged in a circle—and possibly even spinning around—so that I don't get too attached to any specific starting point.

Another important feature of permutations is that disjoint cycles always commute. Since (142) and (35) have no symbols in common, it makes no difference whether we write (142)(35) or (35)(142), the effect on every symbol will be the same. Note also that the permutation (142)(35) is indeed the product of (142) and (35), so our notation is simple and consistent. The only things we need to remember are the redundancy and the noncommutativity. In particular, if σ and τ are any two permutations, then $(\sigma\tau)^{-1}$ is the opposite of doing σ then τ, so it must be $\tau^{-1}\sigma^{-1}$—that is, first we undo τ and then we undo σ, just as with socks and shoes.

The reason we're interested in permutations of symbols is that we intend to permute the indeterminates in our polynomial expressions. Given a polynomial p and a permutation σ, let's write p^σ for the effect of σ on p. (I know this notation conflicts with our usual use of exponents, but since permutations are not numbers, there should be no confusion.) Thus, if $p = X^2 + 3YZ - 2Y$ and $\sigma = (XY)$, then we have $p^\sigma = Y^2 + 3XZ - 2X$. Applying the roll permutation $\tau = (XYZ)$ to p gives us $p^\tau = Y^2 + 3ZX - 2Z$.

We can simplify our notation a bit by ordering our indeterminates as $X_1, X_2, \ldots, X_n$ from the start, and then using the symbol 3 to refer to X_3. In this way, we can write (132) as shorthand for the more cumbersome

$(X_1X_3X_2)$. Thus, if $\sigma = (132)$ and $q = X_1^3X_2 - X_3^2$, then $q^\sigma = X_3^3X_1 - X_2^2$. Notice that if σ and τ are any two permutations and p is a polynomial, then $(p^\sigma)^\tau = p^{\sigma\tau}$, so our notation nicely reflects the associativity of actions.

Now we have an interplay between two very distinct algebraic structures: the domain $\mathbb{C}[X_1, X_2, \ldots, X_n]$ of polynomials in n indeterminates and the system S_n of permutations of n symbols. (The name S_n stands for "symmetries of n objects.") The system S_3 then *acts* on $\mathbb{C}[X_1, X_2, X_3]$, moving polynomials around in various ways depending on their symmetry. Thus, the elementary symmetric polynomial $X_1 + X_2 + X_3$ remains fixed under the action of S_3, whereas X_1X_2 can become X_2X_3 or X_3X_1. Notice that X_1X_2 is fixed by the transposition (12) since multiplication of polynomials is commutative.

The most important feature of the way permutations act on polynomials is that the algebraic structure is preserved. Polynomials form a domain, which means they have a notion of addition and multiplication. Fortunately, permutations do not interfere with these operations: if p and q are polynomials and σ is any permutation, then $(p + q)^\sigma = p^\sigma + q^\sigma$ and $(pq)^\sigma = p^\sigma q^\sigma$. So we not only have an action, we have a *compatible* action—one that preserves the global algebraic structure of the polynomial domain.

This idea is capable of massive, far-reaching generalization. As an illustration, suppose we have a square sitting in a plane.

This is a geometric structure consisting of points—namely, the points forming the perimeter of the square. The relevant structural data are the various distances between each pair of points. (Other geometric information, such as angles and area, can be derived from this.)

What are the symmetries of this structure? In other words, what are the permutations of the points of a square that preserve all of the distance information? We can flip the square upside down (exchanging top and bottom sides) or rotate it a quarter turn. These actions move the points around but leave all the pairwise distances intact. In other words, they are *rigid motions*. We aren't bending or deforming anything; we're just lifting the square off the plane, moving it around in space, and then putting it back. It turns out there are exactly eight symmetries of a square (including doing nothing). These form a closed system of invertible activities.

What are the eight symmetries of a square?

A symmetry combined with a symmetry is always a symmetry. This means that symmetries—that is, structure-preserving permutations—behave in many ways like numbers, in that we can combine and invert them. (The fact that they do not commute is what we have to get used to.)

In this way, every structure in mathematics—collections, shapes, number systems, polynomial domains, geometries, you name it—has an associated *symmetry group* attached to it: the set of all permutations of the elements of the structure that preserve the relevant information. The symmetry group of a structure is in some sense a measure of its beauty—the more symmetries, the more symmetrical. A square has eight symmetries, but a circle has infinitely many—including a continuum of rotations.

What are the symmetries of a rectangle?

This gives us an alternative way to think about S_3, the permutations of three objects. Let's imagine an equilateral triangle in the plane, with vertices labeled 1, 2, and 3:

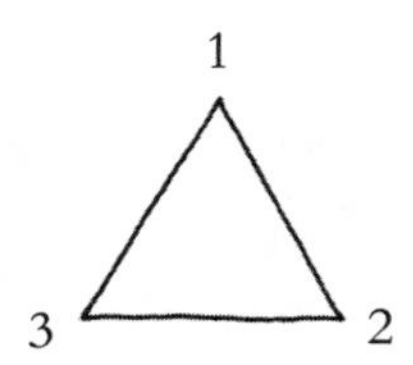

What are the symmetries of this triangle? We can leave it alone, of course. We can also flip it over, so that 1 stays put while 2 and 3 change places. This corresponds to the permutation (23). If we rotate the triangle one-third of a turn clockwise, we obtain the 3-cycle (123).

The symmetry group S_3 can therefore be viewed as the symmetry group of an equilateral triangle. The three swaps become flips, and the rolls are (naturally enough) rotations. The point being that a geometric symmetry of the equilateral triangle is completely determined by its effect on the vertices, so it can be encoded by an element of S_3. An equilateral triangle and a collection of three objects both have six symmetries, so they are equally pretty—at least in this rather simplistic sense.

But the correspondence goes much deeper than that. Not only are we matching rigid motions with permutations, we are doing so in a coherent way that respects the manner in which symmetries combine. The two symmetry groups are not just the same size but are in fact *isomorphic* —that is, there is a multiplication-preserving matching between them. Each symmetry of the triangle corresponds to a permutation of the symbols, and the way that the geometric transformations combine exactly mirrors the way the permutations combine. So not only do these two structures have the same number of symmetries, they have the exact same symmetry *style*.

We could also imagine the triangle as being two sided, with one face painted red and the other blue. The rolls are then the color-preserving symmetries, and the flips change the color. In particular, the two rolls and the do-nothing constitute the symmetry group of the colored, or *oriented* triangle, where we wish to preserve color as well as shape. If we choose to decorate the triangle in a different way, we will obtain a different group of symmetries. For example, if we color the top vertex red and the other two blue, then the symmetry group of this painted triangle would consist only of the identity 1 and the transposition (23).

Show that a regular n-gon has exactly 2n symmetries.
What are they?

The symmetry group of a given structure therefore depends on exactly which structural features we wish to preserve. The more we wish to preserve, the more demands we place on our symmetries. A decorated sphere will thus have fewer symmetries than a plain one. This means that symmetry is a somewhat abstract and relative notion. How symmetrical something is depends on which aspects of it you wish to take into account.

If we like, we can go even further and attempt to capture the notion of symmetry group axiomatically, just as we did with the field concept. In that case, we were able to dispense with the idea of quantity altogether and to reimagine numbers as abstract entities with patterned behavior. We can do the same thing with the notion of symmetry group. Since the symmetries of a structure form a closed system of invertible activities, we can try to capture this idea in the abstract.

Being activities, our multiplication operation is naturally associative: if ρ, σ, and τ are symmetries, then $\rho\sigma\tau$ is the result of performing ρ, then σ, and then τ, regardless of whether you view it as $(\rho\sigma)\tau$ or $\rho(\sigma\tau)$. We also have an obvious identity element (i.e., doing nothing), and every permutation has an inverse. So the plan is to define an *abstract group* to be any system of invertible entities closed under an associative operation. These need not be activities as such, and the operation need not be concatenation of actions. We are generalizing a concept here, replacing our original notion of activities being performed in sequence with abstract entities being combined abstractly.

To be precise, a group G is a set of elements endowed with a binary operation (which we will call $\cdot$) satisfying the following postulates:

- For all a, b, c in G, we have $(a \cdot b) \cdot c = a \cdot (b \cdot c)$.
- There is an element 1 so that for all a, we have $1 \cdot a = a \cdot 1 = a$.
- For every element a, there is an a^{-1} so that $a^{-1} \cdot a = a \cdot a^{-1} = 1$.

Notice how the lack of commutativity requires us to stipulate that 1 is a *two-sided* identity and that a^{-1} is a two-sided inverse. We cannot simply demand that $1 \cdot a = a$ and $a^{-1} \cdot a = 1$ because that tells us nothing about $a \cdot 1$ and $a \cdot a^{-1}$. We're trying to capture the notion of symmetry group abstractly, and two-sided identities and inverses are essential features of the way symmetries combine. (In particular, doing nothing can certainly be done in either order!)

In many ways these demands are similar to our postulates for a field, only we have a single operation instead of two, and it may fail to be commutative. In fact, our field postulates essentially say that a field is a commutative group under addition whose nonzero elements form a commutative group under multiplication. It is the distributive property that binds these two group structures together. Similarly, the units of a domain also form a multiplicative group. The integers $\mathbb{Z}$ form an additive group, but the multiplication fails to be invertible. The units in this case are 1 and -1, and together they form a multiplicative group of two elements.

One important unifying aspect of the modern approach is to view all of our mathematical structures through the lens of symmetry. Every structural environment has its own symmetry group, and when two structures give rise to the same group (in the sense of isomorphism), we feel them to be connected or related. In the same way that we classify plants and animals by their shared features—fused wings or dicotyledonous seeds—we can classify mathematical structures by their symmetry groups. This has the effect of inverting the relationship between objects and their symmetries. Rather than viewing the structure itself as the object of interest—which then happens to have a certain type of symmetry—we instead put the group at the center and ask, what are all the structures with that style of symmetry?

For instance, we can consider the abstract group C_2 consisting of two elements 1 and a, with $a \cdot a = 1$. We have run into this group several times: it is the additive group $\mathbb{F}_2$ as well as the multiplicative group of units in $\mathbb{Z}$. (Just as $1 + 1 = 0$ in $\mathbb{F}_2$, we have $(-1)(-1) = 1$ in the unit group of $\mathbb{Z}$.) Which structures have this style of symmetry? For a structure to have its symmetry group in the C_2 family, it would mean having a unique (nonidentity) symmetry that is self-inverse. One example would be a bilaterally symmetric shape such as the outline of a vase:

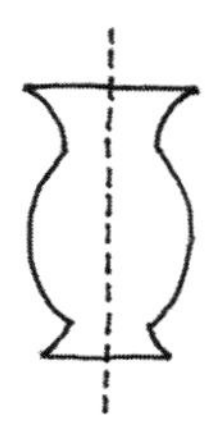

In this case, there are only two activities we can perform that will leave this shape looking the same: do nothing, or flip it over, exchanging the left and right sides.

A somewhat different example would be a Z shape. Here the only symmetry other than the identity is a half-turn rotation. This is also a self-inverse activity, so the symmetry group is again isomorphic to C_2. Now, I'm not saying that Z shapes look like vases, or that reflection is the same as rotation. Rather, it's that these two objects have the same type of symmetry in the abstract, and this can be viewed as a high-level classification. (No one is saying that pigs and peacocks are the same either, but they are both vertebrates.)

Show that the group of spatial rotations of a cube is isomorphic to S_4 acting on the four long diagonals.

This idea of using abstract groups as a classification tool has also been adopted by physicists, chemists, and crystallographers. The symmetry group of a mathematical or physical system can provide valuable guidance as to what constructions and approaches are natural to that environment. For example, the group of rigid motions in the plane controls the possible symmetric wallpaper patterns and mosaic designs and allows us to sort them into symmetry families.

Groups are the simplest and most universal of all algebraic structures and now pervade every corner of mathematics. How funny that they arrived so late in the game! Of course, people have been permuting objects and making symmetrical designs from time immemorial, but the mathematical capture of symmetry by abstract groups did not occur until well into the nineteenth century.

Returning to our problem, the point is that the action of the group S_n on the rational function field $\mathbb{C}(X_1, X_2, \ldots, X_n)$ is structure preserving: if f and g are rational functions (i.e., polynomial fractions) and σ is any permutation, we have

$$\begin{aligned}(f+g)^\sigma &= f^\sigma + g^\sigma,\\ (fg)^\sigma &= f^\sigma g^\sigma.\end{aligned}$$

A permutation σ is thus a field symmetry, moving polynomials and rational functions around while maintaining all of their additive and multiplicative relationships. The symmetrical subfield $\mathbb{C}(s_1, s_2, \ldots, s_n)$ can then be seen as the set of fixed points of this action. These are the rational functions that are unaffected by the symmetry group. In this way, they behave like the center of a circular disk—the rotation group moves the points around, but the center stays put.

Our plan is to construct a sequence of extension fields, starting with the base field $\mathbb{C}(s_1, s_2, \ldots, s_n)$ and working our way up to $\mathbb{C}(X_1, X_2, \ldots, X_n)$. At each stage, we will lose some symmetry, meaning that the relevant symmetry group will be smaller—each new extension will be invariant under fewer permutations than the previous. The symmetry group of each extension will thus be a *subgroup* of S_n—that is, a subset of the permutations that forms a closed system under multiplication and inversion, thereby forming a group of its own. We can also turn this around and say that for each subgroup H of S_n we have a corresponding *subextension*—namely, the subfield of rational functions fixed by H. For example, in the cubic case, we saw that the subgroup of roll permutations has a fixed field generated over $\mathbb{C}(s_1, s_2, s_3)$ by the polynomial δ.

Thus, rather than chaining our way up a tower of field extensions, we can instead chain our way *down* through the various subgroups of S_n until we arrive at the smallest subgroup, consisting only of the identity element 1 itself. This corresponds to the full rational function field, since everybody is fixed by 1. So we have a simple and elegant correspondence between the subgroups of S_n and the subextensions of $\mathbb{C}(X_1, X_2, \ldots, X_n)$ over $\mathbb{C}(s_1, s_2, \ldots, s_n)$. Each subgroup has its fixed field, and every field extension has its symmetry group. The larger the field, the smaller the group and vice versa.

This means we can put symmetry groups at the center of our inquiry. Rather than finding clever polynomial combinations and then seeing how symmetrical they happen to be, we can instead use the symmetry groups to manufacture the field extensions directly. The problem of solving polynomial equations by radicals then becomes the problem of finding subgroups of S_n with the right properties. In the case of quadratic and cubic equations, we succeeded fairly easily because S_2 and S_3 are very small groups and it is not hard to understand all of their subgroups.

To obtain a quartic formula, we'll need to examine the various permutations in S_4 and see what subgroups they form. If we're lucky, we will find a chain of subgroups from S_4 down to 1 whose fixed fields form a *radical tower*—a tower of field extensions each of which is obtained from the previous by adjoining a square, cube, or higher root.

Before we get down to business, I want to introduce some convenient notation. We will be dealing with various groups and subgroups, fields and field extensions, so we could use some handy symbols to convey these relationships. If G is a group, I'll write $H \leq G$ to indicate that H is a subgroup of G. Again, this means that H is not just any old subset of G, but also a group in its own right. If H is a *proper* subgroup (meaning not all of G), then I'll write $H < G$. (I am co-opting the "less than" symbol because groups are not numbers, so there is no risk of confusion. I also think it makes good sense.)

Likewise, if F is a field and K an extension field of F, then we can write $F \leq K$ (or $F < K$) to say that F is a subfield (or proper subfield) of K. This is also written K/F to indicate that K extends or "lies above" F. (I usually pronounce K/F as "K over F.")

Finally, if K is a field and G is a group of field symmetries acting on K, then we'll write K^G for the fixed field—the subfield of K consisting of those elements that are fixed by the action of G. (I know we are once again abusing our exponent notation, but it's quite standard in this context.) For example, the fully symmetric subfield $\mathbb{C}(s_1, s_2, s_3)$ is the fixed field $\mathbb{C}(X_1, X_2, X_3)^{S_3}$.

Now let's try to understand the subgroups of S_4 and the extension fields they generate. To ease notation, let's write K for the field $\mathbb{C}(X_1, X_2, X_3, X_4)$ and F for $\mathbb{C}(s_1, s_2, s_3, s_4)$. (The letter K stands for *Körper*, the German word for field, and of course F is the English version.) Then S_4 acts on the field K, and F is the fixed field K^{S_4}. For each subgroup H of S_4, we then have a corresponding fixed field $E = K^H$. The picture looks like this:

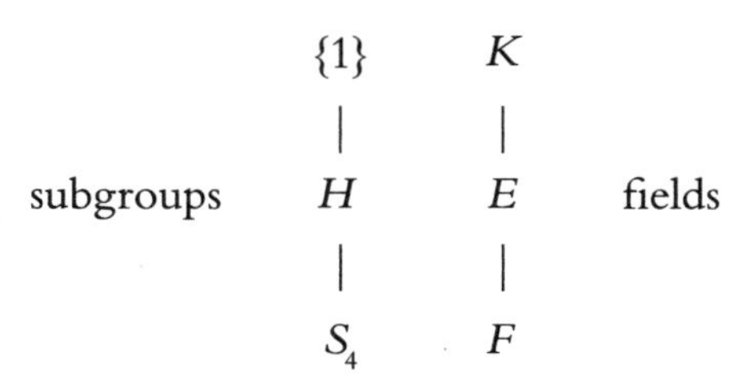

On the left we have the subgroups of S_4, with the trivial subgroup $\{1\}$ at the top and the full group S_4 at the bottom. Then H will run through every possible subgroup in between. On the right we have the corresponding fixed fields, with the rational function field K at the top (corresponding to no symmetry other than the identity) and the fully symmetric subfield F at the bottom. The field $E = K^H$ will then run through the various fixed fields, which are subextensions of K/F (i.e., subfields of K containing F). We can then write $K/E/F$ and $\{1\} \leq H \leq S_4$.

The first thing to do is to list all the elements of S_4. There are twenty-four permutations total, and we can arrange them into families based on their cycle decompositions:

- One identity: 1
- Six transpositions: (12), (13), (14), (23), (24), (34)
- Eight 3-cycles: (123), (132), (124), (142), (134), (143), (234), (243)
- Six 4-cycles: (1234), (1243), (1324), (1342), (1423), (1432)
- Three double-swaps: (12)(34), (13)(24), (14)(23)

What then are the subgroups of S_4? A subgroup is a closed system, so whatever elements we choose to include, we'll have to take all of their various products and inverses as well. For example, if a subgroup contains both (1234) and (213), then it must also contain the products (1234)(213) = (34) and (213)(1234) = (14). These will produce further products (as well as their inverses), and if we are not careful, our subgroup will balloon to all of S_4. We need to find the self-contained symmetry groups hiding inside this larger group of symmetries.

In the quadratic and cubic cases, we managed to construct a polynomial δ with the special property that it negates under all transpositions

—namely, the product of all differences between indeterminates. In the present (quartic) case, this would be

$$\delta = (X_1 - X_2)(X_1 - X_3)(X_1 - X_4)(X_2 - X_3)(X_2 - X_4)(X_3 - X_4).$$

It is not hard to see this δ also has the required property.

Why does δ get negated under any transposition?

As anyone who has ever rearranged books on a shelf will tell you, any given permutation can always be achieved by a sequence of transpositions. For instance, the 4-cycle (1234) can also be written (12)(13)(14). Do you see why? Depending on whether the number of transpositions is even or odd, their product will either leave δ alone or negate it. This means there are two kinds of permutations: the *even* permutations that preserve δ and can be factored into an even number of transpositions, and the *odd* permutations that negate δ and require an odd number of transpositions.

Show that this continues to hold for S_n in general.

In particular, we discover that there is no way for an odd number of swaps to cancel out to nothing because δ cannot be both preserved *and* negated. (This is one of the major annoyances with fields like $\mathbb{F}_2$, where $1 = -1$; we lose *sign* as a detector.)

What are the even and odd permutations in S_4? It turns out that a k-cycle in S_n is even or odd depending on whether k is odd or even. Of course a 2-cycle is odd (being a transposition), and a 3-cycle is always even because $(abc) = (ab)(ac)$. Fortunately, the pattern continues, with odd-length cycles being even and even-length cycles being odd.

Why do cycles alternate this way?

Thus, the even permutations in S_4 consist of the identity, the eight 3-cycles, and the three double-swaps, for a total of 12. Exactly half the permutations are even and half are odd. In fact, this is always true.

Show that exactly half the elements of S_n are even.

The nice thing about the even permutations of S_n is that they form a subgroup. It's not hard to see that even times even is always even and that the inverse of an even is also even. The odd permutations do not offer this luxury. (In fact, an odd times an odd is always even.) The subgroup of even permutations of S_n is denoted A_n, and is known as the *alternating group* on n symbols.

What are the elements of the group A_3?

Just as in the cubic case, our polynomial δ is invariant under the even permutations and antisymmetric under transpositions. This means that its square is fully symmetric and must therefore live in the base field F. Let's write $D = \delta^2$ as usual. (If you are either brave or a masochist, you can work out how to express D explicitly in terms of the ESPs.) We get our first extension field $F(\delta)$ as a quadratic radical extension of F generated by $\sqrt{D}$. The symmetry group of this extension is thus the alternating group A_4.

So far, our chain of subgroups and extension fields consists of $A_4 < S_4$ and $F(\delta)/F$. The next step is to find a suitable subgroup of A_4. Here, we luck out because it turns out that the double-swaps (along with the identity) form their own little group of four elements:

$$W = \{1, (12)(34), (13)(24), (14)(23)\}.$$

Show that W is commutative and isomorphic to the symmetry group of a rectangle.

The subgroup W sits inside A_4 in a very special way: all of the other elements of A_4 are 3-cycles. To generate the fixed field of W we need to construct a polynomial that is invariant under W but not invariant under any 3-cycle. The quadratic polynomial $(X_1 + X_2)(X_3 + X_4)$ is clearly invariant under double-swaps, so it's one possible candidate. If we apply the 3-cycle (123), we obtain $(X_2 + X_3)(X_1 + X_4)$, which is also W-invariant.

Repeating the process, we next get $(X_3 + X_1)(X_2 + X_4)$, and then we return to where we started. So we have three different polynomials that are all symmetric under double-swaps and rotate through each other when we apply a 3-cycle.

Show that this is true no matter which 3-cycle we choose.

Now the clever and beautiful idea is to form the sum of all three of these polynomials but to weight them with the complex cube roots of unity. Thus, we define γ as

$$(X_1 + X_2)(X_3 + X_4) + \omega(X_2 + X_3)(X_1 + X_4) + \omega^2(X_3 + X_1)(X_2 + X_4).$$

As we have seen, the polynomial γ is already W-invariant. Now if we apply any 3-cycle, the terms rotate into each other, thus causing γ to get multiplied or divided by ω. This means that γ^3 is invariant under every 3-cycle as well as under W, so it is therefore fixed by all of A_4. Thus, γ^3 lives in the extension $F(\delta)$, and we can even write down an explicit expression for γ^3 in terms of δ and the ESPs, if we so choose.

The jump from S_4 down to A_4 produced a quadratic radical extension $F(\delta)/F$ (with δ^2 lying in the base field F), and now the drop to W yields a cubic radical extension $F(\delta, \gamma)/F(\delta)$ with γ^3 in $F(\delta)$. Our next step is to find an appropriate subgroup of W to exploit. There aren't many to choose from—in fact, the only nontrivial subgroups of W consist of the identity and one other element. Let's say we choose the subgroup

$$V = \{1, (13)(24)\}.$$

As a group in the abstract, V is naturally isomorphic to C_2.

Now we need to determine the fixed field of V. This field will presumably be an extension of $F(\delta, \gamma)$ generated by a polynomial that is V-invariant but not W-invariant. Perhaps the simplest such object would be the linear polynomial

$$\beta = X_1 - X_2 + X_3 - X_4.$$

This combination is clearly invariant under V, and the effect of the other two elements of W is to negate it. This means that β^2 must be W-invariant, so it belongs to the fixed field $K^W = F(\delta, \gamma)$. Thus, $F(\delta, \gamma, \beta)$ is a quadratic radical extension of $F(\delta, \gamma)$, and with a little perspicacity, we could even obtain an explicit description of β^2 in terms of γ, δ, and the ESPs.

The final step is the drop from V to the identity subgroup $\{1\}$, and this is most easily accomplished using the fourth roots of unity to write

$$\begin{aligned} X_1 = \tfrac{1}{4}(X_1 + X_2 + X_3 + X_4) + \tfrac{1}{4}(X_1 + iX_2 - X_3 - iX_4) \\ + \tfrac{1}{4}(X_1 - X_2 + X_3 - X_4) + \tfrac{1}{4}(X_1 - iX_2 - X_3 + iX_4), \end{aligned}$$

with similar expressions for X_2, X_3, and X_4. Here we are taking the sum of the indeterminates and then weighting the terms with the fourth roots of unity (namely, 1, i, -1, and $-i$) in a cyclic manner. As you can see, most of the terms cancel because the fourth roots of unity total to zero.

Why do the complex n^{th} roots of unity always total to zero?

The first term in parentheses is the elementary symmetric polynomial s_1, and the third term is exactly β. Let's write α and α' for the second and last expressions, respectively. Thus,

$$\begin{aligned} \alpha &= X_1 + iX_2 - X_3 - iX_4, \\ \alpha' &= X_1 - iX_2 - X_3 + iX_4. \end{aligned}$$

These two have the feature that when we perform the 4-cycle (1234), they get multiplied by a square root of -1, either i or $-i$. That means if we do it twice, they must negate. But doing (1234) twice is the same as doing (13)(24), so we find that α and α' negate under this particular double-swap.

The consequence is that α^2, $\alpha\alpha'$, and α^2 are all V-invariant. This means that the field extension $F(\delta, \gamma, \beta, \alpha)$ is quadratic over $F(\delta, \gamma, \beta)$. This field must also contain α', since the product $\alpha\alpha'$ is in the lower field. Since X_1 and the other indeterminates can be expressed as a combination of s_1, α, β, and α', we see that $F(\delta, \gamma, \beta, \alpha)$ must in fact be the entire rational function field K, and so we are done. The chain of subgroups $\{1\} < V < W < A_4 < S_4$ generates a tower of extension fields

$$F < F(\delta) < F(\delta, \gamma) < F(\delta, \gamma, \beta) < F(\delta, \gamma, \beta, \alpha) = K.$$

Once again, we were able to arrange our subgroups so that each corresponding field extension is *radical*—that is, some power of the generator is contained in the previous extension.

The quartic formula thus owes its existence to W, a very special commutative subgroup of A_4. In fact, the fixed field of W—namely, the extension field $F(\delta, \gamma)$—is precisely the field generated by the roots of the resolvent cubic in Ferrari's solution. The remaining extension to K is achieved by a pair of quadratic extensions, and thus represents the splitting of the (modified) quartic into two separate quadratic polynomials. In hindsight, we can now see the work of the sixteenth-century algebraists as a brilliant but naive journey through an underground maze, without the benefit of the light of symmetry.

So did we just get lucky, or does this approach always work? The existence of a general formula for the solution of a polynomial equation of degree n comes down to whether $\mathbb{C}(X_1, X_2, \ldots, X_n)$ is the top of a radical tower over $\mathbb{C}(s_1, s_2, \ldots, s_n)$. We were able to show this for $n \leq 4$ by finding an explicit chain of symmetry subgroups that just happened to work. The question is whether we will continue to get lucky for larger values of n. In particular, we would like to obtain, if possible, a general *quintic* formula.

The French mathematician and physicist Joseph-Louis Lagrange was the first to recognize the possibility of using symmetric polynomials to solve equations, and published a memoir on the subject in 1770. Finally, here was real hope of a new, successful attack on the general quintic equation. Unfortunately, no obvious sequence of subgroups could be found to do the trick. By the end of the century, many algebraists were beginning to feel that a general quintic formula might simply not exist. We can use symmetry groups to help us devise formulas when they do exist; the question is whether a quintic formula can be *ruled out* using these methods.

The answer, as you may have surmised, is yes: we can prove that there does not exist an explicit formula (in the language of radicals) for the solution of the general quintic equation. This was demonstrated by the Italian mathematician Paolo Ruffini, who published an essentially correct but incomplete argument in 1799. The first complete proof was provided by the Norwegian prodigy Niels Henrik Abel in 1823. The insolubility of the general quintic equation is now known as the *Abel-Ruffini theorem*. Both proofs rest on the fact that there is no analog of W in S_5. The chain gets as far as $A_5 < S_5$ and then stops. No subgroup of A_5 has the required property that its fixed field is a radical extension.

In fact, Abel goes on to show that this same property holds for all $n \geq 5$. That is, the field $\mathbb{C}(X_1, X_2, \ldots, X_n)$ can *never* be attained by a radical tower of extensions of $\mathbb{C}(s_1, s_2, \ldots, s_n)$ unless $n \leq 4$. The existence of the quadratic, cubic, and quartic formulas is therefore something of a coincidence, having to do with the unusually small number of symmetries involved. (Seen in this light, the existence of W in A_4 seems especially fortuitous.) It turns out we had no right to think that the language of radicals would be strong enough to unravel every polynomial equation, and in fact it isn't. We were lulled into optimism by our success in the cubic and quartic cases. The truth is that being contained in a radical tower is a very special demand, making such extensions relatively rare. Again, if you want something to have a particular property—such as being rational, algebraic, radical, or sayable at all—then you have to expect it to be uncommon.

The Abel-Ruffini theorem does not spell complete doom for the classical algebra project, however. Abel's result shows that there cannot exist a one-size-fits-all general formula, but that still leaves the door open for each *individual* quintic to be solvable by radicals in its own idiosyncratic way. What is needed is an extension of these methods from the context of the rational function field to *arbitrary* field extensions. Unfortunately, Abel died of tuberculosis at the age of twenty-six and did not live to see the enormous influence of his ideas.

SOLVABILITY

The study of polynomial equations and their solution by radicals was both clarified and boldly reinvented by the brilliant French algebraist Évariste Galois. Young, idealistic, and politically radical, Galois was expelled from college during the Revolution of 1830, jailed for treason, and killed in a duel at the age of twenty. (This is about as romantic as math gets, folks.) Galois is almost single-handedly responsible for the initial development of abstract group theory and even coined the term *group*.

Despite his failures as a political activist, Galois's insight and imagination sparked a revolution in algebra—a sudden and dramatic increase in altitude. His most famous achievement was to determine the precise conditions for a polynomial equation to be solvable by radicals. In particular, he was able to show that certain specific quintics—the equation $x^5 = x + 1$, for example—cannot be explicitly solved. This equation has a real solution $x \approx 1.1673$, but there is simply no way to express it in the restricted language of radicals.

This discovery really did put an end to the classical algebra project and also helped to usher in the modern era. We're not just solving equations anymore; we're studying fields and domains and groups and their interplay. This is what always happens when you climb higher. You no longer see individual plants and animals; you see *ecosystems*. The study of polynomial equations has simply been folded into the larger, more general study of fields and field extensions.

Let's begin by reminding ourselves what it means for a polynomial equation to be solvable by radicals. This means that its solutions can be explicitly written as a hierarchical combination of field operations and radicals—that is, square, cube, and higher roots. For example, the number $\sqrt[4]{5\sqrt{-3} + \sqrt[3]{2}}$ can clearly be expressed in this language. By its very construction, such a number must then be contained in a radical tower:

$$\mathbb{Q} < \mathbb{Q}(\sqrt[3]{2}) < \mathbb{Q}(\sqrt[3]{2}, \sqrt{-3}) < \mathbb{Q}(\sqrt[3]{2}, \sqrt{-3}, \sqrt[4]{5\sqrt{-3} + \sqrt[3]{2}}).$$

In other words, each successive extension is generated from the previous by adjoining a certain radical of a particular element.

Actually, we have to be a little careful here about the meaning of our symbols. As we have seen, an expression like $\sqrt[4]{5\sqrt{-3}+\sqrt[3]{2}}$ is inherently ambiguous. There are two different square roots of -3 and three different cube roots of 2. Which ones are we talking about? And precisely which fourth root are we taking? If we vary our choices, we will produce different numbers. On the other hand, these will all be roots of the same irreducible polynomial equation. So in order to be complete (as well as unambiguous), we should include them *all.* This essentially requires that we throw in all of the relevant roots of unity as well. The effect will be to make our chain of extension fields longer, but that's no problem. Since we are only concerned with whether our numbers are expressible or not, the length of the tower is immaterial.

We can thus improve our radical tower by first adjoining to $\mathbb{Q}$ all of the cube and fourth roots of unity. (The square roots of unity, 1 and -1, are already included). In other words, we can begin our chain with the extensions $\mathbb{Q} < \mathbb{Q}(\omega_3) < \mathbb{Q}(\omega_3, \omega_4)$. Here I am using ω_n to denote a *primitive* complex n^{th} root of unity—that is, one whose powers generate all the others. Thus, ω_3 is simply another name for ω (or its inverse), and we can take ω_4 to be $\pm i$. If we like, we can accomplish this in one go by making the extension $\mathbb{Q} < \mathbb{Q}(\omega_{12})$, since both ω_3 and ω_4 are powers of ω_{12}. Now, when we throw in $\sqrt[3]{2}$, it no longer matters which cube root of 2 we choose because the field $\mathbb{Q}(\omega_{12}, \sqrt[3]{2})$ contains all three.

Can you find a number n so that $\mathbb{Q}(\omega_n) = \mathbb{Q}(\omega_4, \omega_5, \omega_6)$?
What is the general pattern?

In general, given any field F, we will say that a polynomial p in $F[X]$ is *solvable* if its roots can be expressed in the language of radicals. (This is the same as saying that the equation $p(x) = 0$ can be explicitly solved.) Let's write K for the extension field of F generated by the roots of p. Such an extension is called a *splitting field* because p factors completely (or "splits") over K. To say that p is solvable is then to say that the field K is contained in a radical tower over F.

To be precise, this means we would have a sequence of extension fields starting from the base field F and culminating in a field $\tilde{F}$ at the top of the tower:

$$F = F_0 < F_1 < F_2 < \cdots < F_m = \tilde{F}.$$

Here I am indexing the extensions starting with F (which I am calling F_0) and working my way up to $\tilde{F} = F_m$, the mth extension in the sequence. Each link in this chain is then a radical extension—that is, for each index $k = 1, 2, \ldots, m$, the field F_k is obtained from the previous field F_{k-1} by adjoining a radical of the form $\sqrt[n]{a}$ for some $n \geq 2$ and some element a in F_{k-1}. Just as before, we can eliminate any ambiguity in these expressions by first adjoining all of the requisite roots of unity, so that $F_1 = F(\omega_N)$ for some (possibly quite large) integer N. Such an extension is clearly radical, since ω_N can be written as $\sqrt[N]{1}$. The advantage of doing this is that we then don't have to worry about any particular choice of meaning of $\sqrt[n]{a}$.

The question of whether a given equation can be solved by radicals can now be rephrased entirely in the language of field extensions: the polynomial p in $F[X]$ is solvable precisely when the splitting field of p is contained in a radical tower over F. We have managed to capture the linguistic notion of a number being sayable in a specific language as a precise condition on field extensions. Our problem then becomes how to tell if a given field extension is or is not contained in a radical tower.

While we're on the subject, I can't resist mentioning another beautiful and historically important question that (rather surprisingly) lives in this same neighborhood: the problem of *straightedge and compass construction*. First, let me explain what that means. Along with the idealized Euclidean plane, the classical Greek geometers also imagined idealized tools based on those used by architects and carpenters. A *straightedge* is an imaginary perfect stick that can be used to draw imaginary perfect lines. Given any two distinct points in the plane, we can use the straightedge to construct the unique line containing them both. The *compass* allows us to select one point as center and another to determine the radius, thus creating a circle.

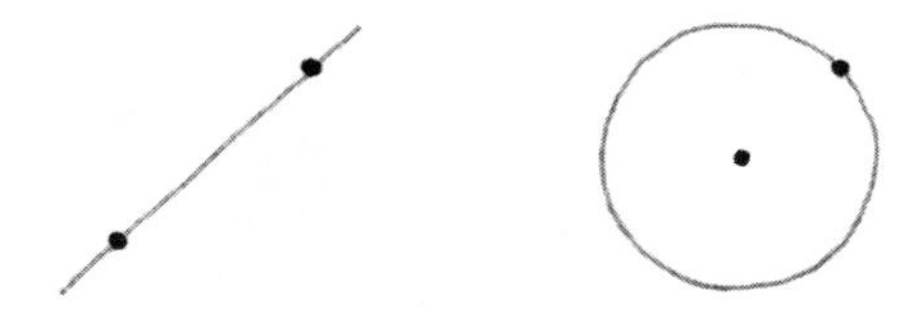

We can think of these as abstract tools, but they are really *permissions* that we allow ourselves: given two points, we get to bring into existence the infinite line containing them, and given a center and another point, we can conjure up the corresponding circle. When lines and/or circles intersect, a new point is then determined. The question is, what shapes can we make with only these tools?

Given two points, can you devise a straightedge and compass construction of their midpoint?

Here's how I like to think of it. We start with two (or more) arbitrary points in the plane. For reasons that will become clear, I want to imagine these as black, with the rest of the plane being white. To use either of our tools, we select two black points. These then determine a line or a circle, to be colored gray. Whenever gray lines and circles intersect, those intersection points then become black. (The idea is that a gray point is not yet "active" and cannot be used until it is determined by intersection.) In this way, we build up the set of black points via straightedge and compass construction. The question is then which points can be built in this way.

As an example, let's consider the classic construction of a square on a given side. Here we are given two adjacent corners of a square, and the goal is to construct the other two corners. (This essentially comes down to constructing a perpendicular to a given segment.) Clearly, it is enough to construct one of these corners since the other can be dealt with analogously. The construction proceeds in several steps:

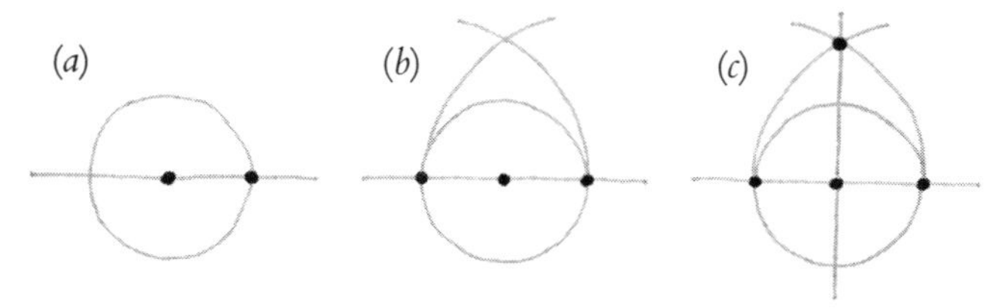

Let's begin by drawing a line through the two given points with the straightedge. Then we can use the compass to draw a circle around one of them, using the other as radius (*a*). This brings a new black point into existence on the other side of the circle. Now we have two black points symmetrically placed with respect to the center. Placing the compass on these two points (*b*), we then produce two large circles. (I have shown only a portion of these.) Since this configuration is symmetrical, their intersection point must lie directly above the center. (There is also a corresponding intersection below the center.) Finally, we use the straightedge to connect this point to the center, forming a perpendicular (*c*). The intersection of this line with the circle is then our desired point. We have shown that a square is indeed constructible, given two of its corners.

The modern way to proceed is to view our plane as the complex numbers $\mathbb{C}$, taking our two initial points to be 0 and 1. (There is no loss of generality here, since we can always shift, rotate, and scale the plane accordingly.) We can now reframe our question by asking which *numbers* are constructible. The problem of constructing a square can then be reimagined as the construction of the complex number i. The interesting translation problem is to determine what "straightedge and compass" means in the language of complex arithmetic.

Is every rational number constructible?

Contrary to what you might expect, straightedge and compass construction is not really a problem in geometry. Rather, it is a linguistic question pertaining to a certain algebraic language—the one generated by these two imaginary tools. As it happens, the straightedge and compass are powerful enough to allow us to add, subtract, multiply, and divide any two complex numbers. (For addition, this essentially comes down to constructing a parallelogram given three of its corners.) We can easily construct the complex conjugate of a given complex number, hence also its real and imaginary parts. In other words, the constructible numbers form a subfield of $\mathbb{C}$ that is closed under conjugation.

Show that if w and z are constructible, then so are $w + z$ and wz.
Show that if z is constructible, then so are $-z$, $1/z$ and $\bar{z}$.

Though the language of these tools is quite expressive, it is still somewhat limited. For example, if a, b, c, and d are complex numbers, and we draw one line through a and b and another through c and d, then their intersection point can be expressed in terms of a, b, c, and d using only field operations and complex conjugation. The consequence is that if our currently constructed points all lie in a certain conjugation-closed subfield, then using the straightedge alone will not allow us to escape.

How exactly does the intersection point depend on a, b, c, and d?

With the addition of the compass, we can do a bit better. For instance, it's not hard to show that by using both tools we can construct the square root of any complex number.

Given a complex number z, how do we construct $\sqrt{z}$?

Our tools thus allow us to perform not only the field operations and conjugation, but square root as well.

This turns out to be the limit of their expressive power, however. The points of intersection of a line and a circle are determined by a quadratic equation, so the newly determined points can always be expressed using square roots. The same goes for the intersection of two circles. The constructible numbers are therefore precisely those complex numbers that can be obtained from 0 and 1 using nothing other than the four field operations, complex conjugation, and square root.

That means we can translate the entire problem of straightedge and compass construction into the simpler language of fields and field extensions. At any given stage in such a construction, we will find ourselves with a certain set of complex numbers—namely, the points that have been constructed so far. These numbers, together with their complex conjugates, generate a field extension of our initial ground field $\mathbb{Q}$ that is closed under conjugation. As the construction proceeds, a new complex number is determined at each stage. If this number is determined using only field operations and conjugation (e.g., as the intersection of two lines), then it already lies in the current field. However, if a square root is required for its description (meaning that the compass is needed), then this number will generate a quadratic extension, and our field will be enlarged. In this way, we build up our extension field until it includes the set of points that we are trying to construct. A complex number is therefore constructible precisely when it lies in an extension field of $\mathbb{Q}$ obtained via a sequence of quadratic radical extensions.

Thus, rather than speaking of a point in the plane being constructed by means of a geometric straightedge and compass, we can instead talk about a complex number being contained in a *quadratic tower* over $\mathbb{Q}$. The number $\sqrt{3 + 2\sqrt{-2}}$ is then automatically constructible because it resides in the quadratic tower $\mathbb{Q} < \mathbb{Q}(\sqrt{-2}) < \mathbb{Q}(\sqrt{3 + 2\sqrt{-2}})$. A given problem in geometry—the construction of an equilateral triangle, say—becomes the question of whether a certain number (in this case ω_3) does or does not lie in such a quadratic tower. Here, since ω_3 is contained in the radical quadratic extension $\mathbb{Q}(\sqrt{-3})$, we are assured that such a construction must exist.

Can you design a straightedge and compass construction of an equilateral triangle?
How about a regular hexagon or octagon?

The trouble with problems like straightedge and compass construction is that it's all well and good when a construction exists—we then simply have to find it. But suppose there is no such construction. How will we ever know? It's one thing to describe an object explicitly in a given language; it's quite another to prove that it cannot be said at all.

And in fact, there are a number of straightedge and compass construction problems that the classical Greeks could not solve, the most famous being the *Delian problem*, or *duplication of the cube*: to construct a cube with exactly twice the volume of a given cube. (Legend has it that the oracle at Delos was consulted during the Athenian plague, and she advised that the altar of Apollo be doubled in size.)

Here we are given the side length of a cube, and we are asked to construct a side of the doubled cube. This basically comes down to constructing the number $\sqrt[3]{2}$, which is the same as asking if the field $\mathbb{Q}(\sqrt[3]{2})$ is contained in a quadratic tower over $\mathbb{Q}$. As we shall see, our modern understanding of fields in the abstract will allow us to prove that no such cubic extension can have this property, and the duplication of the cube is therefore impossible. In fact, it can be shown that if α is a root of an irreducible polynomial of degree n, then the field $\mathbb{Q}(\alpha)$ cannot be contained in a quadratic tower over $\mathbb{Q}$ unless n is a power of 2. Thus, by proving purely algebraic theorems about field extensions and their properties, we can rule out the existence of a straightedge and compass construction without having to draw a single line or circle. It turns out that the Delian problem is not so much about cubes as it is about cubics.

For the same reason, *trisecting the angle* (that is, dividing an arbitrary angle into thirds) is also known to be impossible. One way to see this is to use the fact that we can construct a regular hexagon, and hence an angle of 120 degrees. If a generic trisection method existed, we could then use it to obtain a 40-degree angle (i.e., one-ninth of a full turn). This is tantamount to constructing a complex ninth root of unity ω_9, which happens to satisfy the irreducible sixth-degree equation $x^6 + x^3 + 1 = 0$. Since 6 is not a power of 2, this construction is therefore impossible. As a consequence, we discover that a regular nonagon (nine sides) also cannot be constructed with straightedge and compass.

Show that ω_9 satisfies the equation $x^6 + x^3 + 1 = 0$.

The same goes for the construction of a regular heptagon (seven sides). A complex seventh root of unity ω_7 is also a root of an irreducible polynomial of degree 6, and therefore cannot be contained in a quadratic tower.

Show that ω_7 satisfies the equation
$$x^6 + x^5 + x^4 + x^3 + x^2 + x + 1 = 0.$$

The most difficult of the classical construction problems (and the last to be solved) is *squaring the circle*, the construction of a square with the same area as a given circle. This is equivalent to constructing the number π, a number that we now know to be transcendental and therefore incapable of lying in *any* algebraic extension of $\mathbb{Q}$, let alone a quadratic tower. Therefore, squaring the circle is also impossible. No wonder the Greeks had such a difficult time!

In the same way, many such "linguistic" problems throughout mathematics (e.g., the integration of elementary functions) can be transferred to the field extension context. The question of whether a particular object can be represented in a given language can be rephrased as whether a certain field lies in a certain type of extension tower. The problem of solving polynomial equations leads to radical towers, whereas straightedge and compass construction concerns quadratic towers in particular. Thus, the more we understand about fields and field extensions, the more we can say about the power of these languages.

Can you construct a regular pentagon?

I just love how these two seemingly disparate problems—solving algebraic equations and making geometric constructions—have now been unified. This once again illustrates the power of abstraction and generalization. We can now see both problems as aspects of the larger project of understanding field extensions and radical towers in general. The hope would be that our previously successful symmetry methods could somehow be adapted to this much wider, more general context. In fact, it was precisely this extension of Abel's approach that Galois achieved.

In some ways, Galois's idea is quite simple. We saw before that the permutation group S_n acts on the rational function field $\mathbb{C}(X_1, X_2, \ldots, X_n)$ in such a way that the field operations are preserved. This means we can view each element of S_n as a field symmetry of $\mathbb{C}(X_1, X_2, \ldots, X_n)$. In other words, an element of S_n acts as a permutation of the *entire field*, not just the the n indeterminate symbols. As usual, we may view this field as an extension of the field $\mathbb{C}(s_1, s_2, \ldots, s_n)$, where $s_1, s_2, \ldots, s_n$ are the elementary symmetric polynomials. Galois's breakthrough was to recognize S_n as the group of *field symmetries of* $\mathbb{C}(X_1, X_2, \ldots, X_n)$ *that fix each element of the subfield* $\mathbb{C}(s_1, s_2, \ldots, s_n)$.

Now suppose that F is any field, and K is an extension field of F. Then we can replace the permutation group S_n by the group of field symmetries of K that fix every element of F. This is known as the *Galois group* of the extension K/F and is often written $G_{K/F}$. (Happily, G can stand for both Galois and group.) The Galois group $G_{K/F}$ is not the group of *all* field symmetries of K, but rather the subgroup consisting of those symmetries that leave every element of the base field F untouched. In other words, $G_{K/F}$ is the symmetry group of K "from F's point of view." If we like, we can think of $G_{K/F}$ as the symmetry group of the *extension* K/F. This is the bold new idea, the simplest and most natural way to generalize the action of S_n on $\mathbb{C}(X_1, X_2, \ldots, X_n)$ to arbitrary field extensions.

Let's look at a simple example: the extension $\mathbb{Q}(\sqrt{2})/\mathbb{Q}$. The Galois group of this extension is the set of all field symmetries of $\mathbb{Q}(\sqrt{2})$ that fix every element of $\mathbb{Q}$. Suppose σ is such a symmetry. Then σ permutes the elements of the field $\mathbb{Q}(\sqrt{2})$ while leaving $\mathbb{Q}$ fixed. Let's continue to write x^σ for the effect of σ on x. Then for any elements x and y in $\mathbb{Q}(\sqrt{2})$, we have

$$(x + y)^\sigma = x^\sigma + y^\sigma,$$
$$(xy)^\sigma = x^\sigma y^\sigma.$$

This is what it means for σ to be a field symmetry: a permutation of the elements of a field that leaves the algebraic relationships intact, just as the symmetries of a square permute its points while preserving the geometric structure. We know that every element of $\mathbb{Q}(\sqrt{2})$ has the form $a + b\sqrt{2}$, where a and b are in $\mathbb{Q}$. The effect of σ on such an element is then $(a + b\sqrt{2})^\sigma = a^\sigma + b^\sigma(\sqrt{2})^\sigma$, which equals $a + b(\sqrt{2})^\sigma$, since σ leaves rational numbers alone. The symmetry σ is therefore completely determined by its effect on the generator $\sqrt{2}$. Once we know where σ sends $\sqrt{2}$, then we know everything.

Since σ preserves products, we have $(\sqrt{2})^\sigma \cdot (\sqrt{2})^\sigma = 2^\sigma$, which is just 2. This means $(\sqrt{2})^\sigma$ must also be a square root of 2, so either $(\sqrt{2})^\sigma = \sqrt{2}$ or $(\sqrt{2})^\sigma = -\sqrt{2}$. In the first case, σ is the identity, or do-nothing permutation. In the second case, it is the symmetry that sends $a + b\sqrt{2}$ to $a - b\sqrt{2}$. The Galois group thus consists of only two symmetries, one of which leaves $\sqrt{2}$ and $-\sqrt{2}$ alone, while the other interchanges them. So in this case, the Galois group $G_{\mathbb{Q}(\sqrt{2})/\mathbb{Q}}$ is isomorphic to S_2, the permutations of two objects. Notice that our field $\mathbb{Q}(\sqrt{2})$ is generated over $\mathbb{Q}$ by the roots of the polynomial $X^2 - 2$, and that our field symmetries then act as permutations of these roots.

In general, if p is a polynomial in $F[X]$ and K is its splitting field, then the Galois group $G_{K/F}$ must permute the roots of p. To see this, suppose that σ is an element of the Galois group and α is any element of K. Since σ preserves sums and products and also leaves the coefficients of p fixed, we must have $p(\alpha)^\sigma = p(\alpha^\sigma)$. Thus, if α is a root of p, then so is α^σ.

Since the roots of p generate the extension K/F, an element σ in $G_{K/F}$ is completely determined by its effect on these roots. That means we may regard the elements of the Galois group as finite permutations, as opposed to abstract field symmetries. If the polynomial p has degree n (and thus has exactly n roots, counted with multiplicity), then the Galois group must permute these roots among themselves, just as S_n permutes the roots $X_1, X_2, \ldots, X_n$ in the function field case.

The difference is that not every permutation of the roots necessarily occurs—that is, we can view the Galois group as a *subgroup* of the permutation group S_n, but it needn't be all of S_n. Although a generic polynomial of degree n will allow all possible permutations, a *special* polynomial may have a much smaller Galois group due to coincidental algebraic relationships among its roots. (This was not an issue in the function field case because the roots are indeterminate and therefore independent of each other.)

For example, the polynomial $p(X) = X^3 + X^2 - 2X - 1$ has degree 3, and thus has three complex roots. One might expect the Galois group of its splitting field to be S_3, allowing for any of the six possible permutations. But in this particular case, the Galois group has only three elements and consists only of the roll permutations. The reason for this unexpected loss of symmetry is that if ξ denotes one of the roots, the other two are given by $\xi^2 - 2$ and $\xi^3 - 3\xi$. This is a rather unusual state of affairs. In particular, it means that if a field symmetry fixes ξ, it must fix the other two roots as well.

Show that if ξ is a root of $X^3 + X^2 - 2X - 1$
then so are $\xi^2 - 2$ and $\xi^3 - 3\xi$.

To be somewhat more concrete—and frankly, to avoid some unpleasant complications that arise in the general case—let's assume that our base field F is either $\mathbb{Q}$ or an extension of $\mathbb{Q}$, such as $\mathbb{Q}(\sqrt{2})$ or $\mathbb{Q}(i)$. (Things get a little weird for more exotic fields like $\mathbb{F}_7$.) Now suppose p is a polynomial in $F[X]$ and that K is its splitting field over F. The important thing here is that K contains not just one root of p but *all* of them. In this context, K is said to be a *Galois* extension of F. (You know you've made it big in math when your name gets used as an adjective.)

The extension $\mathbb{Q}(\sqrt{2})/\mathbb{Q}$ is therefore Galois because it contains both roots of $X^2 - 2$. By contrast, $\mathbb{Q}(\sqrt[3]{2})/\mathbb{Q}$ is *not* a Galois extension because it does not contain the other two roots of $X^3 - 2$. That means the symmetry group of this extension reduces to the identity only and fails to provide enough permutations to allow us to apply our previous symmetry techniques. Just as in the rational function field case, we will want to permute the roots of our polynomial, which means we need them all to be present. The analogy therefore works best when the extension is Galois.

So let's suppose that K/F is a Galois extension. An element σ of the Galois group $G_{K/F}$ is called a *conjugation* of K/F. I like to think of a conjugation as a way of moving the elements of K around without F noticing. The reason we can get away with sending $a + b\sqrt{2}$ to $a - b\sqrt{2}$ is that $\mathbb{Q}$ can't tell the difference between the two roots of $X^2 - 2$. We saw this before with the extension $\mathbb{R}(\sqrt{-1})/\mathbb{R}$ and complex conjugation.

For simplicity, let me write $G = G_{K/F}$. The idea, as before, is to set up an association between symmetry groups and fields. If $H \leq G$ is a subgroup of the Galois group, then H acts on the field K, and we can again form the subfield K^H of fixed points. In other words, the field K^H—known as the *fixed field* of H—is the set of all elements of K that are fixed by every conjugation in H. This is clearly an extension field of F, since F is fixed not only by H but by all of G. Thus, to each subgroup of $G_{K/F}$ is associated a certain subextension of K/F:

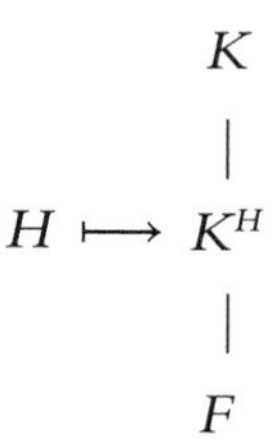

In particular, the fixed field of the full Galois group G is just the base field F, and that of the trivial group $\{1\}$ is the field K itself.

Conversely, suppose E/F is any subextension of K/F—that is, E is a subfield of K containing F. Since K is Galois over F, it must be the splitting field of some polynomial p in $F[X]$. But E is an extension of F, so we can choose to view p as a polynomial in $E[X]$ whose splitting field is K. Thus, K is also Galois over E, and the Galois group $G_{K/E}$ is then a subgroup of G. (If a symmetry of K fixes E, it certainly must fix F, since $F \leq E$.) In this way, each subfield E gives rise to a subgroup:

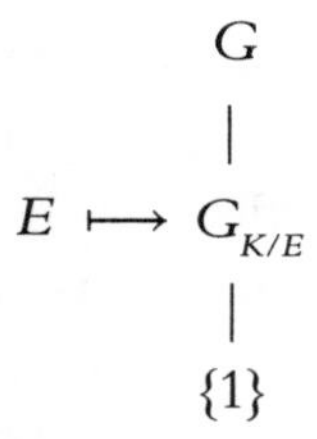

The central result of the subject—which has come to be known as *Galois theory*—is that these assignments are in fact perfect inverses of each other. Each subfield has its own unique subgroup and vice versa. This means the fixed field of the subgroup $G_{K/E}$ is just E, and the Galois group of K/K^H is exactly H. We obtain a complete matching between the subgroups of $G_{K/F}$ and the subextensions of K/F. This matching is known as the *Galois correspondence*.

$$\begin{array}{ccc} \{1\} & K & \\ | & | & E = K^H \\ H & E & H = G_{K/E} \\ | & | & \\ G & F & \end{array}$$

Just as we saw before (with the permutation subgroups acting on the rational function field), this correspondence is *inclusion reversing*—the larger the subfield, the smaller the subgroup, and vice versa. In fact, Abel's results can now be viewed as a special case of the Galois correspondence: $\mathbb{C}(X_1, X_2, \ldots, X_n)$ is simply a Galois extension of $\mathbb{C}(s_1, s_2, \ldots, s_n)$ with Galois group S_n.

The Galois correspondence is the kind of discovery that a modern mathematician dreams of making—an unforeseen reciprocal relationship between two very different types of abstract structures: groups and fields. Here, the symmetry groups are the actors, moving numbers around in patterned ways; the fields provide the stage for them to act upon. Each element of the Galois group knows who it fixes, and each field element knows who fixes it. We obtain a beautiful, perfect matching between subgroups and subfields, but *only* if the extension is Galois to begin with.

Let's take a look at a concrete example: the field extension $\mathbb{Q}(\sqrt{2}, \sqrt{3})$ over $\mathbb{Q}$. This is a Galois extension, being the splitting field of the quartic polynomial $(X^2 - 2)(X^2 - 3)$ over the rationals. Every field symmetry is then determined by its effect on the two generators $\sqrt{2}$ and $\sqrt{3}$, which are either fixed or negated. The Galois group thus consists of four symmetries: 1, the identity conjugation that fixes both $\sqrt{2}$ and $\sqrt{3}$; the conjugation θ that fixes $\sqrt{2}$ and negates $\sqrt{3}$; the conjugation φ that fixes $\sqrt{3}$ and negates $\sqrt{2}$; and $\theta\varphi$ that negates them both. Notice that these symmetries commute and that each of them is self-inverse. The Galois group of $\mathbb{Q}(\sqrt{2}, \sqrt{3})/\mathbb{Q}$ is therefore isomorphic to our old friend *W*, the group of double-swaps of four objects (or the symmetries of a rectangle, if you prefer).

As we have seen, this group has exactly three subgroups other than the identity and the group itself, and each of these consists of two elements. These subgroups are $\{1, \theta\}$, $\{1, \varphi\}$, and $\{1, \theta\varphi\}$. The fixed field of θ is clearly $\mathbb{Q}(\sqrt{2})$, and that of φ is $\mathbb{Q}(\sqrt{3})$. Which elements are fixed by $\theta\varphi$? This conjugation negates both $\sqrt{2}$ and $\sqrt{3}$, which means it must then fix their product $\sqrt{2} \cdot \sqrt{3} = \sqrt{6}$. In fact, it's not hard to show that every element of $\mathbb{Q}(\sqrt{2}, \sqrt{3})$ has the form $a + b\sqrt{2} + c\sqrt{3} + d\sqrt{6}$ for some a, b, c, d in $\mathbb{Q}$. It follows that the fixed field of $\theta\varphi$ is exactly $\mathbb{Q}(\sqrt{6})$. Thus, the complete system of subextensions of $\mathbb{Q}(\sqrt{2}, \sqrt{3})/\mathbb{Q}$ is given by the diagram

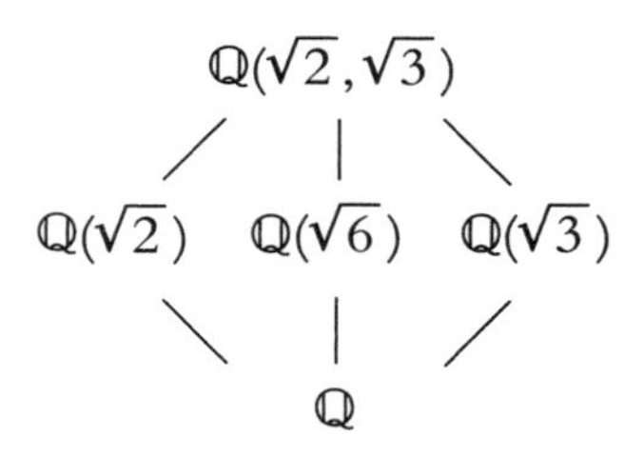

In this case, all three of our subfields are Galois extensions of $\mathbb{Q}$, and they are all radical quadratic extensions as well.

Here we see one of the most important features of the Galois correspondence. The field $\mathbb{Q}(\sqrt{2}, \sqrt{3})$ is infinite, as are all of its subfields.

On the face of it, there's nothing stopping us from forming infinitely many subfields $\mathbb{Q}(\alpha)$ by throwing in whatever element α we like, such as the field $\mathbb{Q}(\sqrt{2} + \sqrt{3})$. However, since each subfield corresponds to a subgroup of the Galois group—which is *finite*—there can be only finitely many intermediate subfields, and in this case, only three.

Show that $\mathbb{Q}(\sqrt{2}+\sqrt{3})$ *is in fact* $\mathbb{Q}(\sqrt{2}, \sqrt{3})$.

So apart from any other conveniences, the main value of the Galois correspondence is that it reduces a potentially infinite problem to a finite one.

Another reason the Galois correspondence is so useful and important is that it allows us to transfer information. Assuming that K/F is Galois (so that the correspondence holds), all of the information regarding this extension and its intermediate subfields is then contained in the Galois group $G_{K/F}$. The question is how to tease it out. We need to understand how the various properties of field extensions are reflected in the corresponding Galois groups. In particular, we want to understand how the Galois group controls whether an extension is radical or contained in a radical tower.

Before we get into these details (which are admittedly rather technical), I want to take a minute to appreciate the larger strategic picture. Our original, naive questions (solving polynomial equations by radicals or producing straightedge and compass constructions) were naturally phrased in terms of certain entities (numbers or points in the plane) and their description within a fixed linguistic framework. In each of these cases, we managed to reframe the problem in the language of fields and field extensions. This means we are shifting our primary focus from numbers themselves to systems of numbers. In other words, we are making a level jump: our individuals used to be numbers within a fixed field environment; now they are the fields themselves. We are asking questions about relative relationships *among* fields—who is contained in whom and how—just as we are used to asking about numbers and their relationships.

And now, having translated our problems into this new, more abstract setting, we are once again shifting our emphasis from the fields themselves to their symmetry groups. With each of these moves—from numbers to fields, from fields to groups—we gain altitude and power, simplicity and elegance. What we lose is concreteness and tangibility and, consequently, intuition. That means we have to work harder to make these new structures more familiar and concrete. This is essentially what it means to live a mathematical life—stretching your mind to reach for the next higher abstraction and doing whatever it takes to gain intuition and understanding so that you can continue to think and create and make progress. (If that sounds exhausting, believe me, it is.)

Our first project is to understand what a radical tower looks like from the symmetry group perspective. Suppose $\tilde{F}/F$ is a field extension arising from a radical tower. Then we have a sequence

$$F = F_0 < F_1 < F_2 < \cdots < F_m = \tilde{F}$$

in which each successive extension F_k/F_{k-1} is radical. The central question is whether such extensions $\tilde{F}/F$ are rare or unusual. If so, then we should expect that fact to be reflected in the corresponding symmetry groups—that is, being a radical tower should impose some sort of condition on the Galois group $G_{\tilde{F}/F}$. If we can figure out what that condition is, then we will have succeeded in translating "radical tower" into purely group-theoretic language.

The trouble is, such extensions $\tilde{F}/F$ usually fail to be Galois in the first place, which means we will not be able to apply the Galois correspondence. For example, the sequence of extensions

$$\mathbb{Q} < \mathbb{Q}(\sqrt{5}) < \mathbb{Q}(\sqrt{2+\sqrt{5}})$$

is certainly a radical tower over $\mathbb{Q}$. And in fact, each successive extension is perfectly Galois. The field $\mathbb{Q}(\sqrt{5})$ is of course the splitting field of $X^2 - 5$ over $\mathbb{Q}$, and the top field $\mathbb{Q}(\sqrt{2+\sqrt{5}})$ is the splitting field of $X^2 - (2 + \sqrt{5})$ over $\mathbb{Q}(\sqrt{5})$. The problem is that the combined extension $\mathbb{Q}(\sqrt{2+\sqrt{5}})/\mathbb{Q}$ is not so well behaved. This extension is generated by $\sqrt{2+\sqrt{5}}$, which is a root of the quartic polynomial $(X^2 - 2)^2 - 5$ over $\mathbb{Q}$. This polynomial has three other roots, however, including the number $\sqrt{2-\sqrt{5}}$. This number is not contained in the field $\mathbb{Q}(\sqrt{2+\sqrt{5}})$, and therefore this extension is *not* a splitting field. Since the complete set of roots is given by $\pm\sqrt{2\pm\sqrt{5}}$, we would need to include them all in order for our extension to be Galois.

Notice that this is the exact same issue that arose earlier when we were worried about the ambiguity of such expressions. We decided to improve the situation by throwing in all of these additional roots as well, thus lengthening our towers but making them more amenable to our methods. So we can replace our original tower by the improved version

$$\mathbb{Q} < \mathbb{Q}(\sqrt{5}) < \mathbb{Q}(\sqrt{2+\sqrt{5}}) < \mathbb{Q}(\sqrt{2+\sqrt{5}}, \sqrt{2-\sqrt{5}}).$$

Now we are in the convenient position of having a sequence of extensions, each of which is radical and Galois, resulting in a total extension that is also Galois.

Similarly, we saw before that the expression $\sqrt[3]{2}$ is somewhat ambiguous because it is not clear which cube root of 2 we are referring to. This is intimately connected with the fact that the extension $\mathbb{Q}(\sqrt[3]{2})/\mathbb{Q}$ fails to be Galois because it lacks the other two roots of the polynomial $X^3 - 2$. To remove the ambiguity, we decided that it would be best to throw in the appropriate roots of unity to obtain the extension $\mathbb{Q}(\omega_3, \sqrt[3]{2})/\mathbb{Q}$. Now we have a Galois extension, and we needn't worry about which cube root of 2 we are talking about because we are including all of them. So it's interesting that our concerns about ambiguity are essentially the same as the desire for our field extensions to be Galois.

Let's assume from now on that all of our radical towers have been improved in this way—that is, we throw in all the alternate roots of the defining polynomials of our generators, as well as the relevant roots of unity needed to make our extensions Galois. There is certainly no harm in doing this, and it will make life a whole lot easier. In fact, let's go ahead and make the additional assumption that our base field F *already* contains all of the necessary roots of unity. I know this will not be true in general, and that we'll eventually need to look at what happens when roots of unity are adjoined to the base field, but I want to sidestep this technicality for the moment and deal with the simplest possible scenario. (The reason this issue did not arise in the function field case is because the base field already contains $\mathbb{C}$ and consequently every root of unity one could ever desire.)

So let's suppose we have a radical tower with the pleasing feature that every extension in the chain is radical and Galois, and the resulting extension $\tilde{F}/F$ is also Galois. Let's write $G = G_{\tilde{F}/F_k}$. Now we can apply the Galois correspondence to obtain a chain of subgroups

$$G = G_0 > G_1 > G_2 > \cdots > G_m = \{1\},$$

corresponding to the chain of extension fields

$$F = F_0 < F_1 < F_2 < \cdots < F_m = \tilde{F}.$$

The group G_k is then the Galois group $G_{\tilde{F}/F_k}$ attached to the kth field in the chain, and the fixed field $\tilde{F}^{G_k}$ is therefore F_k. Our increasing chain of extension fields is reflected in a corresponding decreasing chain of Galois subgroups.

So far, all of this would hold for *any* chain of extension fields culminating in a Galois extension $\tilde{F}/F$. The question is, what additional demands need to be imposed on the subgroups G_k so that the extensions F_k/F_{k-1} are both radical *and* Galois? This question can be broken into two parts. First of all, we need to understand how to tell if a given Galois extension K/F is radical. What feature does the Galois group $G_{K/F}$ possess that indicates whether K/F is a radical extension, as opposed to some ordinary, run-of-the-mill Galois extension? Second, given an arbitrary Galois extension K/F, how can we tell if a subextension E/F is Galois? What condition on the subgroup $G_{K/E} \leq G_{K/F}$ do we need to impose in order to guarantee that E is a Galois extension of F? Let's begin by addressing this question.

Suppose that K/F is Galois, and let E/F be a subextension. If σ is an element of $G_{K/F}$, then σ moves the elements of K around. In particular, it moves the elements of E around inside K. Let's write E^σ for the effect of σ on E—that is, E^σ is the subfield of K obtained by applying the conjugation σ to every element of K.

Why is E^σ a field?

For example, the splitting field of $X^4 - 2$ over $\mathbb{Q}$ is the extension field $\mathbb{Q}(i, \sqrt[4]{2})$. (Here we can take $\sqrt[4]{2}$ to mean the positive real fourth root of 2, just to be definite.) The Galois group includes a conjugation that fixes i and sends $\sqrt[4]{2}$ to $i\sqrt[4]{2}$, an alternate root of $X^4 - 2$. This conjugation then sends the non-Galois subfield $\mathbb{Q}(\sqrt[4]{2})$ to $\mathbb{Q}(i\sqrt[4]{2})$. In this way, the Galois group $G_{K/F}$ not only moves the elements around, it moves the *subfields* around as well.

What it means for a subextension E/F to be Galois is that $E^\sigma = E$ for all conjugations σ in $G_{K/F}$. A conjugation may *permute* the elements of the subfield E, but it will not move E to a different field. If an extension is Galois (meaning it's a splitting field), then it cannot be taken out of itself by any conjugation; it is generated by a complete set of roots, and these can at worst be rearranged. So we now have an alternate way to think about the Galois criterion: an extension is Galois if it is *closed under conjugation*.

If E/F is Galois, it means that every element of $G_{K/F}$ is also a symmetry of E/F. In other words, there is a natural *restriction* from $G_{K/F}$ to $G_{E/F}$. Every conjugation of K/F becomes a conjugation of E/F when we restrict our attention to how it acts on E. In this way, $G_{E/F}$ becomes a shadow, or *image* of $G_{K/F}$. Restriction is always structure preserving (i.e., the restriction of a product is the product of the restrictions), but it usually collapses information. In particular, the set of conjugations in $G_{K/F}$ that restrict to the identity on E is precisely the subgroup $G_{K/E}$, by definition.

As an illustration, let's go back to our previous example: the extension $\mathbb{Q}(\sqrt{2}, \sqrt{3})/\mathbb{Q}$. This is a Galois extension with Galois group $G = \{1, \theta, \varphi, \theta\varphi\}$, where θ is the conjugation that fixes $\sqrt{2}$ and negates $\sqrt{3}$ and φ does the opposite. Let's now consider the Galois subextension $\mathbb{Q}(\sqrt{2})$. Restricting θ to this field, we see that it acts trivially—that is, both 1 and θ do nothing to this field. On the other hand, when φ is restricted to this subfield, it acts as the nonidentity element of $G_{\mathbb{Q}(\sqrt{2})/\mathbb{Q}}$, sending $\sqrt{2}$ to its negative. The element $\theta\varphi$ (which is also $\varphi\theta$) has this same effect as well. In other words, the Galois group G restricts to $G_{\mathbb{Q}(\sqrt{2})/\mathbb{Q}}$, collapsing the subgroup $\{1, \theta\}$ to the identity.

One amusing (though not totally apt) metaphor is to consider a rectangle with a horizontal line through the middle:

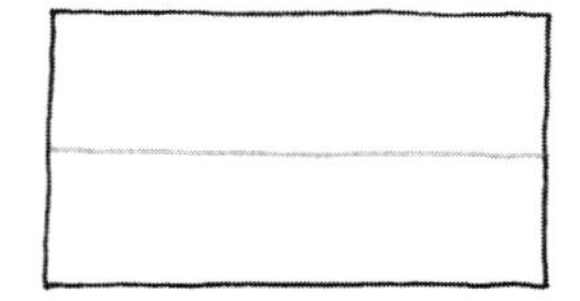

The symmetry group of the rectangle can also be written as $G = \{1, \theta, \varphi, \theta\varphi\}$, only now θ represents the vertical flip that exchanges the top and bottom sides, whereas φ is the horizontal flip. Their product $\theta\varphi$ then acts as a 180-degree rotation. However, if we restrict our attention to the gray horizontal line, then we see that both 1 and θ do nothing at all, whereas both φ and $\theta\varphi$ act on the line as *reflection*. (And of course we can do a similar thing with a vertical line as well.) The reason this analogy is not entirely satisfactory is that here the element $\theta\varphi$ fixes only the center point of the rectangle, whereas in the $\mathbb{Q}(\sqrt{2}, \sqrt{3})/\mathbb{Q}$ case, it fixes an entire subfield $\mathbb{Q}(\sqrt{6})$. But I still like it anyway.

The upshot is that when E/F is a Galois subextension, we have an especially pretty restriction $G_{K/F} \to G_{E/F}$ that collapses the subgroup $G_{K/E}$ to the identity. This is very reminiscent of the modular map that sends $\mathbb{Z}$ to $\mathbb{Z}_n$, collapsing the multiples of n to zero. If we like, we could say that the elements of $G_{E/F}$ are simply those of the parent group $G_{K/F}$ viewed (mod $G_{K/E}$), so to speak. In general, if $H \leq G$ are arbitrary groups, we would like to speak of G (mod H) in the same way that we talk about $\mathbb{Z}$ (mod 7). It would be nice to have the power to collapse subgroups in the same way that we can send all the multiples of 7 to 0.

Unfortunately, due to the noncommutativity of groups in general, this is not quite as simple as it is with $\mathbb{Z}$. We are certainly allowed to call two elements of G equivalent if they are off by an element of H—that is, if one is a multiple of the other by an element of the subgroup. In the case of $\mathbb{Z}$, the group operation is addition, and our subgroup is the multiples of 7. Two integers are equivalent (mod 7) if they differ by an element of this subgroup.

The problem is that now there are two kinds of multiples, left and right. So in order to have a consistent notion, we need the sets of left and right multiples of a given element σ to be the same. Let's write σH for the collection of right multiples $\sigma\tau$, as τ runs through the elements of H, and $H\sigma$ for the left multiples. Then the condition we need for G (mod H) to make sense as a shadow group of G is that $\sigma H = H\sigma$ for all σ in G. This does not mean that $\sigma\tau = \tau\sigma$ for each τ in H but only that the *collections* σH and $H\sigma$ are the same. The point being that when H collapses to the identity, then both σH and $H\sigma$ should be the complete set of elements equivalent to σ (mod H), so they need to be the same. In the case of the (additive) group $\mathbb{Z}$, the subgroup $7\mathbb{Z}$ consisting of the multiples of 7 automatically satisfies this requirement since addition is commutative: $3 + 7\mathbb{Z} = 7\mathbb{Z} + 3$.

Another way to think of it is that in order for G (mod H) to accurately reflect the group structure of G, we need products of equivalent elements to be equivalent. If σ and τ are any two elements of G, and h is any element of H, then we need $\sigma\tau$ and $(\sigma h)\tau$ to be equivalent (mod H). This means $(\sigma h)\tau$ must be equal to $\sigma\tau h'$ for some h' in H. The upshot is that $h\tau = \tau h'$, so left multiples of τ by H must also be right multiples.

Now we can see that for a subextension E/F to be Galois, the corresponding subgroup $G_{K/E}$ needs to collapse coherently, so we must have $\sigma G_{K/E} = G_{K/E}\sigma$ for all σ in $G_{K/F}$. This is the way the Galois criterion is encoded in the symmetry groups. The subfield E is Galois over F precisely when the subgroup $G_{K/E}$ has this two-sided multiples property.

In general, if a subgroup $H \leq G$ has the property that $\sigma H = H\sigma$ for all σ in G, we say that H is a *normal* subgroup of G, written $H \lhd G$. This is the nicest way that one group can sit inside another—the way that allows us to form the so-called *quotient group* G (mod H), the image of G that collapses H to the identity. This group is usually denoted G/H by analogy with division. For example, when we form the quotient group $\mathbb{Z}_7 = \mathbb{Z}/7\mathbb{Z}$, we are essentially chopping the group $\mathbb{Z}$ into seven pieces, or equivalence classes, one for each possible remainder. Similarly, the group G/H consists of the equivalence classes σH, where multiplication of classes is given by $\sigma H \cdot \tau H = \sigma\tau H$. The condition that H be normal is needed to ensure that this definition is consistent.

Show that the subgroup of roll permutations of S_3
is normal and the quotient group is of type C_2.

What happens if you try to form the quotient group S_3/H,
where H *is the non-normal subgroup* $\{1, (12)\}$?

Another way to think about normality is to consider the effect of conjugation on the various subfields of a Galois extension K/F. Suppose H is a subgroup of $G = G_{K/F}$ and E is the fixed field of H. If σ is an element of the Galois group G, then σ moves the subextensions of K/F around, as we have seen. In particular, it will move E to E^σ. But what happens on the group side? Since E is the fixed field of H, it means that H is the Galois group $G_{K/E}$. What group corresponds to E^σ? Which elements of G leave E^σ alone?

The answer is very pretty. The elements of G that do nothing to E^σ are precisely those of the form $\sigma^{-1}\tau\sigma$ for some τ in H. (Essentially, this is saying that if you want to fix E^σ, then you should undo σ, leave E alone, and then put σ back.) Let's write $\sigma^{-1}H\sigma$ for the collection of all such elements—that is, $\sigma^{-1}H\sigma$ is the set of all $\sigma^{-1}\tau\sigma$, as τ runs through the elements of H.

Show that $\sigma^{-1}H\sigma$ is a subgroup of G
and that its fixed field is precisely E^σ.

What this means is that the Galois group G_{K/E^σ} is equal to $\sigma^{-1}G_{K/E}\sigma$, the subgroup obtained by multiplying the elements of $G_{K/E}$ on the right by σ itself and on the left by its inverse, σ^{-1}. This action of σ on subgroups is known as *conjugation*, by analogy with the action of σ on subfields. In other words, the manner in which σ moves the subfields around corresponds exactly to the way it moves the subgroups around: σ moves the field E to E^σ and the corresponding subgroup H to $\sigma^{-1}H\sigma$.

Now we can reinterpret what it means for a subextension to be Galois. If E/F is a Galois subextension of K/F, then $E^\sigma = E$ for every σ in $G_{K/F}$. In other words, being Galois means being a *fixed point* under this action. The condition $E^\sigma = E$ thus corresponds to $\sigma^{-1}H\sigma = H$. A subgroup corresponds to a Galois subextension exactly when it is fixed by conjugation. We could therefore take this as our definition of normal subgroup, if we prefer. (Of course, this is entirely equivalent to saying that $H\sigma = \sigma H$, our previous notion of normal subgroup.)

We have thus succeeded in capturing the Galois property of extensions in purely group-theoretic terms: the subextension E/F is Galois exactly when the subgroup $G_{K/E}$ is normal in $G_{K/F}$. The corresponding Galois group $G_{E/F}$ can then be seen as the quotient group $G_{K/F}/G_{K/E}$, the result of modding out $G_{K/F}$ by the normal subgroup $G_{E/F}$:

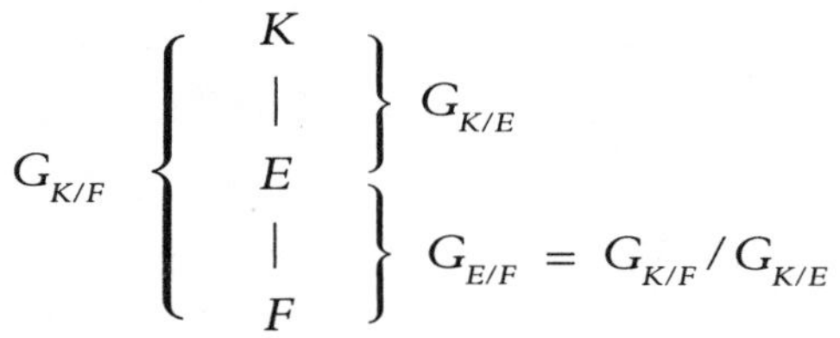

$$G_{K/F} \left\{ \begin{array}{c} \left.\begin{array}{c} K \\ | \\ E \end{array}\right\} G_{K/E} \\ \left.\begin{array}{c} | \\ F \end{array}\right\} G_{E/F} = G_{K/F}/G_{K/E} \end{array} \right.$$

Our next translation project is to capture the notion of a *radical* extension in the language of Galois groups. Suppose F is a field and K is a radical extension of F. Then $K = F(\sqrt[n]{a})$ for some element a in F and some $n \geq 2$. This is a very special situation, being the simplest type of algebraic extension. We should expect something simple on the group side as well. What can we say about the corresponding Galois group $G_{K/F}$?

The first bit of difficulty, as we have seen, is that such an extension need not be Galois. This is typically due to the lack of nth roots of unity in the base field. To make our lives easier, let's consider the special case where the base field F already contains all of the nth roots of 1, so that K is then the splitting field of $X^n - a$ over F. For the sake of clarity, let's choose a particular value of $\sqrt[n]{a}$ and call it β. The complete set of roots of $X^n - a$ is then given by

$$\beta, \beta\omega, \beta\omega^2, \ldots, \beta\omega^{n-1},$$

where $\omega = \omega_n$ is a (primitive) n^{th} root of unity. The Galois group $G = G_{K/F}$ will then permute these n roots in some way. Notice that since $K = F(\beta)$, any conjugation σ is completely determined by its effect on β.

This means that for each conjugation σ, we must have $\beta^\sigma = \beta\omega^k$ for some unique exponent k—that is, each element of the Galois group has the effect of multiplying β by a certain n^{th} root of unity. In this way, we can think of the action of σ on β as a *rotation*. The roots of $X^n - a$ form a regular n-sided polygon in the complex plane, and multiplication by an nth root of unity has the effect of rotating this polygon by a certain angle—that is, some number of n^{th}s of a turn. In other words, our Galois group G can be viewed as a group of rotations in the plane, meaning it's about as simple as a group can possibly be. In particular, there must be a smallest rotation that occurs in G, so that all other conjugations are obtained as repetitions of this one element.

In general, we say that a group G is *cyclic* if it is generated by a single element—that is, if there is some σ in G so that every element has the form σ^m for some integer m. The *order* of a group is just the number of elements in it. Let's write C_n for the abstract cyclic group of order n. Thus, the group of rotations of a regular n-gon is cyclic and isomorphic to C_n. The additive group $\mathbb{Z}_n$ of integers (mod n) is also in the C_n family. (If we like, we could even include the additive group $\mathbb{Z}$ as an instance of the abstract cyclic group C_∞ of infinite order.)

We have discovered that if K/F is a radical extension, then the Galois group $G_{K/F}$ turns out to be cyclic (at least assuming that the base field contains the necessary roots of unity). If K is obtained from F by adjoining the nth root of an element, then the Galois group can be viewed as a set of rotations of a regular n-gon. However, this does not mean that G is necessarily all of C_n. The smallest rotation in G might not be $1/n$ of a turn, but rather a multiple of it—that is, the order of G may only be a *divisor* of n.

For example, suppose we're dealing with the radical quartic extension $\mathbb{Q}(i, \sqrt[4]{4})$ over $\mathbb{Q}(i)$. Here our base field contains the fourth roots of unity (i.e., ± 1, $\pm i$), so we can regard the elements of our Galois group as rotations of a square. However, the polynomial $X^4 - 4$ is reducible and can be factored as $(X^2 + 2)(X^2 - 2)$. Its roots are thus $\pm\sqrt{2}$ and $\pm i\sqrt{2}$, and these are the four corners of our square. But there is no conjugation sending $\sqrt{2}$ to $i\sqrt{2}$, since the latter is not a square root of 2. This means that the 90-degree rotation, or quarter turn, does not occur. The reason is that our field extension $\mathbb{Q}(i, \sqrt[4]{4})$ is really the same as $\mathbb{Q}(i, \sqrt{2})$, so it is actually quadratic, not quartic. The Galois group thus consists only of the identity and the conjugation that sends $\sqrt{2}$ to $-\sqrt{2}$. In other words, the Galois group can only rotate the square 180 degrees.

Thus, we find that the Galois group of a radical extension $F(\sqrt[n]{a})/F$ is always cyclic, and its order divides n. (In the event that the polynomial $X^n - a$ is irreducible over F, the order is exactly n.)

Fortunately, the converse is also true. That is, if our base field F contains the nth roots of unity and K/F is a Galois extension whose Galois group $G = G_{K/F}$ is cyclic of order n, then K must be a radical extension of the form $F(\sqrt[n]{a})$. The idea is that since G is cyclic, it must be generated by a single element σ. The elements of G are therefore $1, \sigma, \sigma^2, \ldots, \sigma^{n-1}$, with $\sigma^n = 1$. The conjugation σ then moves the elements of the field K around, and when this action is repeated n times, it brings everyone back to where they started.

This is certainly reminiscent of the rotations of a regular n-gon, and in fact it can be shown (though not so easily) that there must exist a nonzero element β of K so that $\beta^\sigma = \beta\omega_n$—that is, there is always at least one element of the field that is rotated by the action of σ. The number β^n must then be fixed by this conjugation:

$$(\beta^n)^\sigma = (\beta^\sigma)^n = (\beta\omega_n)^n = \beta^n.$$

Since σ generates G, it means that β^n is fixed by the entire Galois group, and hence it must lie in the base field F. We thus have $\beta^n = a$ for some element a in F, and so $\beta = \sqrt[n]{a}$. This means the extension $F(\beta)/F$ is radical with cyclic Galois group C_n. (We know it's not a smaller group because $X^n - a$ has n distinct roots.) But $F(\beta)$ is a subfield of K, and K/F also has Galois group C_n, so the two fields must in fact be the same. Thus, $K = F(\sqrt[n]{a})$. So the beautiful pattern is that K/F is radical precisely when the Galois group $G_{K/F}$ is cyclic.

In terms of the Galois correspondence, and again assuming that our base field F contains all the necessary roots of unity, we see that a subextension E/F is radical precisely when it is Galois and has a cyclic Galois group $G_{E/F}$. But this means the corresponding subgroup $H = G_{K/E}$ is normal in G and has a cyclic quotient G/H. Another way to say this is that $H \lhd G$ and there is a σ in G such that G decomposes into disjoint sets $H, \sigma H, \sigma^2 H, \ldots, \sigma^{n-1}H$.

Now we are ready to translate "radical tower" into the language of group theory. Suppose we have a radical tower $\tilde{F}/F$ consisting of a sequence of radical extensions F_k/F_{k-1}, for $k = 0, 1, 2, \ldots, m$, where (as usual) we assume that $\tilde{F}/F$ is Galois and that $F = F_0$ contains all of the relevant roots of unity. This gives rise to a corresponding chain of subgroups $G_k < G_{k-1}$, as we have seen. The group G_0 is then the top group $\tilde{G} = G_{\tilde{F}/F}$, and G_k is the Galois group $G_{\tilde{F}/F_k}$ corresponding to the kth extension field F_k.

For each index k, we thus have an extension tower:

$$G_{k-1}\left\{\begin{array}{l}\left.\begin{array}{c}\tilde{F}\\ |\\ F_k\end{array}\right\}G_k\\ \left.\begin{array}{c}|\\ F_{k-1}\end{array}\right\}G_{k-1}/G_k\end{array}\right.$$

Since F_k/F_{k-1} is both normal *and* radical, we know that G_k is then a normal subgroup of G_{k-1} and the quotient group G_{k-1}/G_k must be cyclic. This means that our Galois group $\tilde{G}$ has the remarkable feature that it possesses a decreasing chain of subgroups

$$\tilde{G} = G_0 \rhd G_1 \rhd G_2 \rhd \cdots \rhd G_m = \{1\},$$

where each G_k is normal in G_{k-1} with cyclic quotient G_{k-1}/G_k. This is the way the radical tower concept is reflected in the symmetry groups. Notice that this condition can be stated in purely group-theoretic language, independent of any extension fields or polynomials. This is an abstract property of groups in general and is therefore worthy of investigation in its own right.

Suppose G is any group with the property that there exists a chain of subgroups, each a normal subgroup of its predecessor, with cyclic quotient groups. Such a group is said to be *solvable.* A solvable group is special in that it can be factored, in a sense—we can find a normal subgroup so that when we mod out by it, we get a cyclic group, the simplest kind of group there is. This normal subgroup then contains a normal subgroup of its own, also with a cyclic quotient. This continues all the way down to the trivial subgroup consisting of the identity alone. In some ways, this is reminiscent of prime factorization, although clearly more complicated since we are dealing with groups instead of integers and noncommutative groups at that.

Solvability turns out to be a very useful and interesting property, quite apart from its connection to field extensions and radical towers. When a group is solvable, it means we have a *way in*—the various normal subgroups act as crowbars that allow us to pry open the structure of the group, one cyclic quotient at a time. This makes solvable groups a very nicely behaved species within the larger family of groups in general. In fact, it is not hard to show that solvability is an "inherited" trait:

- If G is solvable, then any subgroup $H \leq G$ is solvable.
- If G is solvable and $H \triangleleft G$, then the quotient G/H is solvable.

Another reason that solvable groups are so convenient is that cyclic groups are always commutative. This means that a solvable group, though not typically commutative itself, can at least be factored (in the above sense) into pieces that are. That makes being solvable the next best thing to being commutative. On the other hand, if our group is commutative to begin with, then every subgroup is normal, and it turns out we can always shatter the group into cyclic quotients. In other words:

- All commutative groups are solvable.

Notice that these are behavioral statements that can be understood (and proven) in the context of abstract groups alone—that is, these properties of solvability do not require that our groups be Galois groups per se, or even symmetry groups. These are theorems of pure group theory, independent of any application they may have to the study of polynomial equations.

This feature is typical of most major works of mathematics, in fact. Longer, more convoluted mathematical arguments tend to break into parts, like movements of a symphony, and some of these parts are capable of wider generalization than others. It's as if every object, pattern, and theorem of mathematics seeks its own natural level of abstraction, like balloons filled with gases of varying density. Some notions must remain close to the ground, weighed down by specific structural demands and constraints, whereas others are lighter and less encumbered and can float up to a higher level. As mathematicians, we are interested in removing whatever obstacles we can so that our ideas can rise as high as possible and extend over the widest range of problems and phenomena. Mostly, this means recognizing those aspects of a problem or argument that are independent of the specific setting. In this way, solvability of groups ceases to be a technical issue relating to radical towers and becomes a larger, more intrinsic feature of groups in the abstract.

We are now in a position to state and prove Galois's main result. Suppose F is now an arbitrary base field and K/F is any Galois extension. Then Galois's theorem states that *K is contained in a radical tower over F precisely when the Galois group $G_{K/F}$ is solvable.* This means we need to prove two things. First, we need to show that if K is contained in a radical tower, then its Galois group must be solvable. Second, we need to show the converse—that is, if K/F is Galois with solvable Galois group, then K must be contained in a radical tower. Fortunately, we have already laid most of the groundwork, so these arguments will be relatively short. The main issue here is that we are no longer making any assumption about the roots of unity being already contained in the base field F. That means we'll need to investigate what happens when we throw them in.

So let's suppose that our Galois extension K is contained in a radical tower $\tilde{F}/F$. As always, we can assume that we have enlarged our tower so that it contains all of the necessary conjugates and roots of unity in order to make $\tilde{F}/F$ a Galois extension. Let's write $\tilde{G} = G_{\tilde{F}/F}$ for its Galois group. We may also assume that the necessary roots of unity are adjoined to the base field from the start, so that our radical tower begins with the extension $F < F(\omega_N)$ for some natural number N. (Here N can be any integer divisible by all of the various n for which we require the nth roots of 1.)

An extension of the form $F(\omega_N)/F$ is known as a *cyclotomic* extension. (This is derived from the Greek for "circle cutting," the complex N^{th} roots of unity producing an equal subdivision of the unit circle). Fortunately, such a cyclotomic extension is always Galois, being the splitting field of the polynomial $X^N - 1$ over F. It is also a radical extension, since we can write $F(\omega_N) = F(\sqrt[N]{1})$. However, in contrast to the radical extensions we considered before, the Galois group of a cyclotomic extension need not be cyclic.

For example, the extension $\mathbb{Q}(\omega_{12})/\mathbb{Q}$ has a Galois group isomorphic to our old friend W, the symmetries of a rectangle. To see this, we first note that ω_{12} is not just any root of the equation $x^{12} = 1$. The number ω_{12} is a *primitive* twelfth root of unity, meaning that it satisfies no equation $x^m = 1$

for any exponent m less than 12. That is, it's a twelfth root of unity and no smaller.

As a notational convenience, let's write $\omega = \omega_{12}$ for short. Since ω is primitive, it means the Galois group must send ω to some other primitive twelfth root of unity, possibly itself. These are just the powers of ω that remain primitive. Thus, the generator ω must get sent to one of ω, ω^5, ω^7, or ω^{11}. (We cannot send ω to ω^3, for instance, because ω^3 is actually a fourth root of unity and ω is not.) The Galois group thus acts on ω by raising it to various powers, namely those relatively prime to 12.

Each of these conjugations is in fact self-inverse: if σ denotes the conjugation taking ω to ω^5, for instance, then we have

$$\sigma^2(\omega) = \sigma(\sigma(\omega)) = (\omega^5)^5 = \omega^{25} = \omega.$$

In other words, the fact that $5^2 = 1 \pmod{12}$ makes $\sigma^2 = 1$. Likewise, we have $7^2 = 11^2 = 1 \pmod{12}$ as well. Similarly, since $5 \times 7 = 11 \pmod{12}$, we find that each of the nonidentity elements of the Galois group is the product of the other two. Thus, we see that this group has the same internal structure as W. Another way to say it is that the Galois group is isomorphic to the multiplicative group of *units* in $\mathbb{Z}_{12}$. In fact, this is true in general.

Show that the Galois group of $\mathbb{Q}(\omega_n)/\mathbb{Q}$
is isomorphic to the unit group of $\mathbb{Z}_n$.

Despite the fact that the Galois groups of cyclotomic extensions are not usually cyclic, they are still quite well behaved. Since we may view an element of such a group as an exponentiation, sending ω to ω^k for some k, we see that these conjugations must always *commute* with each other:

$$(\omega^k)^l = (\omega^l)^k = \omega^{kl}.$$

So the Galois group of a cyclotomic extension need not be cyclic, but at least it's always commutative. (A commutative group is also said to be *abelian*, in honor of Abel, who was among the first to recognize the importance of commutative groups in the theory of equations.)

Returning to our proof, we now have a radical tower $\tilde{F}/F$ that begins with a cyclotomic extension $F < F(\omega)$, where $\omega = \omega_N$ is chosen to provide us with all of the roots of unity needed for the succeeding radical extensions. By our previous results, we know that the radical tower $\tilde{F}/F(\omega)$ therefore has a solvable Galois group, since the base field contains all the required roots of unity.

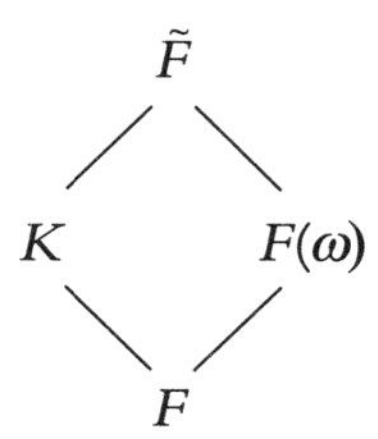

Fortunately, the cyclotomic extension $F(\omega)/F$ is Galois with an abelian (i.e., commutative) Galois group. That means $G_{F(\omega)/F}$ is automatically solvable, since all commutative groups have this property. This tells us that in fact the Galois group $\tilde{G}$ of the entire radical tower $\tilde{F}/F$ must be solvable, since we can join together the chains of subgroups taking us from $\tilde{F}$ to $F(\omega)$ with those going from $F(\omega)$ down to F. Thus, we find that our Galois extension K is contained in a radical tower $\tilde{F}/F$ with a solvable Galois group. But K/F is then a Galois subextension of $\tilde{F}/F$, so we have a restriction mapping from $\tilde{G}$ to $G_{K/F}$, the Galois group of our extension. That means $G_{K/F}$ is an image, or quotient of $\tilde{G}$. But a quotient of a solvable group is always solvable, so $G_{K/F}$ is therefore solvable, and we are done. Being contained in a radical tower forces our Galois extension to have a solvable Galois group.

For the converse, let's suppose that K/F is Galois and that the Galois group $G = G_{K/F}$ is solvable. That means we have a sequence of subgroups, each normal in the preceding, with cyclic quotients of various orders. Let's take N to be a positive integer divisible by all of these orders and write $\omega = \omega_N$ for a choice of primitive N^{th} root of unity. If our base field F happened to contain ω, then we would be all set to apply our previous results. Since it does not necessarily have this property, we must instead adjoin ω to the base field deliberately. So let's first consider the modified extension $K(\omega)/F(\omega)$, obtained by adjoining ω to both our base field and the extension K.

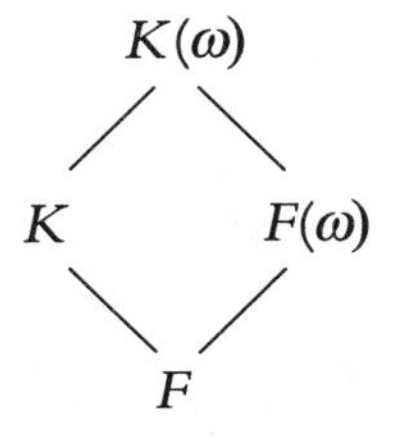

This modified extension is still Galois, since $K(\omega)$ is generated over $F(\omega)$ by the same elements that generate K over F. Let's write G' for the Galois group $G_{K(\omega)/F(\omega)}$. Since K/F is a Galois subextension, we again have a mapping $G' \to G$, sending each conjugation σ of $K(\omega)/F(\omega)$ to its restriction to K/F. (Since σ fixes every element of $F(\omega)$, it certainly fixes F.) Notice that there is no collapsing of information in this case: any σ that restricts to the identity in G must then fix both K and $F(\omega)$, which means it must fix all of $K(\omega)$, making it the identity element of G'. The consequence is that we may regard G' as a *subgroup* of G via restriction. But G was assumed to be a solvable group, which means that G' inherits this solvability. (The chain of normal subgroups for G' may be somewhat shorter than the corresponding chain for G, but the quotient groups are still cyclic, and their orders are at worst divisors of the original orders.)

Now we know that the extension $K(\omega)/F(\omega)$ not only has a solvable Galois group but also contains the necessary roots of unity in the base field. From our preceding work, it follows that $K(\omega)/F(\omega)$ must then be a radical tower. Fortunately, the cyclotomic extension $F(\omega)/F$ is also radical, making the complete extension $K(\omega)/F$ a radical tower as well. And of course this extension contains K. Thus, K is contained in a radical tower over F, and we are finished.

We have now succeeded in translating "solvable by radicals" into the language of group theory. As usual, we'll let F denote our base field, p a polynomial in $F[X]$ whose roots we seek to describe, and K the splitting field of p over F. To say that p is solvable by radicals is to say that the field K lies in a radical tower over F, which is in turn equivalent to the Galois group $G_{K/F}$ being solvable. We thus obtain a classification theorem: *a polynomial is solvable precisely when the Galois group of its splitting field is solvable.*

The problem of determining whether a given polynomial equation can be solved by radicals has now been reduced to two separate steps:

first, determine the Galois group of the splitting field, and second, figure out whether this group is solvable.

The determination of the Galois group associated to a given polynomial is a nontrivial problem. We need to take the information we have—the explicitly given coefficients of the polynomial—and use that information to determine the Galois group of the splitting field. For polynomials of relatively low degree, this can be done using methods similar to those in the previous chapter, forming clever combinations of coefficients that reveal information about the levels of symmetry of the roots. There are still many unsolved problems in this area as well.

The problem of determining whether a given group is solvable is much easier. First of all, such a group is necessarily a finite permutation group. The Galois group of an irreducible quintic, for example, must be a subgroup of S_5, and this group has only $5! = 120$ elements. Since the Galois group must permute the five roots in such a way that each root can be sent to any of the others, there are not actually all that many subgroups of S_5 that can be achieved as Galois groups in this way. In fact, there are only five such groups: the complete permutation group S_5, the alternating subgroup A_5, the cyclic group C_5, and two others of orders 10 and 20. The latter three groups are easily shown to be solvable.

A generic irreducible quintic will have an associated Galois group of S_5, and this group turns out to be *unsolvable*, as is A_5. In fact, as Abel himself showed, the alternating group A_5 is not only unsolvable but *simple*, meaning that it has no nontrivial normal subgroups at all. (In fact, A_5 is the smallest unsolvable group.) This means that any group containing A_5 as a subgroup must also be unsolvable. Thus, the groups A_n and S_n are unsolvable for $n \geq 5$. In particular, since the Galois group of the polynomial $X^5 - X - 1$ happens to be S_5, the equation $x^5 = x + 1$ cannot be solved by radicals.

We can now reinterpret our previous results regarding the function field case. Here, our base field is the field of symmetric rational functions $\mathbb{C}(s_1, s_2, \ldots, s_n)$, and our Galois extension is the full rational function field $\mathbb{C}(X_1, X_2, \ldots, X_n)$. The Galois group is S_n, being the group of all permutations of the roots $X_1, X_2, \ldots, X_n$. The Abel-Ruffini theorem now follows directly from the fact that S_5 is unsolvable. Not only that, but we can now see the quadratic, cubic, and quartic formulas as arising directly from the solvability of the groups S_2, S_3, and S_4. For instance, the sequence of symmetry groups used to obtain the quartic formula is nothing other than the chain of normal subgroups $\{1\} \lhd C_2 \lhd W \lhd A_4 \lhd S_4$ that shows S_4 to be solvable.

As another application of Galois's theorem, we can now classify the constructible complex numbers. We saw before that in order for a number α to be constructible by straightedge and compass, it must be contained in a quadratic tower—that is, a radical tower over $\mathbb{Q}$, all of whose radical extensions are quadratic. Suppose p is the defining polynomial of α, meaning the unique monic irreducible polynomial over $\mathbb{Q}$ that has α as a root. Then α is constructible precisely when the splitting field of p lives in such a quadratic tower, and this means that its Galois group is not only solvable but also has the special additional feature that the quotient groups arising from the chain of normal subgroups are cyclic of order 2. This means the Galois group is quite unusual. In fact, such a group must have order equal to a power of 2, and so must all of its subgroups and quotients. In particular, the degree of p (also known as the degree of α) must then be a power of 2, as I mentioned earlier.

The linguistic question of whether a given polynomial equation can be solved by radicals has now been subsumed into the simpler and more elegant question of whether a given group is solvable. Our investigations begin to move away from the lowly problems of solving this or that particular equation and take on a more abstract, higher-altitude flavor. Which groups occur as Galois groups over which base fields? What are all of the solvable groups of a given order? How does the size of a group control the properties of its various subgroups?

And so classical algebra comes to an end, but it's no cause for dismay. True, we generally can't solve quintic (or higher-degree) equations using the language of radicals, but we've discovered so many beautiful patterns and connections and new points of view that our original questions now seem quaint and naive. The classification of fields and field extensions, the classification of groups and their representations as symmetries, the generalization of all of these ideas to a wider, more encompassing level of abstraction—these are the new paths in the jungle that have been opened up by our explorations and discoveries. We've even managed to make a fair amount of progress on these problems in the past century or so.

Mathematics is always moving forward, and each new generation of mathematicians is tasked with pushing the boundaries a little further. Often, that means climbing higher. Try to imagine the insane levels of abstraction reached in the twentieth century, when the conscious desire for generality and unification had already become a central part of the modern mathematical aesthetic. Galois groups and field extensions were on the cutting edge of mathematics in the 1830s; now they constitute only a small part of the standard background knowledge of any working mathematician.

ALGEBRAIC GEOMETRY

So it turns out that we can't solve most of our mystery number puzzles after all, and we even know why. This means we need to let go of the idea that we can necessarily write down our solutions in some explicit way. Our project has in fact evolved: we're not just solving systems of equations, we're *studying* them—sorting and classifying them, looking for patterns, and trying to understand them as best we can.

The problem then becomes to *describe* the solutions to a given polynomial system, as opposed to naming them explicitly. The number of solutions, for instance, may be knowable even if the solutions themselves cannot be determined. Even when there are infinitely many solutions, they may still satisfy a simple pattern or share certain qualitative features. As always, the question is what patterns are out there and what can be known. What can we prove about systems of polynomial equations and their solutions?

We have already encountered a number of different viewpoints on this matter. Let's consider the simple equation $a + b = 2$. If we like, we can think of a and b as unknowns whose values we seek and the equation as a partial clue to their identities. We could also say that we have a polynomial $X + Y - 2$ in two indeterminates and we are considering its zeros. Of course, these are infinite in number. Yet another viewpoint is that a and b are variables, free to take on a continuum of values, subject to the constraint that they must always add up to 2. In this way, we can regard an equation like $a + b = 2$ as expressing a *relationship* between two otherwise independent numerical variables.

I often think of variables like a and b as sliders (such as on a multichannel mixing board) and an equation like $a + b = 2$ as being the wiring (or programming) that couples them together so that when one moves the other automatically responds. For some reason, I always imagine myself accidentally bumping up against the mixing board, causing the sliders to move. In this case, one would increase and the other would decrease to maintain the constant total.

In general, any algebra problem can be conceived as a set of sliders (the number of them being the dimension of the system), together with wiring instructions provided by a set of polynomial equations. The solutions are simply the legal positions of the sliders, if any. Each equation represents a relationship among continuous variables, and we are demanding that all of these relationships be satisfied.

An entirely new psychological perspective was provided by the French mathematician and philosopher René Descartes in 1637. This was the introduction of *coordinate geometry*—the use of coordinate axes to establish a dictionary between numerical and geometric descriptions. As we have seen, a coordinate system assigns to every point in the plane a numerical label (x, y) for some real numbers x and y. This gives us a way to represent numerical relationships *visually*—the equation $x + y = 2$ can literally be seen as the set of points (x, y) in the plane that satisfy this constraint. We thus obtain the so-called *graph* of the equation $x + y = 2$:

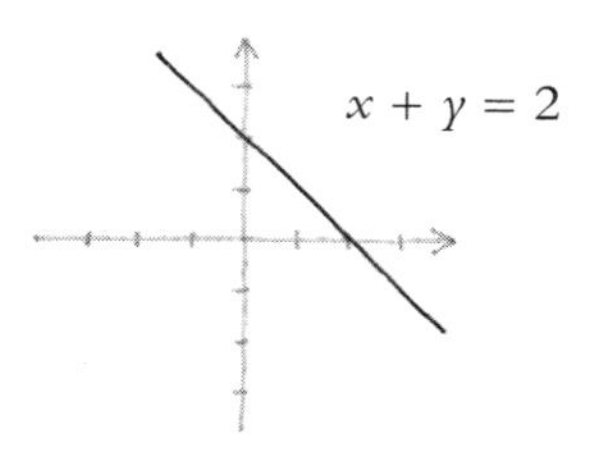

This means that every algebraic equation now has a *look* to it. The hope would be that there is an interesting and intuitively useful correlation between polynomial equations and the shapes of their graphs. This could be valuable in both directions—the geometric imagery informing our algebraic arguments and methods, and the algebraic definiteness helping geometers hold on to pattern information. As you might imagine, there are many deep and powerful (not to mention breathtakingly beautiful) connections to be made and patterns to be discovered in this neck of the jungle.

Sometimes I like to imagine the points of the plane as tiny light bulbs or zero-thickness LEDs. An equation is then an instruction (or incantation) that causes certain points to light up: "All points whose coordinates add up to 2, please come forth!" A graph is neither a path being traced out nor a sequence of points being plotted one at a time, but rather an infinite set of points that appear simultaneously.

Notice that in this case the graph happens to form a *straight line*. This is because the constraint $x + y = 2$ dictates that whenever one variable increases, the other must decrease by a proportional amount. In fact, the same is true for any degree 1 relationship $Ax + By = C$, where A, B, and C are constants (with A and B not both zero). The graph of any such equation is always a straight line, which is precisely why equations of degree 1 are called *linear*. It is especially pleasing that the simplest equations correspond to the simplest shapes.

How do the parameters A, B, and C determine the location and orientation of the line $Ax + By = C$?

What is the equation of the line through (1,1) *and* (2,4)*?*

We now have an entirely new way to think about a system of algebraic equations. We view our variables as *coordinates*—the dimension of the resulting space being the dimension of the system—and our equations as carving out certain subsets of this space. The solutions to our system will then be the points that lie on all of these subsets—in other words, their *intersection*. We thus obtain a geometric interpretation of a polynomial system: we have an ambient space of some dimension, and we are intersecting various curves, surfaces, and other such subspaces inside it.

To take a simple example, suppose we have the system

$$x + y = 2,$$
$$2x - y = 3.$$

If we graph these two equations simultaneously, we obtain a pair of crossed lines in the plane. Their intersection point is then the (unique) solution to the system.

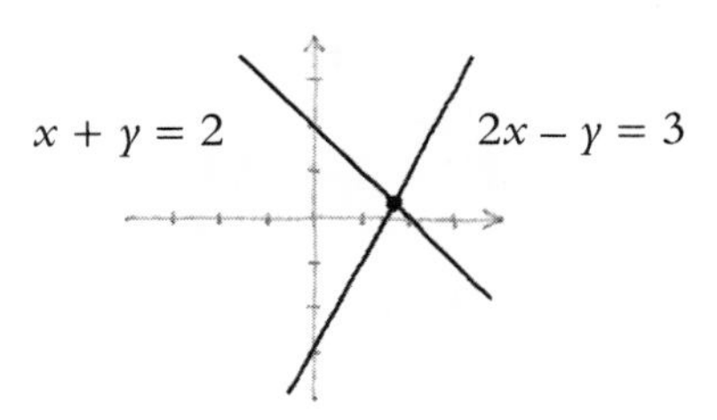

Of course, no crude drawing will tell us the exact location of this point. We still need to do the necessary algebra to determine that $x = \frac{5}{3}$ and $y = \frac{1}{3}$, but now we have the psychological option to think of this pair of values as the single point $(\frac{5}{3}, \frac{1}{3})$ representing the intersection of the two graphs.

What happens when we graph the linear system
$x + y = 2$, $2x + 2y = 5$?

For a somewhat more exciting example, let's look at the system

$$x + y = 2,$$
$$y = x^2.$$

Here we have a linear equation (whose graph is a line) together with a quadratic equation. This carves out a certain shape—a curve, in fact.

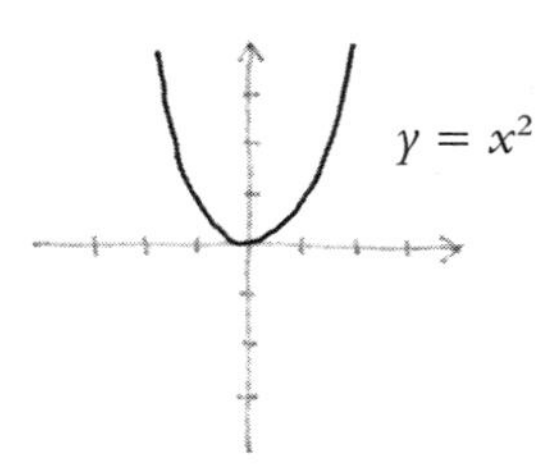

This particular bowl-shaped curve is called a *parabola* and was known to the classical Greek geometers as a *conic section*, or slice of a cone. Specifically, it's the curve obtained by slicing a cone parallel to its slant.

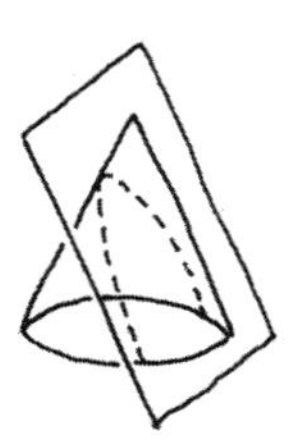

(See *Measurement* for more details.)

Our system is therefore referring to the intersection of a certain line and a certain parabola.

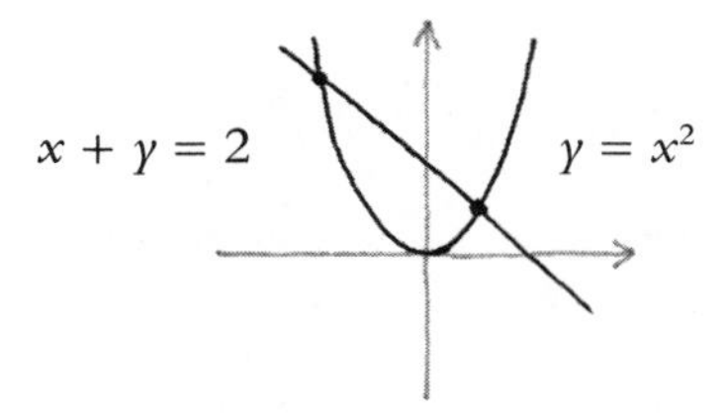

Visually, we see that there should be two solutions to this system, corresponding to the two places where the line crosses the curve. In this case, it's easy to solve the system by substitution: we have $x^2 + x = 2$, so $x = 1$ or $x = -2$. The intersection points are then (1,1) and (–2,4). Algebraically, we know to expect two solutions because we have a quadratic equation; geometrically, this is reflected in the fact that the parabola *bends*, allowing the line to cross it twice.

The system $y = x^2$, $y = 4x - 4$ has only a single solution.
What is the geometric interpretation?

As primates, we are of course always happy to have visual and tactile representations of our ideas. Seeing our equations as shapes and our solutions as intersections is both satisfying and suggestive and may lead us to new ideas and methods we might not otherwise discover. As I said before, we need all the help we can get.

Apart from the intuitive aspect, there is a deeper, more fundamental value to the relationship between equations and their graphs. One of the chief difficulties of working with curves and surfaces and other such geometric objects is our inability to produce mathematically perfect drawings. We can make some crude sketches on paper or even on a computer screen, but this does not really specify a particular shape—at least not in the perfect mathematical sense.

So geometers have always had a description problem. Which curves in the plane, for instance, can be precisely described by language? We can talk about circles and lines and parabolas because they have a pattern to them that can be put into words. Most curves, however, are permanently indescribable due to the fact that there are uncountably many of them. So we're desperate for ways to describe those shapes that can be described.

For geometers, the introduction of coordinates and equations was nothing short of a revolution—a radical and unprecedented expansion of our descriptive powers. Every configuration of points in space is now a collection of number tuples, and if these satisfy an explicit numerical relationship (e.g., a system of algebraic equations), then this relationship essentially becomes the name (or at least one possible name) for the corresponding shape.

The immediate consequence is a profusion of new shapes for us to investigate. Any equation or system of equations in any number of variables is now a geometric object with geometric properties, and all of this shape information must somehow be held in the equations themselves.

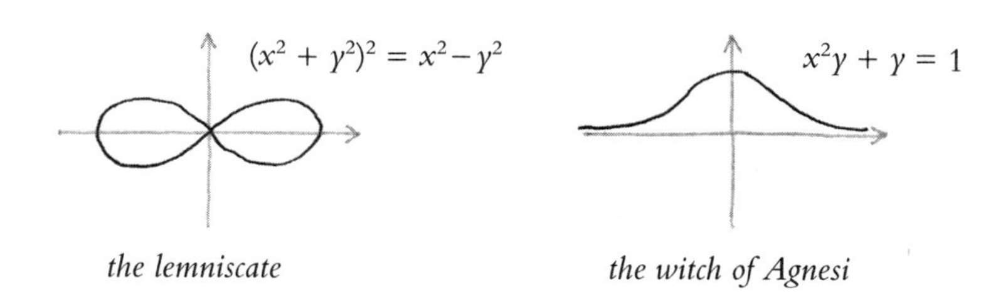

the lemniscate *the witch of Agnesi*

The connection between equations and their graphs deepens and expands the relationship between algebra and geometry. Not only do geometric configurations often lead to algebraic relationships among their measurements, but algebraic equations themselves can now be used to define and specify shapes. We can, for instance, speak of an *algebraic curve* in the plane as one that is carved out by a polynomial equation in two variables. We can then study the relationship between the algebraic features of a given polynomial (e.g., its degree or prime factorization) and the geometric properties of the corresponding curve. We have already seen that degree 1 curves are straight lines, for instance.

What is the shape of the quadratic curve $y^2 = x^2$?

One of my favorite algebraic plane curves is the *nodal cubic* $y^2 = x^3 + x^2$.

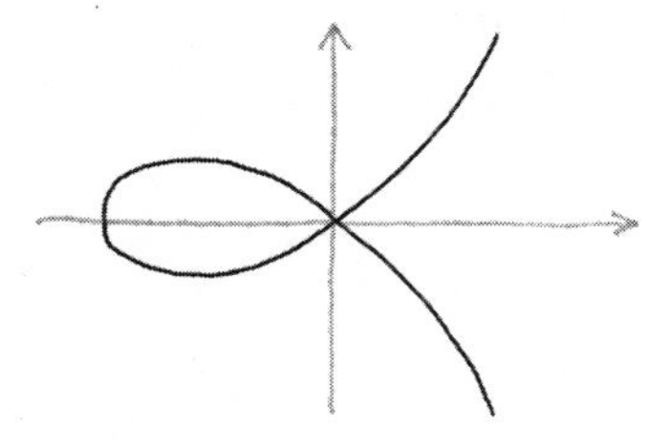

Here we see an amusing example of *self-intersection*. As we trace the curve, it crosses over itself—as if the point (0,0) somehow satisfies the equation twice. Whereas all the other points on this curve feel as though they are sitting on a continuous line, this one is at the center of an X. Such a point, where the behavior of the curve changes suddenly, is called a *singularity*.

Sketch the graph of the cuspidal cubic curve $y^2 = x^3$.
What happens at the singularity (0,0)*?*

Now let's look at a higher-dimensional example. Suppose we have the system of equations

$$x + y + z = 3,$$
$$x^2 + y^2 = 4.$$

The solutions to this system will be certain number triples (x, y, z). These can then be viewed as points in three-dimensional space (also known as 3-space) using a coordinate system with three independent and mutually perpendicular axes. Since these are relatively simple equations, they should correspond to relatively simple shapes in 3-space. The first equation is linear, and its graph turns out to be a *plane*.

Why does a linear equation in three variables describe a plane?

The second equation is quadratic and would normally describe a circle of radius 2 in the xy-plane. But since the z-coordinate can take on any value, the graph in this case becomes an *infinite cylinder* centered along the z-axis. So this particular system is talking about the intersection of a certain cylinder with a certain plane. In this case, the solutions form an *ellipse* in space.

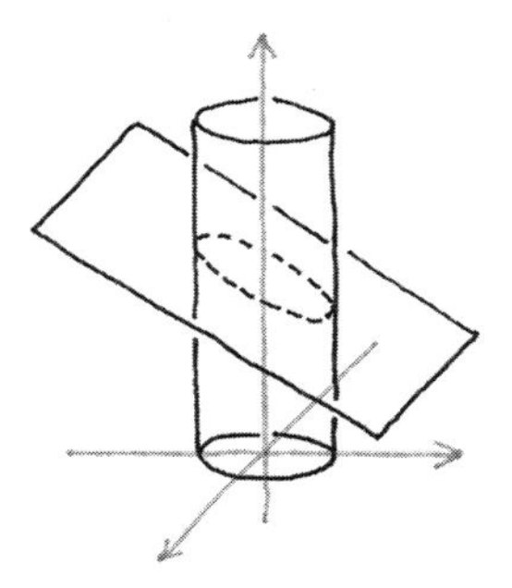

Show that the equation $x^2 + y^2 + z^2 = r^2$ describes a sphere of radius r centered at the origin.

In general, the dimension of our system (and therefore the dimension of our ambient space) will be much higher, well beyond our ability to graph or visualize. Three equations in five variables will typically carve out a curved surface in five-dimensional space. How are we supposed to draw such a thing? I don't know about you, but I find the prospect of trying to visualize seventeen-dimensional space rather daunting, let alone some twisted, self-intersecting nine-dimensional subspace.

So unless we are fortunate enough to be working in very low dimensions, we must abandon the idea of having a visual image in the ordinary sense. Instead, we need to develop our higher-dimensional intuition—to disconnect our mind's eye from our physical eyes, so to speak. This is very much like learning to walk or ride a bicycle. The only way to get good at it is to try (and fail) a lot—to make analogies with one-, two-, and three-dimensional spaces, discover that some of these generalize beautifully whereas others fall flat on their faces, and over many years (if not decades) come to a gestalt understanding of the sorts of things to expect as the spatial dimension increases. Believe me, no one is walking around with the ability to picture in their head what objects in eight-dimensional space look like. It's more a kind of quasi-visual sixth sense informed by experience—one that is always prone to error and always in need of adjustment and improvement.

Show that it's possible for two planes in four-dimensional space to intersect at a single point.
Can you find a way to visualize this scenario?

This means that on the geometric side, we have either crude and approximate physical drawings (when the dimension is low enough) or else vague and unreliable metaphorical imagery (when it isn't). Definite knowledge will have to come from the algebraic side, where the actual information is stored. The value of the geometric viewpoint is not that it solves our problems for us, or even provides new information; it's that it gives us a radically different way to think about what we are doing—a fresh perspective that may inspire new ideas.

One major qualitative difference between algebra and geometry is that algebra is *discrete*. Operations act on finite sets of numbers, and polynomials hold discrete information in their coefficients. Geometric environments, by contrast, offer us the possibility of continuous motion. By treating our unknowns as coordinate variables, we are naturally led to the idea of points moving along curves and surfaces in space, passing through intersections, and so forth—notions we would not have entertained by simply staring at a set of equations.

Suppose we have a system involving n variables. Our ambient space is thus n-dimensional, each variable acting as a coordinate. A given polynomial equation then places a constraint on these coordinates that only certain points will satisfy. Typically, if an equation is at all reasonable (e.g., not $0 = 0$ or $1 = 0$), its graph will then form a certain subspace—a continuum of points having dimension $n - 1$, one fewer than the ambient space itself. As we continue to impose additional polynomial constraints, these will further reduce the dimension of the resulting intersection. In other words, each equation typically knocks the dimension down by one.

Of course, this will fail if the equation is not providing any new information—that is, when it is not algebraically independent of the other equations. The reason we require n independent equations to determine n unknowns is that we need to reduce the dimension of the space of possibilities from n down to zero. Otherwise, there will be at least a curve's worth of solutions.

This means the dimension of the space of solutions is generally equal to that of the ambient space reduced by the number of (independent) constraints. This simple pattern provides us with a rough, qualitative description of the solution space. Thus, a single polynomial equation in three variables will generally describe a surface in 3-space. A pair of such equations will typically result in a curve, being the intersection of two surfaces. If the equations happen to be redundant or incompatible, then the pattern is broken, and we will have either an entire surface of solutions or no solutions at all.

One of the earliest observations about algebraic plane curves concerns the manner in which they bend—specifically, the number of times a given curve intersects a given straight line. We saw before that the quadratic curve $y = x^2$ bends in such a way as to intersect a generic line twice. The reason is that the equation has degree 2, so we are led to finding the roots of a quadratic equation. In general, the pattern is that a plane curve of degree n intersects a generic line exactly n times. This gives us a new, geometric way to think about the degree of a polynomial p in two variables: it's the number of times the plane curve $p(x, y) = 0$ crosses a line.

You may have already noticed some serious problems with this pattern. To take a simple example, let's look at the intersection of the parabola $y = x^2$ with the horizontal line $y = c$, as we vary the parameter c.

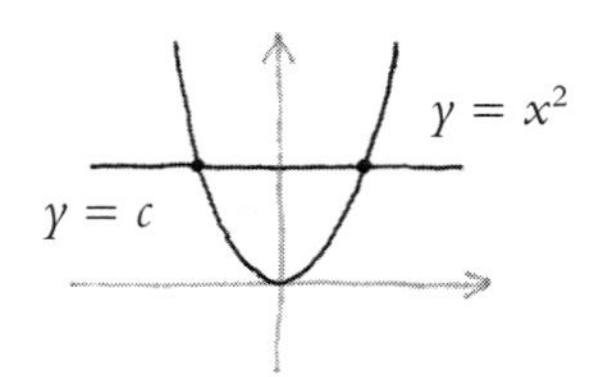

Let's start by imagining c to be a large positive number, so that the line clearly intersects the curve twice. As we continuously lower the value of c, these two intersection points move along the parabola toward each other, and when $c = 0$ they collapse to a single point, breaking the pattern. Even worse, as c becomes negative, we have no intersection at all! There is obviously some sort of miscommunication going on here because the algebra says that since we have a quadratic equation (that is,

$x^2 = c$) we should expect two solutions, but the geometry says that the curve and line simply do not meet if c is negative.

The problem is that we are missing some points. When $c = -1$, for example, the algebra insists that there are solutions $x = \pm i$, $y = -1$. These would correspond to the points $(i, -1)$ and $(-i, -1)$, points we do not have because our coordinates are exclusively *real.* The intersections correspond to the solutions of the quadratic equation $x^2 = c$, and these are real and distinct when c is positive, equal when $c = 0$, and distinct complex numbers when c is negative.

This is reminiscent of our experience with number systems, where we felt an obstruction to a beautiful pattern caused by the absence of certain entities (e.g., negative numbers). Here, we have a very pretty and consistent pattern that all of a sudden gets ruined, simply because we lack the points needed to maintain it. The problem is that $\mathbb{R}$, although a perfectly nice geometric continuum, is not algebraically closed. Some polynomials (e.g., $X^2 + 1$) simply do not have roots, causing our equations to have no solutions and our desired intersection points to not exist.

Naturally, the thing to do is to expand our geometric environment to include not only all pairs of real numbers, but all pairs of *complex* numbers as well. This way, our polynomials are sure to have all of their roots and to factor completely. (This extension of coordinate geometry is part of the general trend of complexification that I was talking about earlier.)

As a simple example, suppose we have the curve $y = q(x)$, where q is a polynomial of degree n. This curve intersects the line $y = 0$ at the points $(\alpha, 0)$, where α is a root of q. Since we are including the complex pairs, every possible α is accounted for.

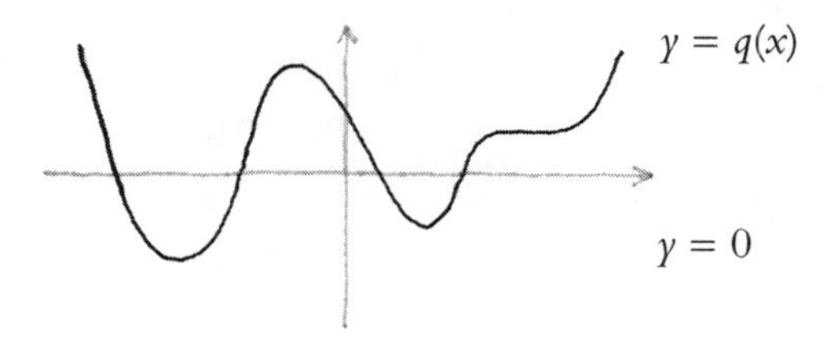

The good news is that the pattern is restored: the total number of intersections is equal to the degree. In fact, we can be a little more precise. Given any plane curve $p(x, y) = 0$, where p is a polynomial of degree n, and any line $Ax + By = C$, substitution will yield an equation of degree n in one variable. This will then factor into n linear pieces over $\mathbb{C}$. If these are distinct, we will then obtain n distinct (complex) intersection points. If not, then some of the intersections will be *multiple*, corresponding to repeated factors. This is precisely what happens with the intersection of $y = x^2$ and $y = 0$, where the two intersection points collapse into each other at (0,0). We obtain the quadratic equation $x^2 = 0$, which has a repeated root at 0. To keep the pattern intact, it is customary to regard this as a *double point*—that is, to count it twice—so that the total number of intersections is always equal to 2. More generally, the resulting polynomial will factor over $\mathbb{C}$ into prime power pieces $(x - \alpha)^m$, where α runs through the distinct roots. We say that the corresponding point has *intersection multiplicity m*. Then the pattern is that the total number of intersections—counted with multiplicity—is always equal to the degree.

The bad news is that we have a hard enough time visualizing two or three real dimensions. How are we supposed to imagine complex two-dimensional space? Since the complex plane is itself two-dimensional, the space of complex pairs is actually *four*-dimensional, and our graphs are actually surfaces! This means that even in the case of plane curves we already need a certain amount of higher-dimensional intuition. Just as the real line is only a thin slice of the complex plane, our graphs—which display only the real points—are merely cross-sections of the *true* graphs, which are surfaces in 4-space. That's pretty frustrating, at least from a visualization point of view.

The truth of the matter is that we're never going to be able to draw these things in their totality. One could argue that the subject is not so much algebraic geometry as it is geometric algebra—that is, when push comes to shove, we're really talking about algebraic structures like numbers and polynomials, only we're using geometric *language*.

Again, the point is that this creates new metaphors and ways of seeing that can lead to new observations and patterns. Often, we find that the new pattern faces an obstacle, which usually takes the form of a lack of some kind—a missing point or number that we wish we had. And you know what comes next: idealization and abstraction, axiomatization and mathematical capture. In this way we create a new, richer environment better suited to the needs of the pattern.

As it happens, our pretty result concerning degrees and intersections turns out to be not entirely true. Notice what happens when we intersect the parabola $y = x^2$ with a *vertical* line such as $x = 2$. Now the bending of the curve becomes irrelevant.

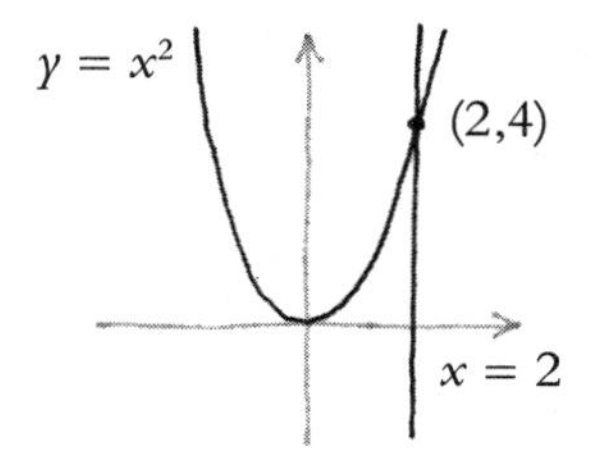

Here the unique intersection point is (2,4). Both the algebra and the geometry agree there is only one solution. Our beautiful pattern—that every line crosses the parabola twice—apparently *still* has exceptions, despite the fact that we expanded our spatial realm to include the complex points.

As a matter of fact, there is an even simpler scenario where our pattern fails miserably: two parallel lines! Since lines are curves of degree 1, our pattern would say that two lines should have exactly one intersection point. This is generically true, but the pattern breaks down when the lines are parallel.

For the sake of definiteness, let's consider a horizontal line and a fixed point below it.

If we look at the various lines through this point, we see that they each intersect the horizontal line at a certain place. As these lines become increasingly horizontal, the intersection points move further and further to the right. On the other hand, when the line goes past horizontal, the intersection point reappears on the far left and then proceeds to get closer.

What seems to be happening is that our intersection point goes off to infinity on the right and then somehow wraps around (as in a video game) and emerges on the left. It's as if our horizontal line is really a loop, with one point "at infinity."

If you are familiar with perspective drawing, this may remind you of the artist's notion of the *vanishing point*, the nonexistent (but apparent) meeting place of parallel lines drawn in perspective:

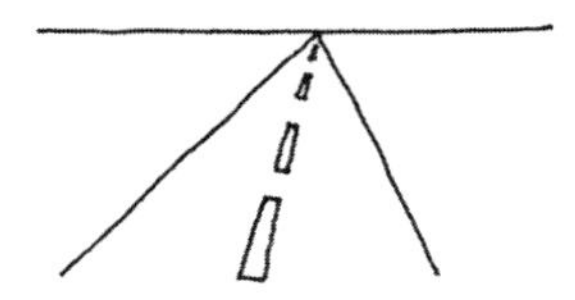

The mathematics of perspective is essentially the study of *projection* — what happens, for instance, when three-dimensional objects are projected onto a plane (such as a canvas or retina) or when objects on one plane are projected onto another. This sort of transformation tends to be fairly destructive, distorting shapes and altering their measurements drastically. Both artists and geometers become interested in those features of an object that are preserved under projection, such as straightness, intersection, and convexity, among others. These are the invariant properties of the object itself, independent of the perspective from which it is viewed.

There are both practical as well as theoretical reasons for wanting to include these points at infinity as part of our geometric environment. The artists want to use vanishing points to get the perspective to look right, and the geometers want them so that our patterns don't break suddenly. If we can somehow get these points at infinity to make sense, then we will have the pretty pattern that any two distinct lines in the plane will *always* intersect at exactly one point, whether they are parallel or not.

Once again, we find ourselves with an extension problem—the need for a new and improved plane that includes our desired points at infinity. As always, the issue is one of mathematical capture. How exactly do we define this new environment so that it is both logically coherent and possesses the properties that we want it to possess?

Just as we extended $\mathbb{R}$ to include new numbers so that our equations would have solutions, we are now hoping to extend the familiar Euclidean plane to include new points so that our lines always intersect. In particular, we want each class of parallel lines—that is, all the lines in a given direction—to meet at infinity. So the idea is to add one point at infinity for each possible direction in the plane, with the additional demand that points at infinity lying in opposite directions (corresponding to the same class of parallel lines) be considered the same. In this way, a class of parallel lines becomes analogous to a set of *concurrent* lines through a given point, only now this point is somehow infinitely far away.

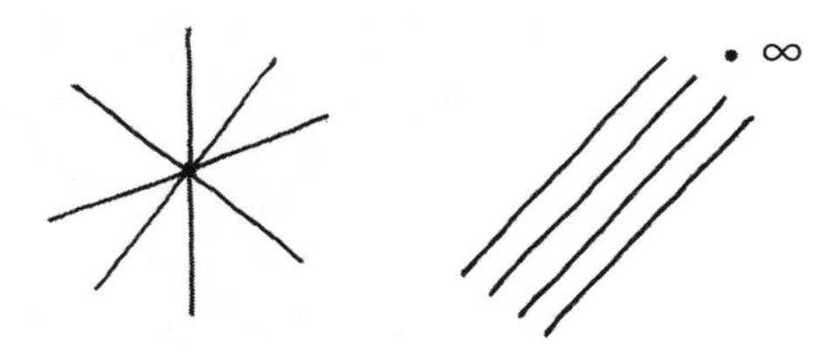

As with algebraic extensions, the goal here is to gain symmetry and uniformity. We don't like it when a pattern has exceptions, so we try to move to an improved structural setting where the pattern can be continued and the exceptions subsumed. This always has the effect of making the original context appear clumsy and naive. We don't want our quadratic equations to break into cases merely because we lack some square roots, and we don't want parallel lines messing up our pretty intersection pattern either.

The plan is to replace the traditional *affine plane* (where parallelism can occur) with the so-called *projective plane* (where it cannot). We need to define this new projective plane in such a way as to include the affine plane as well as the new points at infinity. As usual, the simplest and most elegant way to do this is also the most abstract. We'll start by mimicking what we did before with the line and the point below it. Let's imagine a horizontal plane in space with a fixed point O below it. (The letter O is often used to indicate an origin point.)

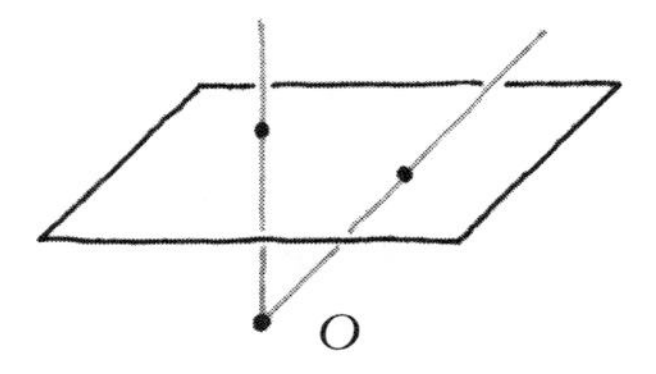

Each point of the plane now corresponds to a line in space through the origin *O*. Under this correspondence, a line in the plane can now be seen as a *plane* in space passing through the origin.

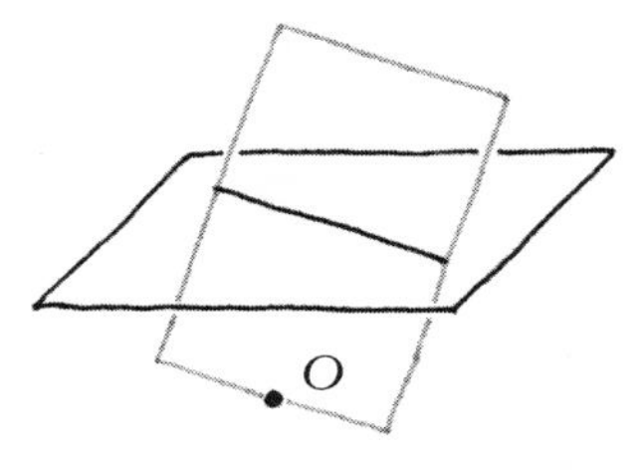

That is, a line of points (in the usual sense) can be reimagined as a plane of lines, all containing *O*. Two distinct lines in the plane then correspond to two planes in space containing the origin.

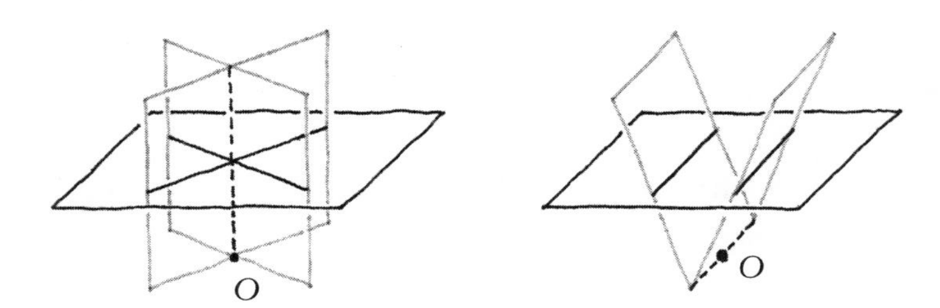

The nice feature here is that any two such planes through *O* *must* intersect in a line through *O*. If the two lines intersect, then the corresponding planes meet in the line through *O* corresponding to their intersection point. If the two lines are parallel, then the planes meet in a horizontal line through *O*, parallel to our original plane. This line is then the embodiment of the point at infinity in that particular direction.

The result is that *every* line in space through O now corresponds to a point in our putative projective plane. If the line is not horizontal, then it hits the plane exactly once and corresponds to an ordinary point in the affine plane. If the line is horizontal, then it instead indicates a direction in the plane corresponding to our desired point at infinity for that direction. In particular, the horizontal plane through O can be regarded as the *line at infinity* consisting of all the points at infinity in every direction.

These observations provide a very clean and symmetrical way to define our new geometric environment: a *projective point* is simply a line in space through the origin O. A *projective line* is a plane through O. Since two lines through O determine a unique plane, we see that any two projective points determine a unique projective line, just as in ordinary plane geometry. Moreover, since any two planes through O must intersect in a line through O, we obtain the beautiful result that any pair of distinct projective lines must meet at a unique projective point, with no exceptions.

Here again we benefit from the enormous flexibility of the abstract viewpoint. Since the only thing we require of points is that two points determine a line and two lines determine a point, it really doesn't matter what points and lines are themselves, so long as they exhibit the desired behaviors. Of course it's weird (and perhaps bewildering) to define points to be lines and lines to be planes, but we've got patterns to hold and investigate, and we need a better container. If we can modify a shoebox to be a terrarium, then I think we can allow some lines in space to play the role of points in an improved, more symmetrical plane.

So let's go ahead and define the *projective plane* to be the set of all projective points—that is, the collection of all lines in space through an arbitrary fixed point O. Notice that this construction is completely symmetrical and emphasizes no particular direction. There is no distinction between ordinary points and points at infinity; instead, all points are created equal. Any arbitrary plane in space (not containing O) could be taken as our affine plane, its infinite points (i.e., directions) then being held by the parallel plane through the origin. This is the line at infinity corresponding to that particular choice of affine plane.

Another way to think about the line at infinity is to imagine the affine plane drawn in perspective, as if it were horizontal and just below eye level:

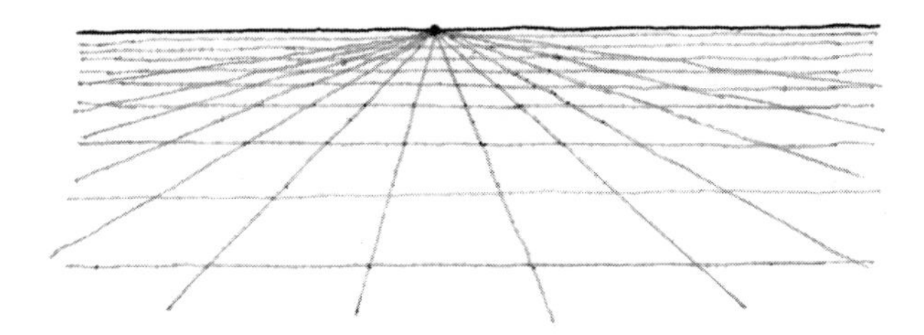

In this view, the line at infinity appears as the *horizon*, each point of which is the vanishing point for a certain class of parallel lines. A choice of affine plane is thus a choice of perspective.

Returning to our example of the parabola and vertical line, we can now see that the quadratic curve $y = x^2$ and the line $x = 2$ do in fact intersect twice—not only at the affine point (2,4) but also at the point at infinity corresponding to the vertical direction. The reason is that the arms of the parabola get increasingly steep, heading toward vertical. This becomes strikingly evident if we draw the graph in perspective:

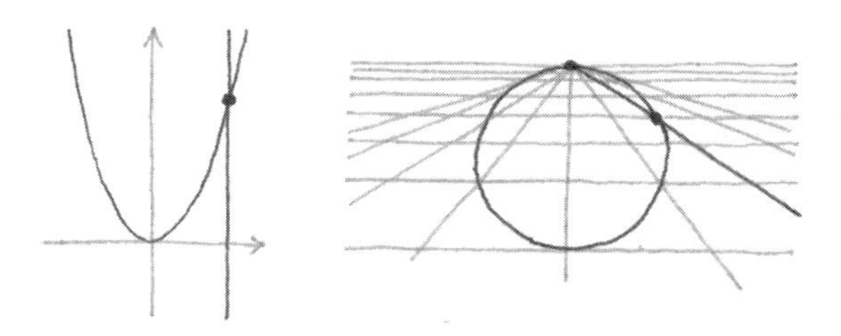

Under this projection, the parabola becomes a circle tangent to the horizon (i.e., the line at infinity), and we can clearly see the two points of intersection. The pattern that was broken has now been restored, at the cost of having to learn to see in a new and more enlightened way—a steep price, but one I am always happy to pay!

Unfortunately, this is still a bit too vague and visual for our purposes. To really understand the interplay between equations and their graphs, we will need to expand our notion of coordinates to the new projective setting. It's one thing to say that a curve appears to be heading in a direction corresponding to a certain point at infinity; it's quite another to name that point precisely and to verify that it satisfies our algebraic constraints.

Fortunately, it's not hard to coordinatize the projective plane; we just need to be a little open-minded about what we mean. Since a projective point is a line in 3-space, we can simply employ our usual three-dimensional coordinate system. We can start by taking our origin O to be the point (0,0,0) in 3-space. A given triple (a, b, c) then determines a unique line—namely, the one through O and the point with those coordinates. We can refer to any line through O in this way, by choosing a random point on it.

As you can see, there is a ton of redundancy in this scheme. Given any triple (a, b, c), which we may regard as a three-dimensional vector, any scaling of it will also refer to the exact same line. So if we want to talk about a line using the coordinates of one of its points, we will have to accept that the lines (1,2,3) and (2,4,6) are one and the same. In other words, it is the *proportion* of the coordinates that matters, not the coordinates themselves.

To clarify the distinction between affine and projective points, we can adopt what are known as *homogeneous coordinates*. The idea is to use ordinary parenthesized triples (a, b, c) to denote points in three-dimensional space, and a bracketed triple $[a, b, c]$ to refer to the line through the origin containing that point. The points (1,2,3) and (2,4,6) are therefore distinct points in space, but [1,2,3] and [2,4,6] are equal and refer to the same projective point. So we can indeed make a coordinate map of the projective plane—we just need triples instead of pairs, and we need to adopt a new, proportional notion of equality.

We can now think of the projective plane as the collection of all homogeneous triples $[a, b, c]$. Each triple then corresponds to a unique line through the origin in space. There is one amusing proviso, however: the homogeneous triple [0,0,0] itself must be excluded, since it does not correspond to any line. Thus, we are essentially defining the projective plane to be three-dimensional space with the origin removed, "modded out" by proportional equivalence to get the dimension back down to two.

In many ways, this construction is analogous to naming fractions by numerator and denominator: it's not the two numbers themselves that matter; it's their ratio that determines the fraction in question. Once again, we have a convenient but highly redundant naming system, and we're just going to have to deal with it. We can use coordinates to navigate our new projective plane, but we need to remember that our points are really lines and can be represented by infinitely many (proportional) triples $[a, b, c]$.

However elegant and symmetrical it may be to define our points to be lines and our lines to be planes, there is something to be said for having an intuitive mental image in which points look and feel like points and lines at least appear one dimensional. The trouble with the projective plane is that it has too many points; we can't fit the whole thing into one flat surface. We can choose an arbitrary plane in space and get most of the points, but we will still miss out on its line at infinity. There simply is no appropriate plane that intersects *every* line through the origin.

One amusing alternative is to use a sphere instead. A sphere (centered at the origin) intersects every line through O, so we aren't missing any projective points. On the other hand, each line through the origin intersects the sphere *twice*, at diametrically opposite points.

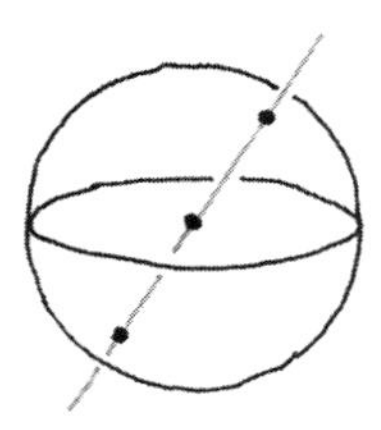

This means we can view the projective plane as the set of *antipodal pairs* of points on a sphere—that is, the projective plane is now a sphere, with opposite points identified. The projective lines would now be *equators* —the circles formed when a sphere is cut in half by a plane. (Since projective lines are planes through the origin, they must cut the sphere exactly in half.) So if you want your points to be point-like and you are willing to accept curved lines, then this is a viable option.

Still another possible model is to view the projective plane as three overlapping affine planes glued together. We know that every projective point $[x, y, z]$ must have at least one nonzero coordinate, since $[0,0,0]$ is not a valid point. That means we can always rescale any point so that one of its coordinates is equal to 1. Thus, every point in the projective plane must live on one or more of the affine planes given by $x = 1$, $y = 1$, and $z = 1$. For example, the point [2,0,3] can be regarded as $[1,0,\frac{3}{2}]$ or $[\frac{2}{3},0,1]$. Any point whose homogeneous coordinates are all nonzero would then be contained in all three of these affine planes.

However we choose to think about it, we now have a perfectly well-defined geometric environment that extends and completes the traditional affine plane, making it more symmetrical and better behaved. In particular, our missing intersection points are now accounted for, and our pretty pattern is restored. The parabola $y = x^2$ and the line $x = 2$ now intersect at two points, one of them corresponding to the vertical point at infinity.

To make this a bit more definite, let's rephrase our equations in projective terms. The equation $y = x^2$ expresses a relationship between the affine coordinates x and y. We can regard any affine point (x, y) as the point $[x, y, 1]$ in homogeneous coordinates. If we write $[X, Y, Z]$ for a generic point in projective space, then the points of the affine plane where Z is nonzero can be rescaled as $[\frac{X}{Z}, \frac{Y}{Z}, 1]$. That means our affine coordinates x and y are simply ratios of projective coordinates: $x = X/Z$, $y = Y/Z$.

Now we can rewrite our equation $y = x^2$ as $Y/Z = (X/Z)^2$. Multiplying both sides by Z^2 to clear denominators, we obtain the equation $YZ = X^2$. This is the "projectivized" version of the affine equation $y = x^2$. Notice that our new equation is *scaling invariant*, meaning that if the triple (X, Y, Z) satisfies this constraint, so do all of its multiples—exactly what we need for the solutions to make sense as projective points $[X, Y, Z]$. The reason our new equation behaves so nicely with respect to scaling is that every term has the same total degree (in this case 2). Such an equation (or polynomial) is said to be *homogeneous*. Thus, the cubic polynomial $X^3 + 3X^2Y + 5Y^3$ is homogeneous, but $X^3 + 2XZ$ is not.

If we restrict our new homogeneous equation $YZ = X^2$ to the affine plane $Z = 1$, we simply recover our original affine equation $Y = X^2$. But now we have the additional possibility of Z being zero, in which case our equation also forces $X = 0$. So along with the traditional affine points $[x, y, 1]$ with $y = x^2$, we also include the points of the form $[0, Y, 0]$ as part of our graph. Here Y must be nonzero, so these are really all the same point, namely [0,1,0]. From the affine point of view, this would be the point at infinity in the vertical direction.

In the same way, we can convert the affine line $x = 2$ to the homogeneous form $X = 2Z$. This says nothing new when Z is nonzero, but when $Z = 0$ we obtain the solution [0,1,0] as well. This means the complete solution to the projectivized system

$$\begin{aligned} YZ &= X^2, \\ X &= 2Z, \end{aligned}$$

consists of the two projective points [2,4,1] and [0,1,0]. We can now see that our original affine interpretation was only one possible perspective view of the *true* graph, which favors no particular coordinate or vantage point.

To an algebraic geometer, the projectivized, homogeneous equations are the *perfected* versions of the original affine equations, and their solutions in the projective plane form the perfected and completed graph. There are no longer any ordinary points or points at infinity; that distinction is merely an artifact created by an arbitrary and unnecessary choice of perspective.

The quadratic curve $xy = 1$ *and the horizontal line* $y = 0$ *do not appear to meet. What's really going on?*

We have now seen two ways in which our original notion of graph turns out to be provincial and naive: our drawings are lacking both the complex points as well as the points at infinity. That means we've been looking only at one particular perspective view of one particular slice of the truth. Of course, it's perfectly natural for us to want to view our variables and coordinates as real numbers and to imagine points in the plane or in three-dimensional space. The trouble is that this is what *we* want; it's not necessarily what *math* wants. Abstract patterns do not conform themselves to our puny imaginations. They do what they do, and it is up to us to listen to what they have to say.

Our algebraic patterns are telling us that it's best to work in the context of an algebraically closed field (such as $\mathbb{C}$) so that we have all the numbers we need to factor polynomials completely. On the other hand, our geometric patterns are saying that there are more points in the plane than we thought, and these should be included for the sake of symmetry. In both cases, we are extending our system—not forsaking any of our prior structures or patterns, but allowing them to be more fully realized. Both complexification and projectivization have become standard methods throughout mathematics (and even modern physics) for precisely this reason.

Surprisingly, the unit circle $x^2 + y^2 = 1$ *includes two points at infinity. What are they?*

It's high time for a massive generalization, wouldn't you say? The notions of affine and projective plane are easily stretched to arbitrary dimensions, and there is no particular algebraic reason to restrict ourselves to a fixed background field. So let's start with any field $\mathbb{F}$ and any natural number n. We can then define *n-dimensional affine space over* $\mathbb{F}$ to be the set of all n-tuples $(x_1, x_2, \ldots, x_n)$ with entries in $\mathbb{F}$. Let's write $\mathbb{A}^n_{\mathbb{F}}$ for this space. (It has become traditional among modern geometers to indicate the dimension of a space as a superscript, analogous to an exponent.) Thus, $\mathbb{A}^2_{\mathbb{R}}$ is a model of the usual Euclidean plane.

Notice that $\mathbb{A}^1_{\mathbb{F}}$ is simply $\mathbb{F}$ itself—that is, we are allowing ourselves to say that any field is a "line" over itself. The complex number field, which we often regard as a plane when we are being $\mathbb{R}$-centric, is also capable of being viewed as $\mathbb{A}^1_{\mathbb{C}}$ and is therefore a line from a complex standpoint. Unfortunately, this then makes the expression "complex plane" somewhat problematic, as it could refer to $\mathbb{C}$ itself or to $\mathbb{A}^2_{\mathbb{C}}$, the affine plane over $\mathbb{C}$ consisting of all pairs (z_1, z_2) of complex numbers. (This is the kind of thing that always happens once you start expanding the meanings of your words and ideas.) As a more extreme example, the affine plane over $\mathbb{F}_2$ consists of only four points!

What is the graph of $y = x^2$ *in the affine plane over* $\mathbb{F}_7$*?*

By analogy with our construction of the projective plane, we can define *n-dimensional projective space over* $\mathbb{F}$—written $\mathbb{P}^n_{\mathbb{F}}$—to be the set of all lines through the origin in affine $(n+1)$-dimensional space. As before, we can view this as the space of all homogeneous $(n+1)$-tuples $[X_0, X_1, X_2, \ldots, X_n]$ with entries in $\mathbb{F}$, not all zero. (Here I prefer indexing from 0 to n to avoid having to write the subscript $n+1$.) Of course, two homogeneous coordinate descriptions are to be considered equal if they are proportional—that is, off by a nonzero scaling factor in $\mathbb{F}$. The classical projective plane is then $\mathbb{P}^2_{\mathbb{R}}$.

How many points are in the projective plane over $\mathbb{F}_2$?

Just as before, we can also view $\mathbb{P}^n_{\mathbb{F}}$ as being covered by overlapping copies of $\mathbb{A}^n_{\mathbb{F}}$, each affine subspace corresponding to a particular coordinate being nonzero. For instance, any point $[X_0, X_1, X_2, \ldots, X_n]$ of $\mathbb{P}^n_{\mathbb{F}}$ with nonzero X_0 can be written in the rescaled form $[1, \frac{X_1}{X_0}, \frac{X_2}{X_0}, \ldots, \frac{X_n}{X_0}]$, corresponding to a point with coordinates $(\frac{X_1}{X_0}, \frac{X_2}{X_0}, \ldots, \frac{X_n}{X_0})$ in affine n-space. Each of the $n+1$ projective coordinates thus provides its own affine patch, and together they cover all of $\mathbb{P}^n_{\mathbb{F}}$. In this way, projective space is analogous to the globe, and the affine patches are like flat maps, no one of which can encompass the entire surface of the planet.

For this reason, geometers tend to think of projective space and homogeneous equations as *global* objects, whereas affine spaces and equations are *local*. If we are concerned with the behavior of our solutions at or near a particular point, for instance, then we can restrict to an affine patch containing that point and treat the problem locally. Often, as with intersection multiplicity, we can define various structures locally at each point, then assemble them into a global statement that is symmetrical and projective, independent of any particular reference frame.

In a way, we now have a new operation we can perform. Given any system of polynomial equations in n variables—which we can regard as defining a certain subset of n-dimensional affine space—we can choose to projectivize it, recasting our equations in homogeneous terms via the substitutions

$$x_1 = \frac{X_1}{X_0},\ x_2 = \frac{X_2}{X_0},\ \ldots,\ x_n = \frac{X_n}{X_0}.$$

After clearing denominators, each new equation will then be a homogeneous polynomial, meaning that all of its terms will have the same total degree.

Why are projectivized equations always homogeneous?

The graph of the projectivized system will include the original graph as well as possible additional points at infinity. Instead of curves and surfaces that sail off indefinitely, our graphs are now closed and bounded, the points at infinity serving to tie up any loose ends. Projectivization thus creates a feeling of completion.

We can also go the other way, from global to local. Given any homogeneous system, we can always choose a particular coordinate and restrict to the corresponding affine patch. For instance, we can simply set $X_0 = 1$, and our system then becomes affine. The only potential loss are the points with $X_0 = 0$, but these can be handled separately.

As a simple illustration of these ideas, lets consider the two-dimensional system

$$y = x^2,$$
$$y^2 = x^3.$$

Geometrically, this corresponds to the intersection of a parabola with a cuspidal cubic.

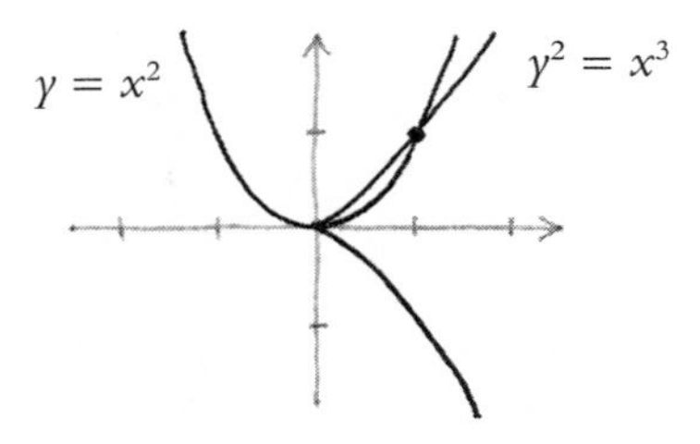

Here we can easily see the intersection points (0,0) and (1,1). But we also understand that there's a lot more to this picture than meets the eye. The complete graph contains not only points at infinity but an entire surface of complex points as well. On the other hand, if we are mainly concerned with intersection points and their various multiplicities, we may be able to get away with a more limited view. Just looking at the affine real picture, we would expect an intersection at the vertical point at infinity, since both curves seem to be bending in that direction.

Solving the equations algebraically (e.g., using substitution), we get $x^4 = x^3$, so we are looking for the roots of the quartic polynomial $x^4 - x^3$, which factors as $x^3(x - 1)$. The first factor corresponds to the intersection (0,0) counted with multiplicity 3, and the second factor indicates a simple intersection (that is, multiplicity 1) at the point (1,1). These are the only intersection points in the complex affine plane.

To include possible points at infinity, we can projectivize our equations to obtain the homogeneous system

$$YZ = X^2,$$
$$Y^2Z = X^3.$$

The points with nonzero Z have already been accounted for, so now we set $Z = 0$, which forces $X = 0$, and we get only the vertical point at infinity [0,1,0], as predicted.

To understand this intersection point in more detail, we can restrict to a local affine neighborhood. Setting $Y = 1$, we obtain the affine system

$$z = x^2,$$
$$z = x^3,$$

and our intersection point is now at (0,0) in xz-coordinates. The system reduces to $x^3 = x^2$, so we look to the factorization $x^3 - x^2 = x^2(x - 1)$ to find that the multiplicity of the intersection point [0,1,0] (corresponding to $x = 0$) is 2.

We now have a grand total of six intersections: the point [1,1,1] counted once, the point [0,0,1] counted three times, and the point [0,1,0] counted twice. We started with two polynomial equations, one quadratic and the other cubic. Our usual rule of thumb would then predict (after elimination) an equation of degree $2 \times 3 = 6$, with six solutions. Now we see what we need to do to make this pattern really work: we need to count *all* the solutions, real and complex, finite and infinite, and each with the proper multiplicity.

This result is known as *Bézout's theorem*: the total number of intersections of any two algebraic curves in the complex projective plane is equal to the product of their degrees. In other words, as long as we work in $\mathbb{P}^2_{\mathbb{C}}$, our rule of thumb works perfectly. This is where the pattern likes to live, so we must follow it to its lair—because that's what pattern hunters do.

There is actually one important exception to Bézout's theorem. Naturally, we would not expect the pattern to work if the two curves were identical because there would be infinitely many intersection points. But they needn't be the exact same curve for this to happen. One example would be the line $y = x$ intersecting the pair of crossed lines $y^2 = x^2$. The trouble here is that the second curve, corresponding to the zeros of the polynomial $y^2 - x^2 = (y - x)(y + x)$, actually *contains* the first curve entirely. So we need to add a proviso that the two curves be independent. This means that the two polynomials have no common factors, so the resulting curves have no components in common. With this clause in place, the Bézout theorem is perfectly valid.

This result provides a nice example of local-global interaction. The projective curves and their degrees are the global objects; the intersection points and their multiplicities are the local data. To examine a particular intersection point and to determine the type of crossing that occurs does not require a global view of the entire projective space, only a local affine subspace. So the Bézout theorem is saying that when two curves intersect, the various local data—that is, the intersection multiplicities—total to a global number that is independent of the details of the curves themselves and depends only on their degrees. The product of the degrees is one global measurement, and the total intersection number is another, and they happen to always be equal. What's more, we can prove that this is so.

Of course, any such proof must consist of purely algebraic statements and deductions involving polynomials and factorization and so on. The question really comes down to finding the right algebraic definition of intersection multiplicity that works for all possible cases. This is not an easy problem. Weird things can happen with high-degree curves, and it's not always so clear what we want intersection multiplicity to mean.

This problem becomes even more difficult when we attempt to generalize these notions to higher dimensions and to arbitrary fields. The general problem is to understand the relationship between systems of homogeneous polynomial equations and their solution sets in projective space. As we've already seen with elimination, it's simplest to replace the original polynomials (whose zeros we seek) by the ideal they generate. That way, we don't have to keep track of a specific set of generators.

So let's suppose that α is a *homogeneous ideal*—that is, an ideal in the domain $\mathbb{F}[X_0, X_1, \ldots, X_n]$ generated by homogeneous polynomials. To any such ideal, we can then assign its *zero set* $Z(\alpha)$, consisting of all the points $[X_0, X_1, \ldots, X_n]$ in $\mathbb{P}^n_{\mathbb{F}}$ that satisfy $p(X_0, X_1, \ldots, X_n) = 0$ for every homogeneous polynomial p in α. The subset $Z(\alpha)$ of projective space then provides us with a *picture* of the ideal α, as represented by its simultaneous zeros. In other words, $Z(\alpha)$ can be regarded as the graph of α. For example, we saw that when α is the homogeneous ideal $(YZ - X^2, Y^2Z - X^3)$, then $Z(\alpha)$ consists of three points.

Conversely, given any subset A of the projective n-space $\mathbb{P}^n_{\mathbb{F}}$, we can consider the ideal $I(A)$ generated by all homogeneous polynomials in $\mathbb{F}[X_0, X_1, \ldots, X_n]$ that vanish at every point of A. We thus establish a two-way correspondence between homogeneous ideals and subsets of projective space, analogous to the Galois correspondence between groups and field extensions. In particular, the larger the ideal α, the more is being asked of its simultaneous zeros, and hence the smaller the zero set $Z(\alpha)$; the larger the subset A of projective space, the more demands are placed on the ideal, and consequently the smaller $I(A)$ must be.

That's about where the similarities end, however. The relationship between homogeneous ideals and projective sets is vastly more subtle and complicated than the Galois correspondence. Two different ideals can have the same zero set, and distinct subsets of projective space can have the same ideal. We need to understand the possible subsets that can occur as zero sets, and the various ideals that can create them. Algebraic geometry becomes the study of polynomial ideals through geometric means.

This is where our insatiable curiosity about number puzzles has led us: to homogeneous ideals in abstract polynomial domains and the point sets they determine in n-dimensional projective space. And of course, the story does not stop there. Classical algebraic geometry is mostly concerned with the behavior of curves and surfaces in $\mathbb{P}^n_{\mathbb{C}}$, but many of these patterns are capable of much wider generalization. Consequently, they can be understood more deeply—as well as more simply—by moving to a more abstract, higher-altitude setting.

In particular, problems concerning the classification and resolution of singularities, the study of deformations and smoothly varying families, and the development of intersection theory in higher dimensions have led to the need for new, more flexible structures, which in turn inspire new questions. The challenge becomes not only to discover exciting new patterns and truths, but also to devise increasingly powerful and abstract ways to hold these patterns so they can be revealed and understood as simply as possible and at the greatest possible level of generality.

In recent years—meaning the last century and a half or so—algebraic geometry has become a veritable poster child for abstraction and generality. It would be difficult to name any area of mathematics that has undergone as many fundamental paradigm shifts and increased levels of abstraction as algebraic geometry. And of course, each new generalization brings with it new points, new spaces, and a host of new patterns and problems.

This is the way of all mathematics. The search for beauty and understanding always leads to higher ground and a more encompassing view. We can't help being attracted to pretty patterns, and we can't stop ourselves from sorting and classifying them. Our brains are biological pattern-recognition machines; making connections and analogies is what they do. When I am daydreaming about math, I am giving my brain its favorite drug in its purest form. Mathematics is a means of constantly evolving and changing myself—becoming reborn as a person who sees in a new way and cannot unsee. I feel so grateful! I get to have this amazing playground in my head, and so do you.

We have come to the end of our journey together, and now it's time to pack up and return to civilization. We've seen a lot of exotic creatures on this rafting trip, and we've had to get used to a lot of strange and unexpected behavior. The transition from physical reality to Mathematical Reality can be dizzying. Our eyes and hands are of no use, and nothing in our evolutionary history prepares us for such an airy and intangible landscape. There are even limitations on our ability to hold and communicate information. We were forced to accept the law of the jungle: if a creature can be described at all, it's unusual; if it can be described in a particular way, it's a rare bird indeed.

We've had to face a number of intellectual comedowns on this adventure, not the least of which is that the jungle and its inhabitants do not care about us—they do what they do, not what we want them to do. Yes, we get to explore and forge paths in Mathematical Reality, but we must follow its inclinations more than our own. For this reason, doing math essentially *requires* us to stretch and bend our ideas and thereby expand our consciousness. I hope this has been a fun and rewarding challenge for you.

The challenge for me was to explain these ideas to you in an entertaining and understandable way—to exhibit Mathematical Reality as a beautiful and mysterious jungle and Algebra as one of its most majestic rivers. I wanted to tell the story of algebra in a way that would convey both the power of the modern abstract viewpoint and the profound joy of mathematical epiphany. Mostly, I wanted to see if I could get across the way mathematicians like me think and feel about our subject and why it is that we are willing to spend our entire lives trying to understand and explain abstract imaginary patterns. The truth is, we can't help it; we simply need to know.

And so we enter the jungle, rowing our weak vessels against the current, pursuing each river upstream to its source, in the pristine snowy peaks above the clouds.

INDEX